PRACTICAL GUIDE TO ELECTRONIC AMPLIFIERS

Other books by John D. Lenk:

COMPLETE GUIDE TO COMPACT DISC (CD) PLAYER
TROUBLESHOOTING AND REPAIR

COMPLETE GUIDE TO DIGITAL TELEVISION
TROUBLESHOOTING AND REPAIR

COMPLETE GUIDE TO ELECTRONIC POWER SUPPLIES

COMPLETE GUIDE TO VHS CAMCORDER TROUBLESHOOTING
AND REPAIR

PRACTICAL ELECTRONIC TROUBLESHOOTING

VCR TROUBLESHOOTING AND REPAIR

PRACTICAL GUIDE TO ELECTRONIC AMPLIFIERS

Basics • Simplified design •
Audio frequency • Radio frequency •
Intermediate frequency •
Video frequency • Direct-coupled •
Compound • Differential •
Op-amp/OTA • Tests •
Troubleshooting

JOHN D. LENK

Consulting Technical Writer

PRENTICE HALL, Englewood Cliffs, New Jersey 07632

Library of Congress Cataloging-in-Publication Data

Lenk, John D.
    Practical guide to electronic amplifiers : basics, simplified
  design, audio-frequency, radio-frequency, video-frequency,
  intermediate frequency, direct-coupled, compound, differential, Op
  -amp/OTA, tests, troubleshooting / John D. Lenk.
      p.   cm.
    ISBN 0-13-690843-8
    1. Amplifiers (Electronics)   I. Title.
  TK7871.2.L386   1991
  621.381'535--dc20   90-32721
                                                            CIP

Editorial/production supervision
  and interior design: *Ellen Denning*
Manufacturing buyer: *Kelly Behr*

Cover photograph courtesy
  of Hewlett-Packard Company

© 1991 by John D. Lenk
Published by Prentice-Hall, Inc.
A Division of Simon & Schuster
Englewood Cliffs, New Jersey 07632

The publisher offers discounts on this book when ordered
in bulk quantities. For more information write:
  Special Sales/College Marketing
  Prentice Hall
  College Technical and Reference Division
  Englewood Cliffs, New Jersey 07632

Printed in the United States of America

10 9 8 7 6 5 4 3 2 1

ISBN   0-13-690843-8

PRENTICE-HALL INTERNATIONAL (UK) LIMITED, *London*
PRENTICE-HALL OF AUSTRALIA PTY. LIMITED, *Sydney*
PRENTICE-HALL CANADA INC., *Toronto*
PRENTICE-HALL HISPANOAMERICANA, S.A., *Mexico*
PRENTICE-HALL OF INDIA PRIVATE LIMITED, *New Delhi*
PRENTICE-HALL OF JAPAN, INC., *Tokyo*
SIMON & SCHUSTER ASIA PTE. LTD., *Singapore*
EDITORA PRENTICE-HALL DO BRASIL, LTDA., *Rio de Janeiro*

Greetings from the Villa Buttercup!
To my wonderful wife, Irene, Happy Anniversary. Thank you
for being by my side for all these years.
To my lovely family, Karen, Tom, Brandon, Justin, and Suzzie;
and to our Lambie, be happy wherever you are!
To my special readers, thank you for buying my books
and making me a best seller! This is book number 69.

CONTENTS

8   OPERATIONAL TRANSCONDUCTANCE AMPLIFIERS   332

9   AMPLIFIER TEST PROCEDURES   357

PREFACE

This is a "something for everyone" amplifier book. No matter where you are in electronics, this book provides experimentation/troubleshooting/test/design information on electronic amplifiers that can be put to immediate use.

If you are an experimenter, student, or serious hobbyist, the book provides sufficient information to design and build electronic amplifiers from scratch. The design approach here is the same as used in all the author's best-selling *books on simplified and practical design.*

Design problems start with approximations or guidelines for the selection of all components on a trial-value basis, assuming a specific design goal and a given set of conditions. Then, using these approximate values in experimental test circuits, the desired results (power output, frequency range, rolloff characteristics, and the like) are produced by varying the test component values.

Although the operation of all circuits is described thoroughly where needed, mathematical theory is kept to a minimum. No previous design experience is required to use the design data and techniques described here. The reader need not memorize elaborate theories or understand abstract math to use the design data, thus making the book ideal for the practical experimenter.

With any solid-state/IC amplifier, it is possible to apply certain guidelines for the selection of component values. These rules can be stated in basic equations, requiring only simple arithmetic for their solution.

If you are a service technician or field-service engineer, there are two entire chapters devoted to advanced test and troubleshooting for amplifiers: The amplifier circuits from a cross-section of present-day electronic equipment are chosen as examples. Not only do the chapters cover the operation of the amplifier circuit, but they also describe test and troubleshooting approaches for that type of amplifier in step-by-step detail.

If you are a working engineer responsible for design and/or selection of amplifiers, or anyone who simply wants an all-around source book for solid-state/IC amplifiers, the great variety of circuits and configurations described in this book should fill the gap. The book describes all types of amplifiers in current use: audio, RF, VF, IF, DC, compound, differential, and op-amps, covering both discrete circuits and ICs. Although no book can cover all amplifier circuits in existence, this book describes the *time-tested, yet up-to-date amplifier circuits*.

Chapter 1 is devoted to a review and summary of basic electronic amplifier design from a practical standpoint.

Chapter 2 is devoted to mounting techniques and thermal considerations for amplifier components, particularly transistors and ICs. The main concern is with power amplifiers, both audio and RF, where thermal problems can arise in design and service.

Chapters 3 through 8 describe theory and simplified, step-by-step design for AF, RF, IF, VF, direct-coupled, compound, and differential amplifiers, as well as op-amps and OTAs.

Chapter 9 is devoted to test procedures for amplifiers. These procedures can be applied to a complete amplifier (such as a stereo system) or to specific circuits (such as the audio and RF circuits of a radio transmitter or receiver). The procedures can also be applied to amplifier circuits at any time during design or experimentation.

Chapter 10 is devoted to troubleshooting procedures for a cross-section of amplifier circuits and covers basic problems common to all amplifiers.

Many professionals have contributed their talent and knowledge to the preparation of this book. The author gratefully acknowledges that the tremendous effort to make this book such a comprehensive work is impossible for one person and wishes to thank all who have contributed, both directly and indirectly.

The author wishes to give special thanks to the following: Robert Green, Everett Sheppard, Ron Smith, Walt Herrin, and Jeff Harris of Mitsubishi; Dick Harmon, Nancy Teater, Mike Arnold, Joel Salzberg, and Ross Snyder of Hewlett-Packard; Bob Carlson and Martin Pludé of B&K-Precision Dynascan Corporation; Joe Cagle and Rinaldo Swayne of Alpine/Luxman; Theodore Zrebiec of Sony; John Taylor, Matthew Mirapaul, and Susan Lewis of Zenith; Thomas Lauterback of Quasar; Donald Woolhouse of Sanyo; Judith L. Fleming and J. W. Phipps of Thomson Consumer Electronics (RCA); Tom Roscoe and Terrance Miller of Hitachi; Pat Wilson, Theresa Long, and Ray Krenzer of Philips Consumer Electronics; and Barbara Zeiders of Service to Publishers.

The author extends his gratitude to Doug Sohm, Greg Doench, Paul Becker, Diane Spina, Bernard Goodwin, Jerry Slawney, Ellen Denning, Amanda McGibney, Barbara Cassel, Karen Fortgang, Nancy Menges, Jewel Harris, Rudy

Drelb, Armond Fangschlyster, Barbara Alfieri and Jacques DuBox of Prentice Hall and Ann Marie Norris of Prentice-Hall International. Their faith in the author has given him encouragement, and their editorial/marketing expertise has made many of the author's books international best sellers. The credit must go to them.

A special thanks to Matt Fox, Master Publisher, whose influence on the author has been (and still is) of tremendous benefit. He is the author's oldest ally and friend in the book publishing field!

The author also wishes to thank Joseph A. Labok of Los Angeles Valley College for his help and encouragement throughout the years.

And to my wife Irene, my research analyst, I wish to extend my thanks. Without her help, this book could not have been written.

JOHN D. LENK

PRACTICAL GUIDE TO ELECTRONIC AMPLIFIERS

1

AMPLIFIER BASICS

This chapter is devoted to a review and summary of basic electronic amplifier design. For our discussion, the purpose of an amplifier is to increase the amplitude of a voltage, current, or power. Secondary functions are to isolate signal sources from other circuits and to provide impedance matching.

In both simple and complex amplifiers, an input signal (consisting of a voltage or current having a definite waveform) produces a corresponding output signal with the same waveform, but in amplified form. Thus, an amplifier is a circuit that develops a voltage or current (or power) that has an amplitude greater than the control factor (or input).

1-1 AMPLIFIER CLASSIFICATIONS

Electronic amplifiers are classified in many ways. One method is to classify amplifiers as to *basic circuit connections.*

Depending on the circuit configuration, three basic classifications are applied to two-junction or bipolar transistors: *common emitter, common base,* and *common collector.*

Note that the word *grounded* can be substituted for *common* in describing the three amplifier circuit configurations. Also note that (in this book) the word transistor implies a two-junction or bipolar transistor. Field-effect transistor amplifiers are identified as FET amplifiers (Sec. 1–11).

Amplifiers can also be classified in terms of *operating point* (or *bias* that establishes the operating point). There are four commonly used operating-point classifications: *class A, class B, class AB, and class C.* A class D is also used to indicate the operating point of a square-wave amplifier, but this is so rare that we are not concerned with class D in this book.

Another method of amplifier classification is based on the function to be performed by the amplifier. Using this system, there are *voltage amplifiers* and *power amplifiers.* When more than one transistor is used for either function, the circuit is sometimes referred to as a *compound amplifier* (such as the Darlington compound discussed in Chapter 6).

Sometimes amplifiers are grouped according to frequency of operation: AF (audio frequency), RF (radio frequency), VF (video frequency), and IF (intermediate frequency). The frequency system is further broken down into *narrowband* (where amplification is limited to a specific frequency or narrow range of frequencies), *wideband* (where a wide range of frequencies is amplified at or about the same level, *tuned* (where amplification is limited to a specific frequency, or to a narrow range, by tuning controls), and *untuned* (where the frequency range is set by nonadjustable component values).

Amplifiers can also be classified by the *type of signals* being amplified. For example, there are a-c (alternating-current), d-c (direct-current), and pulse amplifiers. From a practical standpoint, the d-c amplifier also amplifies a-c signals. (However, the reverse is not true.) Likewise, many a-c and d-c amplifiers pass pulse or square-wave signals without difficulty.

In addition to the various classifications, amplifiers can be assigned some specific title that is descriptive of the function or circuit configuration. The most important of these are *d-c (direct-coupled) amplifier, differential amplifier,* and *operational amplifier (op-amp).*

Note that the term d-c amplifier can mean either *direct-coupled amplifier* or *direct-current amplifier.* This confusion arises from the fact that all direct-current amplifiers must be direct coupled. (That is, there must be no capacitors between amplifier stages.) However, note that direct-coupled amplifiers are not limited to direct-current amplification. In general, direct-coupled amplifiers operate with alternating current and pulse signals, as well as direct current.

1-2 BASIC AMPLIFICATION PRINCIPLE

Figure 1-1 shows NPN and PNP transistors connected in a theoretical common-emitter circuit. Figure 1-2 shows a practical common-emitter circuit operated from a single power supply. In both the theoretical and practical circuits, the base–emitter circuit is forward biased, while the base–collector circuit is reverse biased.

In the theoretical circuit, an input signal is applied across resistor R1, while the output is taken from across R2. In the practical circuit, the output is taken from across RL. Under *no-signal* (or *quiescent*) conditions, current flows in the input circuit, causing a steady value of current to flow in the output circuit. When one-half of an alternation of the input signal is applied to the PNP cir-

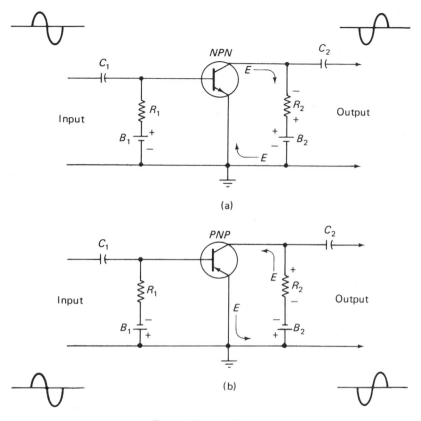

(a)

(b)

Typical Characteristics
Input resistance 500 - 1500 Ω
Output resistance 30 - 50 kΩ
Voltage gain 300 - 1000
Current gain 20 - 50
Power gain 25 - 40 dB

FIGURE 1-1 NPN and PNP transistors connected as theoretical common-emitter amplifiers.

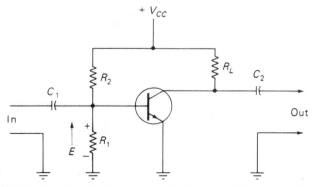

FIGURE 1-2 Practical common-emitter amplifier circuit operated from a single power supply.

cuit of Fig. 1–1b, a voltage is developed across R1. This voltage, positive at the base end of R1, subtracts from the bias voltage provided by B1, causing the base-to-emitter voltage (sometimes called Vbe) to become less negative.

Assume that the B1 battery voltage is 3 V, and the signal (at peak) is 0.5 V. Since the signal voltage drop opposes the B1 battery voltage, the net voltage between the base and emitter decreases to 2.5 V from the original 3 V. The input current also decreases, and less current is available to the collector. Less current flows through R2, and the voltage drop across R2 decreases.

The voltage from collector to emitter (Vce) increases, with the collector becoming more negative than during the no-signal condition. As the signal voltage increases in a positive direction, the output voltage (taken from collector to ground) increases in a negative direction. This action illustrates *phase reversal of the signal,* a characteristic of the common-emitter amplifier. Amplification occurs because the collector current (Ic) is many times greater than the base–emitter current (Ibe).

When the second half of the alternation is applied to the input of the Fig. 1–1b circuit, the voltage is again developed across R1. However, this voltage is negative at the base and adds to the B1 battery voltage so that the net voltage between base and emitter increases to 3.5 V from the original 3 V. The input current increases, and more current is available to the collector. More current flows through R2, and the voltage drop across R2 increases. This increase remains out of phase with the input and is amplified by the same amount as the first half of the alternation.

The total voltage Vbe is a combination of the input signal and B1. The total voltage Vce is a combination of the output signal and B2. In the practical circuit of Fig. 1–2, the input is measured from base to ground, with the output measured from collector to ground. The NPN circuit of Fig. 1–1a operates in the same way as the PNP circuit of Fig. 1–1b, except that polarities of the voltages are reversed.

1-3 RULES FOR LABELING, BIASING, AND POLARITIES IN TRANSISTORS

The following general rules can be helpful in a practical analysis of how transistor amplifiers operate. The rules apply primarily to a class A amplifier, but also remain true for many other transistor amplifier circuits. The rules are included here primarily for those totally unfamiliar with bipolar transistors (and for those who have forgotten!).

1. In the NPN transistor, electrons flow from the emitter to the collector, so the collector must be *positive relative to the emitter.* In the PNP transistor, holes flow from the emitter to collector, so the collector must be *negative relative* to the emitter.

2. The *middle letter* in PNP or NPN applies to the base.

3. The *first two letters* in PNP or NPN refer to the *relative bias polarities of the emitter* with respect to either the base or collector. For example, the letters PN (in PNP) indicate that the emitter is positive with respect to either the base or collector. The letters NP (in NPN) indicate that the emitter is negative with respect to both the base and collector.

4. The *d-c electron current flow* is always against the direction of the arrow in the emitter.

5. If electron flow is into the emitter, electron flow is out from the collector.

6. If electron flow is out from the emitter, electron flow is into the collector.

7. The collector–base junction is always reverse biased.

8. The emitter–base junction is generally forward biased.

9. A *base-input voltage that opposes* or decreases the forward bias also decreases the emitter and collector currents. For example, a negative input to the base of an NPN or a positive input to a PNP base decreases both currents.

10. A *base-input voltage that aids* or increases the forward bias also increases the emitter and collector currents (positive to NPN, negative to PNP).

1-4 BASIC COMMON-EMITTER AMPLIFIER

The common-emitter (or CE) circuit shown in Fig. 1-2 is the most widely used amplifier configuration. The emitter is common to both the input and output circuits and is frequently called the grounded element (although the emitter is not always connected to ground).

One reason for the popularity of the common-emitter amplifier is because both *current gain* and *voltage gain* are possible (as shown by the characteristics of Fig. 1-1). This combination results in a large *power gain*. For example, if the output impedance is 10 times higher than the input impedance, there is a power gain of 10, even if the base–emitter (input) and emitter–collector (output) currents are equal (no current gain). Since it is possible to control a large output current with a small input current in a practical CE amplifier, there is also current gain.

C1 is a coupling capacitor used to block out any d-c components in the input signal developed across R1. Note that R1 also provides a closed circuit for current flow in the base–emitter circuit. This is essential in any transistor amplifier since a transistor is a current-operated device. Capacitor C2 blocks the steady d-c component, but passes the a-c output component developed across RL. As shown by the waveforms of Fig. 1-1, there is a 180° phase reversal between input and output signals.

To summarize the CE amplifier, the input signal is applied between base and emitter, and the output signal appears between emitter and collector. This provides a moderately low input impedance and a very high output impedance, with a phase reversal between input and output. The CE amplifier produces the highest power gain of all three amplifier configurations. The voltage and current gains are fairly high. CE amplifiers are the most often used since there is current gain, voltage gain, resistance gain, and power gain.

1-4.1   Expressing Amplifier Gain

Many terms are used to express the gain of CE amplifiers, as well as other amplifier configurations. For example, there are the terms *alpha* and *beta,* in addition to *current gain, voltage gain, resistance* (or *impedance*) *gain, power gain, voltage amplifier,* and *power amplifier.* The terms are interrelated and are often interchanged (properly and improperly). To minimize this confusion, the following is a summary of how these terms are used throughout this book.

Alpha and Beta.   The terms alpha and beta are applied to transistors connected in the common-base (CB) and common-emitter (CE) configurations, respectively. Both terms are a measure of current gain for the transistor (but not necessarily for the circuit). Alpha is always less than 1 (typically 0.9 to 0.99). Beta is more than 1 and can be as high as several hundred (or more). The relationships between alpha and beta are:

$$\text{alpha} = \frac{\text{beta}}{\text{beta} + 1}, \qquad \text{beta} = \frac{\text{alpha}}{\text{alpha} - 1}$$

The terms alpha and beta do not necessarily represent the current gain of the amplifier circuit in which the transistors are used. Instead, the current gain of the circuit *cannot be greater* than the alpha or beta of the transistor.

Current Gain.   The term current gain can be applied to either the transistor or to the amplifier circuit; it is a measure of change in current at the output for a given change in current at the input. For example, when a 1-mA change in input current produces a 10-mA change in output current, the current gain is 10. Since alpha (CB) is always less than 1, there is no current gain in CB amplifier circuits. (Instead, there is a slight loss.)

Resistance Gain.   The ratio of output resistance (or impedance) divided by input resistance (or impedance) is the resistance gain. For example, if the input resistance is 1 k and the output resistance is 15 k, the resistance gain is

15. The input and output resistances (or impedances) depend on circuit values, as well as transistor characteristics. As discussed throughout this book, many amplifier characteristics depend directly on the relationship between transistor characteristics and the values of circuit components.

Resistance gain, by itself, produces no usable gain for the amplifier circuit. However, resistance gain has a direct effect on the voltage and power gains. For example, using the previous values (current gain of 10, resistance gain of 15), it is possible to have a voltage gain of 150. Assume that the input resistance is 1 k, and a 1-mV signal is applied at the input. This results in an input current change of 1 μA. With a current gain of 10, the output current is 10 μA. This 10-μA current passes through a 15-k output resistance to produce a voltage change of 150 mV. As a result, the 1-mV input signal produces a 150-mV output signal (a voltage gain of 150).

Voltage and Power Gain.   Voltage gain is equal to the difference in output voltage divided by the difference in input voltage. Power gain is equal to the difference in output power divided by the difference in input power. Except in CB amplifiers, power gain is always higher than voltage gain, since power is based on the square of voltage (power $= E^2/R$), as well as the square of current (I^2/R). Using the same values, the input power of the stage is 1×10^{-9}, the output power is 1.5×10^{-6}, and the power gain is 1500.

As in the case of current gain, both voltage gain and power gain depend on *both* transistor characteristics and circuit values.

Voltage and Power Amplifiers.   The function of a *voltage amplifier* is to receive an input signal consisting of a small voltage of definite waveform and produce an output signal consisting of a voltage with the same waveform, but much larger in amplitude. For example, as a radio wave cuts across the antenna of a radio receiver, the wave induces a fluctuating voltage in that antenna, usually on the order of microvolts. The voltage amplifier of the receiver amplifies this voltage to produce a similarly fluctuating voltage that is large enough to operate a *power amplifier,* which, in turn, operates the power-consuming loudspeaker.

These transistors designed for voltage amplification usually have high betas, with small current-carrying capability. On the other hand, power transistors have large current-carrying capacity, but relatively low betas.

Typically, there is at least one (and usually two) voltage-amplifier circuit ahead of the power-amplifier circuit. This permits a low-voltage input signal (say from an antenna, tape head, or industrial transducer) to operate a power-consuming device (radio or stereo loudspeaker or industrial servo motor). If the power involved exceeds about 1 W, the transistor must be operated with *heat sinks* (as discussed in Chapter 2).

1-5 BASIC COMMON-BASE AMPLIFIER

The common-base (CB) amplifier shown in Figs. 1–3 and 1–4 has no current gain, but does provide power gain. The base is common to both the input and output circuits and is frequently called the grounded element (although the base is not always connected to ground).

The input signal is applied between the base and emitter, and the output signal appears between the base and collector. This provides the lowest possible input impedance and a very high output impedance. The output signal is in phase with the input.

C1 and C2 are blocking or coupling capacitors. In the practical amplifier of Fig. 1–4, both transistor junctions are biased from a single supply. Resistor

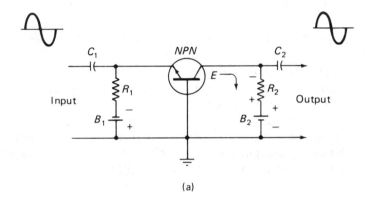

(a)

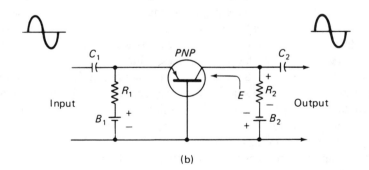

(b)

Typical Characteristics

Input resistance 30 - 150 Ω
Output resistance 300 - 500 kΩ
Voltage gain 500 - 1500
Current gain less than 1
Power gain 20 - 30 dB

FIGURE 1-3   NPN and PNP transistors connected as theoretical common-base amplifiers.

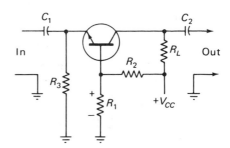

FIGURE 1-4 Practical common-base amplifier circuit operated from a single power supply.

R3 provides a closed circuit for the base–emitter circuit. Bias for the base–collector junction is provided by the R1/R2 voltage divider.

To summarize the CB amplifier, the input signal is applied between base and emitter, and the output signal appears between base and collector. This provides an *extremely low* input impedance and a very high output impedance. The output signal is in phase with the input. The CB amplifier produces high voltage gains and modest power gains, even though there is no current gain. This is possible because of the resistance gain, as discussed in Sec. 1–4.1.

As an example, assume that the input resistance is 100 Ω, the output resistance is 15 k, the current gain is 0.9 (less than 1), and the input signal is 1 mV. With 1 mV across 100 Ω, there is a 10-μA current change. With a current "gain" of 0.9, the output current is 9 μA. This 9-μA current passes through a 15-k output resistance to produce a voltage change of 135 mV. Thus, a 1-mV input signal produces a 135-mV output signal (a voltage gain of 135).

1-6 BASIC COMMON-COLLECTOR AMPLIFIER

The common-collector (CC) amplifier shown in Figs. 1–5 and 1–6 is also known as an *emitter follower*, since the output is taken from the emitter resistance and the output follows the input (in phase relationship).

The input signal is applied at the base, and the output signal appears at the emitter. This provides a high input impedance and a very low output impedance. The output signal is in phase with the input.

C1 and C2 are blocking or coupling capacitors. In the practical amplifier of Fig. 1–6, both transistor junctions are biased from a single supply. Resistor R2 provides a closed circuit for the base–emitter circuit. Bias for the base–collector junction is provided by the R1/R2 voltage divider.

To summarize the CC amplifier (or emitter follower), the input signal is applied to the base, and the output signal appears at the emitter. This provides *extremely high* input impedance and a very low output impedance (usually set by the value of RL). The output signal is in phase with the input. The CC amplifier (or emitter follower) produces modest current and power gains, even though there is no voltage gain. In general, the CC current gain (and hence the power gain) is limited by the current gain (beta) of the transistor.

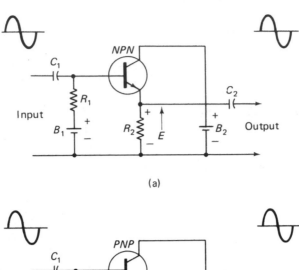

(a)

(b)

Typical characteristics

Input resistance 20 - 500 kΩ
Output resistance 50 - 1000 Ω
Voltage gain less than 1
Current gain 25 - 50
Power gain 10 - 20 dB

FIGURE 1-5   NPN and PNP transistors connected as theoretical common-collector (emitter-follower) amplifiers.

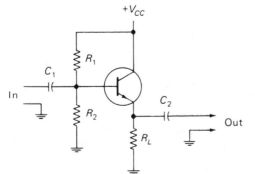

FIGURE 1-6   Practical common-collector (emitter-follower) amplifier operated from a single power supply.

1-7 BASIC AMPLIFIER BIAS NETWORKS

All solid-state amplifiers require some form of bias. As a minimum, the collector–base junction of any solid-state amplifier must be reverse biased. That is, current should not flow between collector and base. Any collector–base current that does flow is a result of leakage or breakdown.

Breakdown must be avoided by proper design. Leakage, usually listed as Icbo, is an undesirable (but almost always present) condition that must be reckoned with in practical use. (Note that the *c* and *b* indicate current flow between collector and base. The *o* indicates that flow is measured with the emitter disconnected, or open.)

Under no-signal conditions, the emitter–base circuit of a solid-state amplifier can be forward biased, reverse biased, or zero biased (no bias). However, emitter–base current must flow under some condition of operation. For example, some current flows all the time in class A and class AB amplifiers. In class B and C amplifiers, current flows only in the presence of an operating signal. With any class of operation, the emitter–base circuit must be biased so that current can flow under some conditions. (Classes of amplifiers are discussed in Sec. 1–8.)

The desired bias is produced by applying voltages to the corresponding transistor elements through bias networks, usually composed of resistors. The following paragraphs describe several basic bias networks. These circuits (or variations of them) represent most of the bias methods used in solid-state amplifiers.

The bias networks (or the resistors used to form the networks) serve more than one purpose. Typically, the bias network resistors (1) set the operating point, (2) stabilize the circuit at the operating points, and (3) set the approximate input/output impedances of the circuit.

Operating Point of Amplifier.   The basic purpose of the bias network is to establish collector–base–emitter voltage and current relationships at the operating point of the amplifier circuit. (The operating point is also known as the *quiescent point, Q point, no-signal point, idle point,* or *static point.*)

Since transistors rarely operate at the Q point, the basic bias networks are generally used as a reference or starting point for design. The actual circuit configuration and especially the bias network values are generally selected on the basis of dynamic circuit conditions (desired output voltage swing, expected input signal level, and so on).

Amplifier Bias Stabilization.   Once the desired operating point is established, the next function of the bias network is to stabilize the amplifier circuit at this point. Although there are many bias networks, each with advantages and disadvantages, one major factor must be considered for any network. The basic bias network must maintain the desired current relationships in the presence of temperature and power-supply changes and possible transistor replacement. In some cases, frequency changes and changes caused by component aging must also be offset by the bias network. This process is generally referred to as *bias stabilization.*

Two undesirable conditions can result when adequate bias stabilization is not provided. First, any changes in temperature, power-supply voltage, and (possibly) frequency also produce changes in collector–emitter and/or base–emitter current. For example, an increase in temperature or supply voltage increases current. In turn, this shifts the operating point of the amplifier circuit. As discussed in the following paragraphs, a shift in operating point can produce distortion and a change in frequency response, as well as other undesired effects. In any event, the amplifier circuit is no longer at the operating point for which the circuit is designed.

Thermal Runaway.   The other undesirable effect of inadequate bias stabilization has to do with power dissipation limits of transistors. When a transistor is operated at or near the maximum power dissipation limits, the transistor is subject to thermal runaway.

When current passes through a transistor junction, heat is generated. If not all this heat is dissipated by the case or heat sink (often an impossibility), the junction temperature rises. This, in turn, causes more current to flow through the junction, even though the voltage, circuit values, and so on, remain the same. In turn, this causes the junction temperature to increase even further, with a corresponding increase in current flow. If the heat is not dissipated by some means, the transistor burns out.

Adequate bias stabilization prevents any drastic change in junction currents, despite changes in temperature, voltage, and so on. Thus, proper bias stabilization maintains the amplifier circuit at the desired operating point (within practical limits) and prevents thermal runaway.

Input/Output Impedances.   The resistors used in bias networks also have the function of setting the input and output impedances of the amplifier circuit. From a theoretical standpoint, the input/output impedances of a circuit are set by a wide range of factors (transistor beta, transistor input/output capacitance, and so on). However, for practical purposes the input/output impedances of a resistance-coupled amplifier (operating at frequencies up to about 100 kHz) are set by the bias network resistors. For example, the output impedance of a CB or CE amplifier is about equal to the collector resistor (between the collector and power supply).

1-7.1 Basic Bias-Stabilization Techniques

There are several methods for providing the bias stabilization of solid-state amplifiers. All methods use a form of *negative feedback* or *inverse feedback.* That is, any change in transistor currents produces a corresponding voltage or current change that tends to offset the initial change. There are two basic methods for producing inverse or negative feedback: inverse-voltage feedback and inverse-current feedback (also known as *emitter feedback*).

Current feedback is more commonly used than voltage feedback in present-day solid-state amplifiers. This is because transistors are primarily current-operated devices, rather than voltage-operated devices. *Note that any form of negative or inverse feedback in an amplifier tends to oppose all changes, even those produced by the signal being amplified.* Thus, inverse or negative feedback tends to reduce and stabilize gain, as well as undesired change. This principle of stabilizing gain by means of feedback is used in virtually all types of amplifiers, as is discussed throughout remaining chapters.

Typical Emitter-Feedback Bias Network. Figure 1-7 shows a typical inverse-current (emitter-feedback) bias network using an NPN transistor. Note that this circuit is essentially the same as the basic CE amplifier shown in Fig. 1-2, but with an emitter resistor to provide bias stabilization. Other bias networks using the same principle are discussed in the following sections. The use of an emitter-feedback resistance in any bias circuit can be summed up as follows.

Base current (and, consequently, collector current) depends on the *differential in voltage* between base and emitter. If the differential voltage is lowered, less base current (and, consequently, less collector current) flows. The opposite is true when the differential is increased. All current flowing through the collector (ignoring collector–base leakage, Icbo) also flows through the emitter resistor. The voltage across the emitter resistor therefore depends (in part) on the collector current.

Should the collector current increase (for any reason), emitter current and the voltage drop across the emitter resistor also increase. This negative feedback tends to decrease the differential between base and emitter, thus lowering

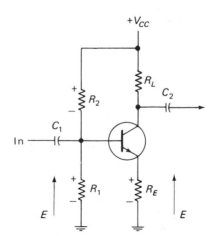

FIGURE 1-7 Typical inverse-current (emitter-feedback) bias network.

the base current. In turn, the lower base current tends to decrease the collector current and offset the initial collector-current increase.

1-7.2 Some Representative Amplifier Bias Networks

Figures 1–8 through 1–14 illustrate typical bias schemes used in solid-state amplifiers. All the bias networks use inverse-current feedback (emitter feedback) of some form.

In addition to providing the required voltage relationships, the bias-network resistors also set the *approximate* input and output impedances of the amplifier circuit, as shown by the equations on the illustrations. Likewise, the

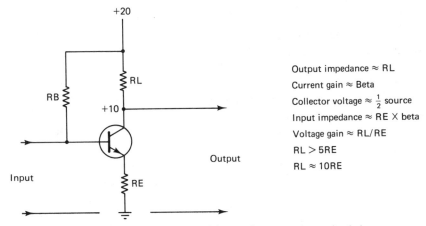

Output impedance $\approx$ RL

Current gain $\approx$ Beta

Collector voltage $\approx \frac{1}{2}$ source

Input impedance $\approx$ RE $\times$ beta

Voltage gain $\approx$ RL/RE

RL $>$ 5RE

RL $\approx$ 10RE

FIGURE 1-8   Bias network with maximum gain and minimum stability.

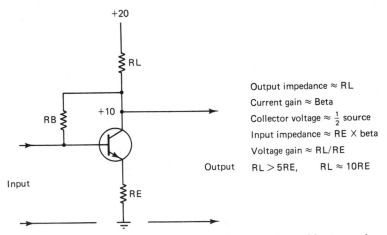

Output impedance $\approx$ RL

Current gain $\approx$ Beta

Collector voltage $\approx \frac{1}{2}$ source

Input impedance $\approx$ RE $\times$ beta

Voltage gain $\approx$ RL/RE

RL $>$ 5RE,   RL $\approx$ 10RE

FIGURE 1-9   Bias network with maximum gain and improved stability.

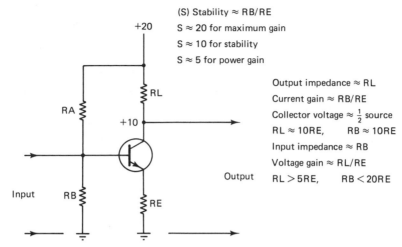

FIGURE 1-10   Bias network with improved stability (circuit characteristics depend on circuit values).

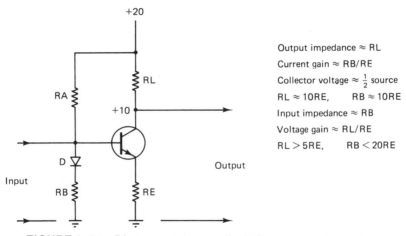

FIGURE 1-11   Bias network with diode for improved temperature stability.

approximate voltage and current gains are set by resistance ratios in many of the circuits. The equations on Figs. 1–8 through 1–14 hold true for operation up to about 100 kHz (and higher in many instances). As operating frequencies increase beyond about 100 kHz, other factors enter into gain and impedance relationships. These factors are discussed in the appropriate chapters.

To help understand the relationship between the bias networks and amplifier circuits, *design examples* are given for the various networks in the following paragraphs. Before we get into the individual bias networks, let us consider some basic bias-design problems.

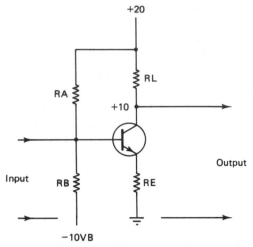

FIGURE 1-12   Bias network with positive and negative supply voltages for control of base current.

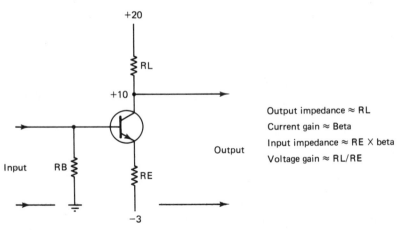

FIGURE 1-13   Bias network with positive and negative supply for emitter–collector currents.

1-7.3   Basic Bias-Design Considerations

The first step in bias design is to determine the characteristics of *both* the circuit and the transistor. For example, is the circuit to be used as an amplifier, oscillator, or switch? What class of operation (A, AB, B, or C) is required? How much gain (if any) is required? What power-supply voltages are available? What transistor is to be used? Must input and/or output impedances be set at some arbitrary value?

In all the following basic bias-circuit examples, it is assumed that some gain is required, that a 2N332 NPN transistor is used, that the power-supply

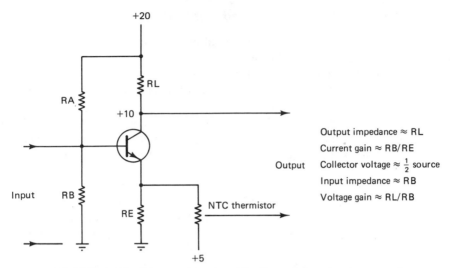

FIGURE 1-14 Bias network with thermistor for improved temperature stability.

voltage is 20 V, and that collector current must flow at all times (class A operation, Sec. 1-8). Note that the same basic bias circuits can be used (but with different values) to produce other classes of operation.

Once these preliminary factors are decided, the design information should be recorded. This requires reference to the transistor datasheet, as well as some arbitrary decisions. (The datasheet information for the 2N332 transistor is given in Chapter 3.)

Available (or Desired) Power-Supply Voltage.   An arbitrary 20 V has been selected. The voltage must be below the collector–emitter breakdown voltage (BVceo). The rated collector–emitter breakdown voltage for a 2N332 is 45 V, so the arbitrary 20 V is well within tolerance.

Maximum Collector-Current Rating.   The maximum rated collector current for a 2N332 is 25 mA (from the datasheet).

Maximum Power Dissipation.   The maximum rated power for a 2N332 is 150 mW at 25 °C and 50 mW at 125 °C. *When the exact operating temperature is not known, use from one-half to one-fourth of the 25 °C rating.* An arbitrary 50 mW is chosen for all the following bias-circuit examples.

Leakage Current (Icbo).   The maximum rated leakage current for a 2N332 is 2 μA at 25 °C and 50 μA at 150 °C. *When the exact operating temperature is not known, use 10 times the 25 °C rating.* An arbitrary 20 μA is chosen for the examples.

1–7.4 *Operating Load Current and Resistance*

Once the basic design characteristics are tabulated, the next step is to calculate the operating load current and resistance. In some amplifier circuits, arbitrary load resistance must be used (to get a given circuit output impedance, for example). In other circuits, an arbitrary load current is required (to get a given power output from the amplifier).

In the following examples, it is assumed that both load current and load resistance are chosen solely to get a given operating point (no-signal collector voltage). Usually, the no-signal collector voltage is one-half the supply voltage. In the examples, this is 10 V (20/2 = 10).

Two major factors determine operating load current for a transistor: *leakage current* and *maximum rated current*. Obviously, the load current cannot exceed the maximum rated current for the transistor (25 mA). Likewise, the load current must not be less than the leakage current (20 μA), or current cannot flow.

If the amplifier is to be operated from a battery, an operating load current near the low end should be selected to minimize battery drain. However, *the load current should not be less than 10 times the leakage current*. Since the leakage current is 20 μA, the minimum load current should be 0.2 mA.

At the high end, the maximum load current should not exceed the *maximum power-dissipation voltage*. Using the arbitrary 50 mW, divided by 10 V ($I = P/E$), the maximum load current is 5 mA.

The operating load current should be midway between these two extremes: 5 − 0.2 = 4.8, divided by 2, or 2.4 mA. The rule for selecting an operating load current midway between the two limits should be followed in all design work, unless there is specific need for a given collector voltage with a given load resistance.

Once the operating load current is established, the next step is to calculate the load resistance value. When operating at frequencies up to about 100 kHz, the load resistance value can be calculated on a d-c basis. At higher frequencies, it may be necessary to select a load resistance on the basis of impedance. This is discussed in the relevant chapters.

When the operating load current is flowing at the selected no-signal point, the collector load resistance should drop the collector voltage to one-half the supply voltage. This is typical for all class A circuits. In the example, the source of 20 V is dropped by 10 V through the load resistance. The desired 10 V divided by 2.4 mA ($R = E/I$) produces a load resistance of 4166 Ω. The nearest standard value is 4300 Ω.

1–7.5 *Determining Base Current*

If the exact base current required to produce a given collector current is known, bias-circuit design is a simple matter. Unfortunately, the exact relationship between base and collector currents (or gain) is never known. Nor can datasheets be trusted for exact gain information. Gain depends on temperature, frequency, and circuit values.

There are two basic ways to find the *approximate* base current that produces a given operating point. The first method involves a *load line* drawn on static collector characteristic curves, as shown in Fig. 1-15. As indicated, a base current of about 0.2 mA produces a collector current of 2.4 mA. This indicates a beta of about 12. Note that the load line is drawn between the source voltage (20 V) and the maximum permitted current (5 mA).

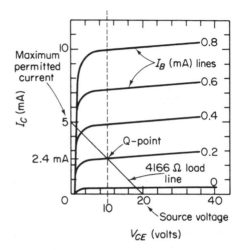

FIGURE 1-15 Load line drawn on static characteristic curves to find approximate base current and Q point.

The second method is to divide the desired collector current by the nominal beta shown on the datasheet. In this case, the nominal beta is 15. Using the 2.4-mA collector current, the base current is 2.4/15, or 0.16 mA.

No matter what method is used, keep in mind that the values are approximate. Therefore, the selected bias resistance values are just as approximate. Also, all resistors have some tolerance for their values (usually 5% or 10%).

In practice, the calculated bias resistance values can be tried with an experimental circuit. Then the resistance values are trimmed to produce the desired results. In the case of a bias circuit, "desired results" are a given collector voltage, current, and load. When the basic bias circuit is used in another circuit, "desired results" can be overall circuit functions (such as gain for an amplifier or output voltage for an oscillator). This aspect of bias adjustment is discussed in the relevant chapters.

1-7.6 Bias Network with Maximum Gain and Minimum Stability

The circuit of Fig. 1-8 offers the greatest possible gain, but the least stability, of all the bias circuits described here. The basic characteristics for this circuit are also shown on Fig. 1-8.

The value of RL is determined by the desired collector voltage and current or by an arbitrary need for a given output impedance, as discussed in Sec. 1–7.4.

The value of RE depends on a trade-off between stability and gain. An increase in the value of RE in relation to RL increases stability and decreases gain. The value of RE should be between 100 and 1000 Ω and should not be greater than one-fifth of RL. An RE that is one-tenth of RL can be considered as typical. Assuming that RL is 4300 Ω (Sec. 1–7.4), RE should be 430 Ω.

The value of RB is selected to provide a given base current at the Q point (no-signal point). For an NPN silicon transistor, the base voltage is about 0.5 V *more positive* than the emitter. (The base of a PNP silicon transistor is 0.5 V *more negative* than the emitter.)

The emitter voltage is found by noting the drop across RE ($E = IR$). Both the base current and collector current flow through RE (although the base current can usually be ignored). Therefore, using our example, the current through RE is 2.4 + 0.2, or 2.6 mA. The drop across RE is 2.6 × 430, or about 1.2 V. With RE at +1.2 V, the base voltage should be about 1.7 V (1.2 + 0.5).

Since RB is connected to the source of 20 V, the drop across RB is 20 − 1.7, or 18.3 V. With a base current of 0.2 mA through RB, a resistance of 91.5 k is required to produce a drop of 18.3 V ($R = E/I$). The nearest standard resistor is 91 k. This is close enough for a trial value of RB.

Transfer Characteristic Curves.   Some transistor datasheets show the collector current for a given base current or a given base–emitter voltage by means of transfer characteristic curves, such as shown in Fig. 1–16. The use of curves make it easier to calculate a more accurate value of RB. (The curves of Fig. 1–16 show a collector current of 2.4 mA, with a base current of 0.2 mA, and with a base–emitter differential of 0.5 V.)

In practical design, the calculated value of RB is only a trial value. In the absence of suitable curves, the value of 0.5 V for the base–emitter differential is a good guideline for all silicon transistors (with 0.2 to 0.3 V for germanium transistors).

1–7.7 Bias Network with Maximum Gain and Improved Stability

The basic characteristics for the bias circuit of Fig. 1–9 are the same as for the circuit of Fig. 1–8 (Sec. 1–7.6), except that stability is increased. The increase in stability is brought about by connecting base resistance RB to the collector rather than to the source.

In the circuit of Fig. 1–9, if the collector current increases for any reason, the drop across RL increases, lowering the voltage at the collector. This lowers the base voltage and current, thus reducing the collector current. The feedback effect is combined with that produced by the emitter resistor to offset any varia-

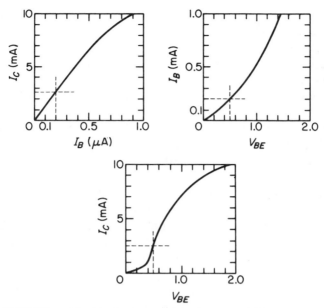

FIGURE 1-16 Typical transfer characteristic curves.

tion in collector current. However, gain for the circuit of Fig. 1-9 is only slightly less than for the Fig. 1-8 circuit.

The values of RL and RE in the Fig. 1-9 circuit are the same as for the Fig. 1-8 circuit. The value of RB is found in the same way, but is a different value. Since RB is connected to the collector (at a theoretical 10 V), the drop across RB is 10 − 1.7, or 8.3 V. With a base current of 0.2 mA through RB, a resistance of 41.5 k is required to produce a drop of 8.3 V. The nearest standard resistor is 43 k.

1-7.8 Bias Network with Improved Stability

The bias circuit shown in Fig. 1-10 offers more stability than either the Fig. 1-8 or 1-9 bias circuits, but with a trade-off of lower gain and lower input impedance. As shown by the characteristics on Fig. 1-10, the input impedance is about equal to RB (at frequencies up to about 100 kHz). Technically, the input impedance is equal to RB *in parallel* with RE × (beta + 1).

In practice, unless the beta is very low, the RE × (beta + 1) factor is much greater than RB. As a result, the value of RB (or slightly less) can be considered as the stage or circuit input impedance (and is so considered when the Fig. 1-10 bias circuit, or any of the variations, appears in examples throughout this book).

The value of RL is determined by the desired collector voltage and current or by an arbitrary need for a given output impedance, as discussed in Sec. 1-7.4.

The value of RE depends on a trade-off between stability and gain. An increase in the value of RE, in relation to RL, increases stability and decreases voltage gain. An increase in the value of RE, in relation to RB, increases stability and decreases current gain. The value of RE should be between 100 and 1000 Ω and should not be greater than one-fifth of RL. An RE that is one-tenth of RL can be considered as typical. Assuming that RL is 4300 Ω (Sec. 1–7.4), RE should be 430 Ω.

The value of RB depends on trade-offs between the value of RE, current gain, stability, and the desired input impedance. As a basic rule, RB should be about 10 times RE. A higher value of RB increases current gain and decreases stability.

The input impedance of the circuit is about equal to RB (actually slightly less). As a result, if the input impedance is of special importance in the circuit, the value of RB must be selected on that basis. This may require a different value of RE to maintain stability and current–gain relationships. Of course, any change in RE results in a change of the voltage gain (assuming that RL remains the same).

The value of RA is selected to provide a given base voltage at the Q point. Assume the same base–emitter relationships as in the previous example (base at 1.7 V, emitter at 1.2 V, 0.5-V differential). Also assume that RB is 10 times RE (430 × 10 = 4300 Ω).

Since RB is connected to the base, the drop across RB is 1.7 V. Therefore, the current through RB is 1.7/4300, or 0.4 mA (approximately). This current combines with the base current (0.2 mA) to produce a total current of 0.6 mA through RA.

Since RA is connected to the source of 20 V, the drop across RA must be 20 − 1.7, or 18.3 V. With a combined current of 0.6 mA through RA, a resistance of 30.5 k is required to produce a drop of 18.3 V. The nearest standard resistor is 30 k.

In addition to stability, the major advantage of the Fig. 1–10 bias circuit is that the input and output impedances, as well as voltage and current gain, do not depend on transistor beta. Instead, *circuit characteristics depend on circuit values.*

1–7.9 Bias Network with Diode for Improved Temperature Stability

The basic characteristics for the bias circuit of Fig. 1–11 are the same as for the Fig. 1–10 circuit, except that temperature stability is increased. The increased stability is produced by diode D connected between the base and RB. The diode (forward biased) is of the same material (silicon) as the base–emitter junction and is maintained at the same temperature. As a result, the voltage drops across diode D and the base–emitter junction are the same and remain the same with changes in temperature.

The values for RL, RE, and RB in the Fig. 1–11 circuit are the same as for the Fig. 1–10 circuit. The value of RA is slightly different. Since the drop across diode D is the same as across the base–emitter junction (0.5 V), the drop across RB is the same as across RE (1.2 V). Therefore, the current through RA is 1.2/4300, or 0.28 mA. This current combines with the base current (0.2 mA) to produce a total current of 0.48 mA through RA.

Since RA is connected to the source of 20 V, the drop across RA must be 20 − 1.7, or 18.3 V. With a combined current of 0.48 mA through RA, a resistance of 38.125 k is required to produce a drop of 18.3 V. The nearest standard resistor is 39 k.

As in the case of the Fig. 1–10 bias circuit, the characteristics of the Fig. 1–11 circuit depend on transistor beta. In practice, diode D is mounted near the transistor so that the base–emitter junction and diode D are at the same temperature.

1-7.10 Bias Network with Positive and Negative Supply Voltages for Control of Base Current

The basic characteristics for the bias circuit of Fig. 1–12 are the same as for the Fig. 1–10 circuit. However, the circuit of Fig. 1–12 is used in those special applications that require a negative and positive voltage, each with respect to ground, to control base current.

The values for RL, RE, and RB in the Fig. 1–12 circuit are the same as for the circuit of Fig. 1–10. The value of RA is different because of the large amount of current through RB. Since RB is connected to a negative source VB, the drop across RB is VB, plus the drop across RE (1.2 V), plus the base–emitter drop (0.5 V). Assume that VB is − 10 V. Then the drop across RB is 10 + 1.2 + 0.5, or 11.7 V. Therefore, the current through RB is 11.7/4300, or 2.7 mA. This current combines with the base current (0.2 mA) to produce a total current of 2.9 mA through RA.

Since RA is connected to the source of 20 V, the drop across RA must be 20 − 1.7, or 18.3 V. With a combined current of 2.9 mA through RA, a resistance of 6310 Ω is required to produce a drop of 18.3 V. The nearest standard resistor is 6200 Ω.

1-7.11 Bias Network with Positive and Negative Supply for Emitter-Collector Currents

The circuit of Fig. 1–13 is used in those special applications where it is necessary to supply collector–emitter current from both a positive and negative source. Since the transistor is NPN, the collector is connected to the positive source through RL, while the emitter is connected to the negative source through RE.

If both sources are about equal, it is difficult to design a circuit that produces any voltage gain (unless emitter-bypass techniques are used, as described in Chapter 3). The collector and emitter currents are about equal (ignoring base current). As a result, if RL drops the positive source to half (say from 20 to 10 V), then RE must drop the entire negative source (from 20 to 10 V), and RE is about twice the resistance of RL. This produces a voltage loss, all other factors being equal.

The basic characteristics for the Fig. 1–13 circuit are essentially the same as for Fig. 1–8, except that voltage gain is low because of the lower RL/RE ratio.

In the following example, it is assumed that the conditions are the same as for the previous bias-network examples, except that RE is returned to a − 3 − V source, rather than ground. Base current is 0.2 mA, collector–emitter current is 2.4 mA, positive source is 20 V, RL drops the positive 20 V to about 10 V, RL is 4300 Ω, and the base and emitter are to be maintained at 1.7 and 1.2 V, respectively, at the Q point.

The voltage drop across RE must equal the desired + 1.2 V, plus the − 3 V from the negative source, or 4.2 V total. Both collector and base current flow through RE, resulting in a total current of 2.6 mA, as in previous examples. A resistance of 1615 Ω is required for RE to drop 4.2 V with 2.6 mA flowing.

The voltage drop across RB must equal the desired + 1.7 V. With a base current of 0.2 mA, a resistance of 8500 Ω is required for RB.

1–7.12   Bias Network with Thermistor for Improved Temperature Stability

The basic characteristics for the bias circuit of Fig. 1–14 are the same as for the Fig. 1–10 circuit, except that the NTC (negative temperature coefficient) characteristics of a *thermistor* provide temperature compensation. The resistance of a thermistor decreases with increases in temperature, and vice versa. For best results, the thermistor is mounted near the transistor so that both devices are at the same temperature.

In the circuit of Fig. 1–14, the thermistor varies the emitter voltage when temperature variations occur. This minimizes the effects of these variations on the emitter current. Resistors RA and RB form a voltage divider to apply a portion of the collector supply voltage in a direction to forward bias the emitter-base junction. Resistor RE and the thermistor form a second voltage divider across the emitter supply voltage.

The direction of the voltage drop across RE places a reverse bias on the base–emitter junction. However, since the forward bias applied to the base is larger than the reverse bias applied to the emitter, the net result is that the emitter–base junction is forward biased.

Should the temperature rise, the emitter and collector currents tend to rise. The same rise in temperature reduces the resistance of the thermistor. This permits more current to flow through the voltage divider. The increase in current flow increases the voltage drop across RE, thus increasing the reverse bias being applied to the emitter–base junction. As a result, the net forward bias of the junction is reduced, thus reducing the emitter and collector currents toward their normal values. Should the temperature decrease, the action is reversed, preventing a decrease in emitter and collector currents.

1-8 AMPLIFIER CLASSIFICATIONS BASED ON OPERATING POINT

As discussed, amplifiers are often classified as to *operating point*, or the amount of current flow under no-signal conditions. The following is a brief summary of the four basic operating-point classifications.

Note that in all four classifications the base–collector junction is always reverse biased at the operating point, as well as under all signal conditions. No base–collector current flows (with the possible exception of reverse leakage current, Icbo). On the other hand, the base–emitter junction is biased so that base–emitter current flows under certain conditions, and possibly under all conditions. When base–emitter current flows, emitter–collector current also flows.

1-8.1 Class A Amplifier

As shown in Fig. 1–17, a class A amplifier operates only over the *linear portion* of the transistor characteristic curve. (The curve represents the relationship between base voltage, or input, and collector current, or output.) At no point of the input signal cycle does the base become so positive or negative that the transistor operates on the nonlinear portion of the curve. The transistor collector current is never cut off nor does the transistor ever reach saturation.

The main advantage of the class A amplifier is the relative lack of distortion. The output waveform follows that of the input waveform, except in amplified form. However, with any class of amplifier there is some distortion, as is discussed throughout this book.

The main disadvantages of class A amplifiers are relative inefficiency (lower power output for a high power input dissipated by the transistor) and the inability to handle large signals. Rarely is a class A amplifier over about 35% efficient. If the power input to a class A amplifier is 1 W (generally, the maximum power dissipation capability of a single transistor), the output is less than 0.3 W.

The peak-to-peak output signal voltage swing of a class A amplifier is limited to something less than the total supply voltage. Since the output voltage must swing both positive and negative, the peak output is less than one-half the supply voltage. For example, assume that the supply is 20 V and the amplifier is biased so that the Q-point collector voltage is one-half the supply, or 10 V. (Such a Q point is generally typical for a class A amplifier.) Under these conditions, the output voltage cannot exceed ±10 V. If distortion is to be at a minimum (the usual reason for class A amplifiers), the output is usually about ±5 V (with a 20-V supply). This keeps the transistor on the linear portion of

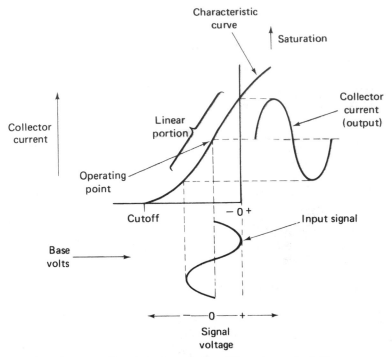

FIGURE 1-17   Typical class A amplifier characteristic curve.

the curve. (Typically, the curve becomes nonlinear near the cutoff and saturation points. However, this can be determined only from an actual test of the amplifier circuit, as described in the appropriate chapters.)

The input voltage swing of a class A amplifier is limited by the output voltage swing capability and the voltage amplification factor. For example, if the output is limited to ±10 V and the voltage amplification factor is 100, the input is limited to ±0.1 V (100 mV).

Because of these limitations, class A amplifiers are generally used as *voltage amplifiers,* rather than power amplifiers. Typically, a class A amplifier stage is used ahead of a power amplifier stage.

1-8.2   Class B Amplifier

As shown in Fig. 1-18, a class B amplifier operates only on one-half of the input signal. Class B operation is produced when the base–emitter bias is set so that the operating point coincides with the transistor cutoff point. For an NPN transistor, this means making the base more negative than for class A operation. (For PNP transistors, class B is produced when the base is more positive than for class A.) Either way, the base–emitter reverse bias is increased for class B operation.

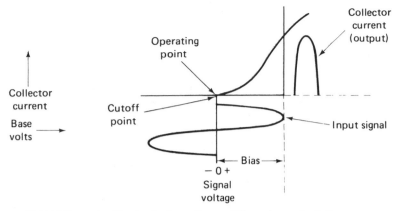

FIGURE 1-18   Typical class B amplifier characteristic curve.

As shown in Fig. 1–18, when the input signal voltage is zero, there is no flow of collector current. During the positive half-cycle of the signal voltage (Fig. 1–18 is for an NPN transistor), the collector current rises to the peak and then falls back to zero in step with the variations of that half-cycle. During the negative half-cycle of the signal voltage, there is no collector current since the base–emitter reverse bias is always greater than the transistor cutoff voltage. Collector current flows only during half the input signal cycle.

There is considerable distortion if a single transistor is operated as class B. This is because the waveform of the resulting collector current resembles that of the positive half-cycle of the input signal and, consequently, does not resemble the complete waveform of the input. Class B is generally used when two transistors are connected in *push–pull.* This makes it possible to reconstruct an output waveform that resembles the full waveform of the input.

The peak output voltage swing of a class B amplifier is slightly less than the supply voltage. Since the output appears only on half-cycles, it is possible to operate class B amplifiers at a higher current (or power) rating than class A, all other factors being equal. For example, if a transistor is capable of 0.3-W dissipation (without damage) as class A, the same transistor can be operated at 0.6 W, class B, since the transistor is conducting collector current only half the time. (This is a theoretical example. In practice, there are factors that limit class B power dissipation to something less than twice that of class A.)

Also note that the peak output of a class B amplifier is equivalent to the peak-to-peak output of a class A amplifier. So, if two transistors are connected in push–pull and operated as class B, the output voltage can be twice that of class A.

Because of these voltage and power factors, class B amplifiers are generally used as *power amplifiers,* rather than voltage amplifiers. In a typical audio amplifier using discrete components (they still exist!), two push–pull transistors are operated in class B, preceded by a single class A amplifier stage. The class

A stage provides voltage amplification, whereas the class B stage produces the necessary power amplification.

1-8.3   Class AB Amplifier

Class B is the most efficient operating mode for audio amplifiers since the least amount of current is drawn. That is, the transistors are cut off at the Q point and draw collector current only in the presence of an input signal. However, true class B operation often results in *crossover distortion.*

The effects of crossover distortion can be seen by comparing the input and output waveforms of Fig. 1–19. In true class B operation, the transistor remains cut off at very low signal inputs (because transistors have a low current gain at cutoff) and turns on abruptly with a large signal. As shown in Fig. 1–19b, there is no current flow when the base–emitter voltage is below about 0.65 V (for a silicon transistor). During the instantaneous pause when one transistor stops conducting and the other starts conducting, the output waveform is distorted.

Distortion of the signal is not the only bad effect of this crossover distortion condition. The instantaneous cutoff of collector current can set up large voltage transients equal to several times the size of the supply voltage. This can cause the transistor to break down.

Minimizing Crossover Distortion.   Although crossover distortion cannot be completely eliminated, it is possible to minimize the effects by operating the output stage as class AB (or somewhere between B and AB). That is, the transistors are forward biased just enough for a small amount of collector current to flow at the Q point. Some collector current flows at the lowest signal levels, and there is no abrupt change in current gain.

The effects of class AB operation are shown in Fig. 1–19b. The combined collector currents produce a *composite curve* that is essentially linear at the crossover point, resulting in a faithful reproduction of the input (at least as far as the crossover point is concerned). Of course, class AB is less efficient than class B, since more current must be used.

Some designers of hi-fi amplifiers use an alternative method to minimize crossover distortion. This technique involves putting diodes in series with the collector or emitter leads of the push–pull transistors. Because the voltage must reach a certain value (typically 0.65 V for silicon diodes) before the diode conducts, the collector current curve is rounded (not sharp) at the crossover point.

1-8.4   Class C Amplifier

Figure 1–20 shows the characteristic curve of a typical class C amplifier. Note that the transistor is reverse biased considerably below the cutoff point.

As shown in Fig. 1–20, during the positive half-cycle of the input signal, the signal voltage starts from zero, rises to the positive peak value, and falls

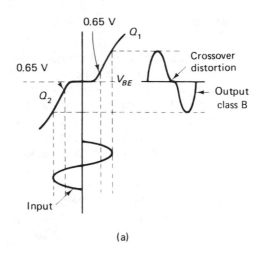

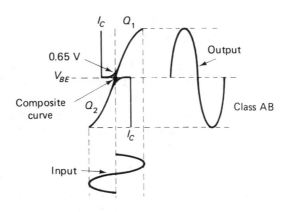

FIGURE 1-19 Effects of crossover distortion.

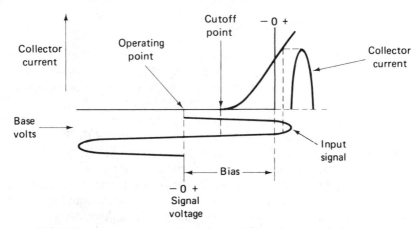

FIGURE 1-20 Typical class C amplifier characteristic curve.

back to zero. (Figure 1–20 is for an NPN transistor.) Note that a *portion of the input signal* causes the base–emitter junction to be forward biased. There is a flow of collector current for a portion of half the input cycle. The negative half-cycle of the input signal lies well below the cutoff point of the transistor.

Collector current flows only during that portion of the positive half-cycle of the input signal between the cutoff point and the peak. The resulting collector current is a pulse, the duration of which is considerably less than a half-cycle of the input signal.

Obviously, the waveform of the output signal cannot resemble that of the input signal. Nor can this resemblance be restored by the push–pull method (as with class B or AB). Class C is limited to those applications where distortion is of no concern. Generally, class C is limited to use in RF amplifiers (Chapter 4).

1–9 AMPLIFIER DISTORTION BASICS

For the purposes of this book, distortion is defined as that condition when the output signal of an amplifier is not identical to the waveform of the input signal. A small amount of distortion is generally present in all amplifiers. However, amplifiers are usually designed to keep such distortion within acceptable limits. In some special cases, amplifiers contain circuits that introduce a form of distortion. This is generally to offset or compensate for distortion already present in the signal.

There are many specific types and causes of distortion in amplifiers. These include *crossover distortion* (Sec. 1–8), *intermodulation distortion,* and *harmonic distortion*. (Both harmonic and intermodulation distortion are discussed in Chapter 9.) However, three basic types of distortion are found in amplifiers: *amplitude distortion, frequency distortion,* and *phase distortion*. Any of these, either separately or in combination, may be present in an amplifier of any type.

1–9.1 Amplitude Distortion Basics

Amplitude distortion occurs in the transistor and is the result of operating the transistor over the nonlinear portion of the characteristic curve. The usual remedy is to use a base–emitter bias that places the operating point well within the linear portion of the curve, preferably at the center of the linear portion. Also, the amplitude of the input signal must be small enough so that the positive and negative half-cycles do not drive the transistor beyond the linear portion. Generally, a low input signal and proper operating point (at the center of the linear portion) mean that gain must be sacrificed.

To sum up, an overdriven amplifier (used to get maximum gain) almost always results in some amplitude distortion. A low-distortion amplifier generally requires two (or more) stages to get the same gain as an overdriven amplifier.

1-9.2 Frequency Distortion

Frequency distortion occurs because the input signal rarely, if ever, is at a single frequency. Instead, the input signal usually contains components of several frequencies, making the signal waveform somewhat complex.

In addition to a transistor, a solid-state amplifier circuit is composed of resistors, capacitors (unless the amplifier is direct coupled, Chapter 6), and possibly inductances (coils and transformers). Capacitors and inductances have reactance. Since reactance is a function of frequency, the different frequencies of the signal encounter different reactances. The high and low frequencies of the signal can be impeded in *different degrees or amounts*. This produces distortion of the signal waveform from the original.

For example, assume that the input signal is a complex waveform composed of three frequencies: 10, 100, and 1000 Hz, all of the same amplitude. The reactance of coupling capacitors between stages is different for each of the three frequencies. Capacitive reactance increases with a decrease in frequency. The 10-Hz signal is attenuated more than the 100-Hz signal and much more than the 1000-Hz signal. Even though the transistor amplifies all three frequencies equally, the signals are no longer of equal amplitude, and the output waveform is different (distorted) from the input.

To minimize the effect of frequency distortion, amplifiers are usually designed to eliminate unwanted capacitance and inductance. Likewise, compensating components may be introduced into the circuit. Such procedures are discussed in relevant chapters.

1-9.3 Phase Distortion Basics

The fact that the input signal contains components of different frequencies is also responsible for phase distortion in an amplifier. When a signal flows through a capacitor or an inductor, the signal encounters a shift of phase. The degree of this phase shift is a function of frequency. The high- and low-frequency components of the signal are phase shifted by different amounts. These different phase shifts cause a distortion of the signal waveform.

As in the case of frequency distortion, the phase distortion in an amplifier may be minimized by proper design to eliminate unwanted capacitance and inductance. Such procedures are discussed in relevant chapters.

1-10 DECIBEL MEASUREMENT BASICS

The decibel, or dB, is widely used in amplifier work to express logarithmically the *ratio* between two power or voltage levels (and less commonly the ratio between two current levels). A decibel is one-tenth of a bel. (The bel is too large for most practical applications.)

Although there are many ways to express a ratio, the decibel is used in amplifiers for two reasons: (1) the decibel is a convenient unit to use for *all types* of amplifiers, and (2) the decibel is related to the *reaction of the human ear* and is thus well suited for use with *audio amplifiers*. The human ear does not hear sounds in direct power ratio. Humans can listen to ordinary conversation quite comfortably, and yet be able to hear thunder (which is taken to be 100,000 times louder than conversation) without damage to the ear. This is because the response of the human ear to sound waves is approximately *proportional to the logarithm of the sound-wave energy* and is not proportional to the energy.

The common logarithm ($\log_{10}$) of a number is the number of times 10 must be multiplied by itself to equal that number. For example, the logarithm of 100 (that is 10×10, or 10^2) is 2. Likewise, the logarithm of 100,000 (10^5) is 5. This relationship is written

$$\log_{10} 100,000 = 5$$

In comparing two powers, it is possible to use the bel (which is the logarithm of the ratio of the two powers). For example, in comparing the power of ordinary conversation with that of thunder, the increase in sound is equal to

$$\log_{10} \frac{\text{power of thunder}}{\text{power of conversation}} \quad \text{or} \quad \log_{10} \frac{100,000}{1}$$

Using the more convenient decibel, the *increase in sound* from ordinary conversation to thunder is equal to

$$10 \log_{10} \frac{100,000}{1} \quad \text{or} \quad 50 \text{ decibels (or 50 dB)}$$

For convenience, the same method is used in measuring the increase in amplifier power, whether the amplifiers are used with audio frequencies or not. The increase in power of any amplifier can be expressed as

$$\text{gain in dB} = 10 \log_{10} \frac{\text{power output}}{\text{power input}}$$

This relationship can also be expressed as

$$\text{gain in dB} = 10 \log_{10} \frac{P2}{P1}$$

Usually, $P2$ represents power output and $P1$ represents power input. If $P2$ is greater than $P1$, there is a power gain, expressed in positive decibels ($+$ dB). With $P1$ greater than $P2$, there is a power loss, expressed in negative decibels

(− dB). Whichever is the case, the ratio of the two powers (*P*1 and *P*2) is taken, and the *logarithm of this ratio* is multiplied by 10. As a result,

$$\text{power ratio of } 10 = 10\text{–dB gain}$$

$$\text{power ratio of } 100 = 20\text{–dB gain}$$

$$\text{power ratio of } 1000 = 30\text{–dB gain}$$

and so on.

1-10.1 Doubling Power Ratios

Doubling the power of an amplifier produces a power gain of + 3 dB. For example, if the volume control of an amplifier is turned up so that the power rises from 4 to 8 W, the gain is up + 3 dB. If the power output is reduced from 4 to 2 W, the gain is down − 3 dB.

If the original 4 W is increased to 8 W, the power gain is + 3 dB. Increasing the power output further to 16 dB produces another gain of + 3 dB, with a total power gain of + 6 dB. At 40 W, the power is increased 10 times (from the original 4 W), and the total power gain is + 10 dB, and so on.

1-10.2 Adding Decibels

There is another convenience in using decibels for amplifier work. When several amplifier stages are connected so that one works into another (stages connected in *cascade*), the gains in each stage are multiplied. For example, if three stages, each with a gain of 10, are connected, there is a total power gain of 10 × 10 × 10, or 1000.

In the decibel system, the decibel gains are added. Using our example, the decibel power gain is 10 + 10 + 10, or + 30 dB. Similarly, if two amplifiers or stages are connected, one of which has a gain of + 30 dB and the other a loss of − 10 dB, the net result is + 30 − 10, or + 20 dB gain.

1-10.3 Using Decibels to Compare Voltages
and Currents

The decibel system is also used to compare the voltage input and output of an amplifier. (Decibels can be used to express current ratios. However, this is generally not practical in amplifiers.) When voltages (or currents) are involved, the decibel is a function of

$$20 \log \frac{\text{output voltage}}{\text{input voltage}} , \quad 20 \log \frac{\text{output current}}{\text{input current}}$$

The ratio of the two voltages (or currents) is taken, and the logarithm of this ratio is multiplied by 20.

It is important to note that, although power ratios are independent of source and load impedance values, *voltage and current ratios in these equations hold true only when the source and load impedances are equal.*

In circuits where input and output impedances differ, voltage and current ratios are calculated as follows:

$$20 \log \frac{E1 \sqrt{R2}}{E2 \sqrt{R1}}, \qquad 20 \log \frac{I1 \sqrt{R1}}{I2 \sqrt{R2}}$$

where $R1$ is the source impedance and $R2$ the load impedance. ($E1 \sqrt{R2}$ and $I1 \sqrt{R1}$ are always higher in value than $E2 \sqrt{R1}$ and $I2 \sqrt{R2}$.)

As is true for the power relationship, if the voltage output is greater than the input, there is a decibel gain (+ dB). If the output is less than the input, there is a voltage loss (− dB).

Note that doubling the voltage produces a gain of + 6 dB. Conversely, if the voltage is cut in half, there is a loss of − 6 dB. To get the net effect of several voltage amplifiers working together, add the decibel gains (or losses) of each.

1-10.4 Decibels and Reference Levels

When an amplifier has a power gain of + 20 dB, this has no meaning in actual power output. Instead, it means that the power output is 100 times as great as the power input. For this reason, decibels are often used with specific reference levels.

The most common reference levels in use are the *volume unit,* or VU, and the *decibel meter,* or dBm.

When the volume unit is used, it is assumed that the zero level is equal to 0.001 W (1 mW) across a 600-Ω impedance. Therefore,

$$VU = 10 \log \frac{P2}{0.001} = 10 \log \frac{P2}{10^{-3}} = 10 \log 10^3 P2$$

where $P2$ is the output power.

Both the dBm and VU have the same zero level base. A dBm scale is (generally) used when the signal is a sine wave (normally 1 kHz), whereas the VU is used for complex audio waveforms.

1-11 FIELD-EFFECT TRANSISTOR AMPLIFIERS

Before going into specific considerations for field-effect transistor (FET) amplifier circuits, we review basic FET operating characteristics. This is necessary since FET characteristics are quite different when compared to the two-junction or bipolar transistors discussed this far.

As an example, a FET is often biased at the *zero-temperature-coefficient* point when used as an amplifier. This is an operating point where the FET drain–source current does not vary with temperature. Furthermore, the characteristics shown on FET datasheets do not correspond to those of two-junction transistors. It is necessary to analyze these datasheet characteristics as they apply to amplifier circuits.

1-11.1 Advantages and Disadvantages of FETs as Amplifiers

The FET has several advantages over a two-junction transistor in amplifier applications. The FET is relatively free of noise and is more resistant to the degrading effects of nuclear radiation. The FET is also inherently more resistant to burnout than the two-junction transistor. There are additional advantages for certain amplifier circuits.

As an example, *high input impedance* (typically several megohms) is very useful in impedance transformations and where the amplifier must be matched to a high-input-impedance signal source. Since the FET is a *voltage-controlled device,* in contrast to current-controlled two-junction transistors, the FET can readily be *self-biased.* This frequently makes for a more simple circuit than is possible with a bipolar transistor. The FET also has a *nonlinear region of operation,* but this is generally of small value for amplifiers, except where automatic gain control is used.

The junction field-effect transistor (or JFET, Sec. 1–11.2) has a very high output resistance, making the JFET useful as a constant-current source. Figure 1–21 shows a comparison of FET and two-junction transistors.

When compared with two-junction transistors, the major shortcoming of the FET is a relatively small gain–bandwidth product. Although the JFET is

Characteristic	JFET	MOSFET	Bipolar (Two-junction)
Input impedance	High	Very high	Low
Noise	Low	Unpredictable	Low
Aging	Not noticeable	Noticeable	Not noticeable
Bias voltage temperature coefficient	Low, predictable	High, not predictable	Low, predictable
Typical gate current	0.1 nA	10 pA	—
Gate-current change with temperature	Medium, predictable	Low, predictable	—
Sensitivity to overload	Good	Poor	Good

FIGURE 1-21 Comparison of FET and two-junction transistors.

free from carrier-transit-time limitations (which set high-frequency limitations for two-junction transistors), parasitic capacitances limit the FET at higher frequencies. This is discussed further in Chapter 4.

1-11.2 FET Operating Modes in Amplifier Applications

Two types of FETs are commonly used in amplifiers, junction (JFET) and metal oxide silicon (MOSFET). As the names imply, JFET uses the characteristics of a reverse-biased *junction* to control the drain–source current, whereas with a MOSFET the gate is a metal deposit on an oxide layer. The gate is insulated from the source and drain. Because of this *insulated gate,* the MOSFET is sometimes referred to as an insulated-gate FET, or IGFET. The terms MOSFET and IGFET are used interchangeably (although MOSFET is preferred by the author).

Both JFETs and MOSFETs operate on the principle of a *channel current* controlled by an electric field. The control mechanisms for the two are different, resulting in considerably different amplification characteristics. The main difference in the two is in the *gate characteristics.* The input to a JFET amplifier acts like a reverse-biased diode, whereas the input of a MOSFET amplifier is similar to a small capacitor.

In addition to the two basic types, there are two fundamental modes of operation for FET amplifiers: *depletion* and *enhancement.* These modes are illustrated in Figs. 1–22 and 1–23, which show the transfer characteristics and basic test circuits for each mode, respectively.

In the depletion mode, maximum drain current (IDSS) flows when the gate–source voltage (VGS) is zero and decreases for increasing VGS.

Enhancement mode is just the opposite in that minimum drain current flows at VGS = 0. With enhancement mode, the drain current increases with increasing VGS.

FETs designated as type A operate in the depletion mode only. Type B FETs operate in both depletion and enhancement modes. Type C FETs operate in the enhancement mode only.

The test circuits of Fig. 1–23 show the biasing for N-channel FETs. Note that VDS is always positive for the three N-channel types. In the useful range of amplifier operation, VGS is negative for a type A FET, positive for type C, and either polarity for type B. For a P-channel FET, all polarities must be reversed.

1-11.3 Basic FET Amplifier Operating Regions

The FET has three distinct characteristic regions, only two of which are operational for amplifiers. Figure 1–24a, the output transfer characteristics, illustrates the different regions. Below the pinchoff-voltage VP, the FET operates in the

VA = Avalanche voltage
VP = Pinchoff voltage
VDS = Drain–source voltage
VP < VDS < VA

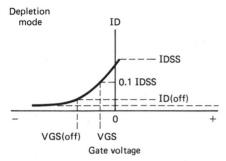

Characteristic	
IDSS @ VGS = 0, VGS(off) @ 0.001 IDSS VGS @ ID = 0.1 IDSS	Type A

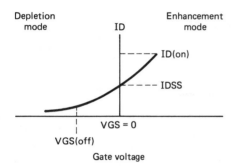

Characteristic	
ID(on) @ VGS > 0 IDSS @ VGS = 0 VGS(off) @ ID = 0.001 IDSS	Type B

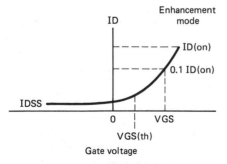

Characteristic	
ID(on) @ VGS > 0 VGS @ 0.1 ID(on) VGS(on) @ ID = 0.001 ID(on) or less IDSS @ VGS = 0	Type C

FIGURE 1-22 FET transfer characteristics.

Test circuit for IDSS

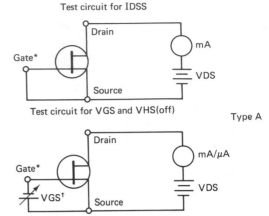

Description

Zero-gate-voltage drain current.
Represents maximum drain current.

Gate voltage necessary to reduce ID to
some specified negligible value at the
recommended VDS (that is, cutoff).

Gate voltage for a specified value of
ID between IDSS and IDS at cutoff
(normally 0.1 IDSS).

Test circuit for VGS and VHS(off)

Type A

*Gates internally connected.
†Adjust for desired ID.

Description

An arbitrary current value (usually
near max rated current) that locates
a point in the enhancement mode.

Zero-gate-voltage drain current.

Voltage necessary to reduce ID to
some specified negligible value at
the recommended VDS (that is,
cutoff).

Type B

*Gates internally connected.
†Adjust for desired ID, normally near
maximum-rated ID.

Test circuits for IDSS and VGS(off)
same as for type A.

Description

An arbitrary current value (usually
near max rated current) that locates
a point in the enhancement mode.

Gate–source voltage for a specified
drain current of 0.1.

Gate cutoff or turn-on voltage.
Leakage drain current.

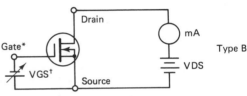

ID(on) test circuit same as for type B.
VGS test circuit same as for ID(on).

Type C

VGS(th) test circuit same as VGS(off)
in type A, except reverse VGS battery
polarity.
IDSS test circuit same as for type A.

FIGURE 1-23 FET basic test circuits.

ohmic or *resistance* region. (The ohmic region is not generally used for amplifiers, except in special cases.) Above VP, up to the drain-source breakdown voltage V(BR)DSS, the FET operates in the *constant-current region,* which is the region most used for amplifier circuits. The third region, above V(BR)DSS, is the *avalanche region,* where the FET is not operated in practical amplifier circuits.

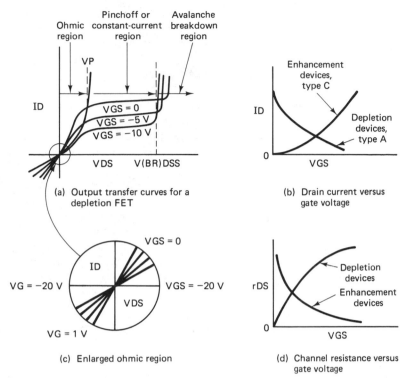

(a) Output transfer curves for a
    depletion FET

(b) Drain current versus
    gate voltage

(c) Enlarged ohmic region

(d) Channel resistance versus
    gate voltage

FIGURE 1-24 FET operating regions.

The drain–source resistance rDS at any point on these curves is given by the slope of the curve at that point. Above VP, changes in VDS produce small changes in drain current ID. The results in a very high rDS and is characteristic of a constant-current source. Also, the actual operating ID is variable and depends on the gate–source voltage. This produces a voltage-controlled current source.

The ID–VGS curve, shown in Fig. 1–24b (and often found on a typical FET datasheet) illustrates how the drain current varies with changes in VGS. For depletion FETs, ID decreases as VGS is increased. For enhancement FETs, ID is enhanced or increased as VGS is increased.

If the FET is operated with a VDS below VP, the curve slope varies considerably as VGS is varied. This is shown in Fig. 1–24c. Since the slope varies, rDS varies. This is considered as operation in the ohmic region. In effect, the drain–source channel is a voltage-variable or voltage-controlled resistor. As shown in Fig. 1–24d, rDS decreases with increasing VGS for enhancement FETs, and vice versa for depletion FETs.

Note that the curves near the origin (Fig. 1–24c) are relatively symmetrical. This means that a-c as well as d-c signals can be handled. In other words, the drain–source channel is bilateral, not unilateral.

Note that a FET is generally operated in the pinchoff region for linear devices (which includes most amplifiers), whereas the ohmic region is used primarily for voltage-variable applications.

1–11.4 *Zero-Temperature-Coefficient Point*

An important characteristic of all FETs is the ability to operate at a zero-temperature-coefficient (0TC) point. This means that if the gate–source is biased at a specific voltage and is held constant the drain current does not vary with changes in temperature. This characteristic makes for very stable amplifier circuits.

The ID–VGS curves of Fig. 1–25 show that the various curves at different temperatures intersect at a common point. If the FET is operated at this value of ID and VGS (shown as IDZ and VGSZ), zero-temperature-coefficient, or 0TC, operation results.

The 0TC point varies from one FET to another and depends on IDSS, the zero-gate-voltage drain current, and VP. The equations shown in Fig. 1–25 provide good approximations of the 0TC point. For example, if VP is 1 V, the 0TC mode is obtained if VGS is 0.37 V (1 − 0.63 = 0.37).

Typically, JFETs show the 0TC characteristic over a wide range of temperatures, approximately 150 °C. MOSFETs are limited to a much narrower range (about 50 °C).

It is sometimes assumed that the forward transadmittance (Yfs or Y21) of the FET does not vary with temperature, particularly if the FET is biased at the 0TC point. (Note that y parameters are discussed in Chapters 3 and 4.) In practical amplifier applications, the transadmittance of a FET is the slope of the ID–VGS curve. The curve of Fig. 1–25 shows that the slope varies with temperature at every point on the curve.

Figure 1–26 shows the temperature coefficients for a typical JFET.

Remember that *it is not always practical to operate a FET at the zero-temperature-coefficient point*. For example, assume that the required VGS to

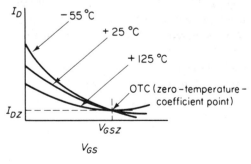

$$I_{DZ} \approx I_{DSS}\left(\frac{0.63}{V_P}\right) \approx \frac{0.4\,I_{DSS}}{V_P^2}$$

$$V_{GSZ} \approx V_P - 0.63$$

FIGURE 1-25 FET zero-temperature-coefficient (0TC) point.

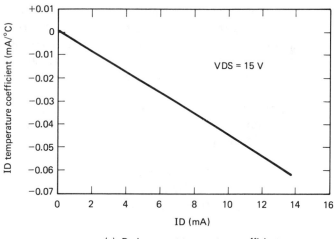

(a) Drain-current temperature coefficient
versus drain current

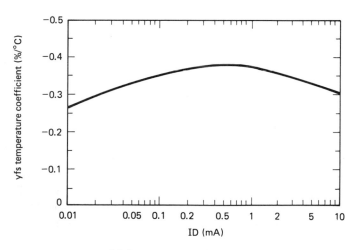

(b) Forward-transadmittance temperature
coefficient versus drain current

FIGURE 1-26   FET temperature-coefficient curves.

produce 0TC is 0.37 V, and the FET is to operate as an amplifier with 0.5-V input signals. A part of the input will be clipped. Or assume that the circuit is to be self-biased with a source resistor (Sec. 1–11.5). An increase in bias resistance to produce 0TC could reduce gain.

Practical Methods for Finding 0TC of FETs.   The values of ID and VGS that produce 0TC can be found using datasheet curves or by means of equations, as shown in Fig. 1–25. However, these values are typical approximations.

A more practical method for determining IDZ requires a soldering tool, coolant (a can of Freon), a transistor curve tracer, and a 1-k resistor (across the base and emitter of the curve-tracer test socket). The 1-k resistor converts the constant-current base drive to a relatively constant voltage drive for the FET gate. Then adjust the curve tracer to display the ID–VGS output family of curves (Fig. 1–24a).

Alternately, bring the soldering tool near the FET and then spray the FET with Freon. Note the voltage step on the curve tracer of VGS that *remains motionless* in the presence of temperature changes. The ID at this voltage step is IDZ.

Typically, FETs with an IDSS of about 10 to 20 mA have an IDZ of about 0.5 mA. Usually, IDZ increases as IDSS increases (but not always, and not in proportion). For example, the IDZ of 300-mA FETs is often on the order of 1 or 2 mA.

1–11.5 Bias Methods for FETs

In linear amplifier applications, the FET is biased by an external supply, by self-bias, or by a combination of both techniques. This applies to all FETs, whether biased at 0TC or at some other operating point.

Figure 1–27 shows the familiar common-source drain characteristic curves of a JFET (as the curves might appear on a typical curve tracer). For a constant level of VDS, ID can be plotted versus VGS as shown in Fig. 1–27b. This curve is generally referred to as a *transfer characteristic*.

From a practical standpoint, the curves of Fig. 1–27 show the amount of current that flows through the FET for a given VGS. For example, either curve shows that if a $-1-V$ bias is applied between gate and source, approximately 10 mA flows. If VDS is 10 V, and a 500-Ω resistor is connected between drain and supply, there is a 5-V drop across the resistor. Of course, this reduces the VDS down to 5 V and possibly changes the characteristics. (In the example of Fig. 1–27a, there is very little change in characteristics.)

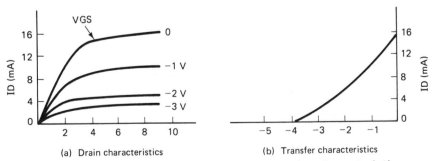

(a) Drain characteristics   (b) Transfer characteristics

FIGURE 1-27 FET common-source drain characteristic curves.

The following paragraphs show how similar curves can be used to find the correct value of bias for a given no-signal FET operating point and provide basic or theoretical methods for finding bias values. In Chapter 3, we discuss step-by-step procedures for finding FET bias values, using actual datasheet information.

External Bias for FETs. Figure 1–28a shows a FET biased by an external voltage source. The input portion of this circuit is redrawn in Fig. 1–28b so that a graphical analysis may be used to determine the no-signal drain current. The graphical analysis consists of plotting the *V–I* (voltage–current) characteristics looking into the source terminal and *V–I* characteristics looking into the supply–voltage terminal.

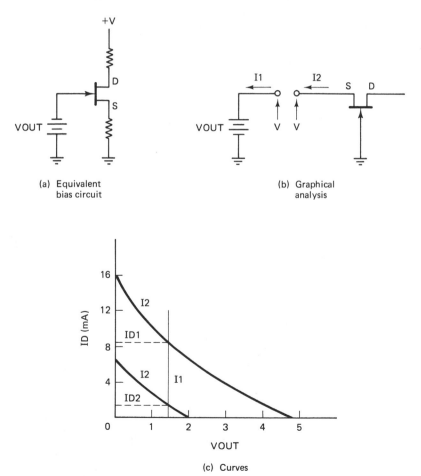

(a) Equivalent bias circuit

(b) Graphical analysis

(c) Curves

FIGURE 1-28 FET biased by external supply voltages.

When the source terminal is connected to VOUT, currents I1 and I2 are equal. Consequently, the no-signal level of source (and drain) current is determined by the point of intersection for the *V–I* plots. For example, about 9 mA of current (ID1) flows when VOUT (now the VGS) is 1.5 V. Figure 1–28c shows this graphical analysis.

Note that two I2 curves are given. These two curves show a typical spread of transfer characteristics among FETs of the same family or type. For example, with the same VOUT of 1.5 V, the lower I2 curve shows that the current is approximately 1 mA (ID2). Thus, if it is desirable that ID be maintained at some level, a form of self-bias must be used.

Self-Bias for FETs.   Self-bias of a FET amplifier reduces (but does not eliminate) variations in no-signal levels of ID. Figure 1–29a shows the use of a source resistor RS to develop a gate–source reverse-biased voltage. As ID increases, VGS becomes more negative, thus tending to prevent and increase in ID. The input portion of the self-bias circuit is redrawn in Fig. 1–29b and is analyzed graphically in Fig. 1–29c.

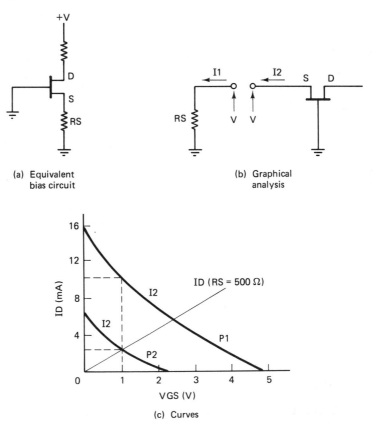

(a) Equivalent
bias circuit

(b) Graphical
analysis

(c) Curves

FIGURE 1-29 FET self-bias characteristics.

The *V–I* characteristic for the resistor is a straight line, having a slope equal to the reciprocal of RS, or 1/RS. The line intersects the two I2 plots at points P1 and P2. No-signal levels of ID are somewhat closer for the circuit of Fig. 1–29 than the circuit of Fig. 1–28. However, it is still possible to have a wide variation in ID.

As an example, ID can vary from about 2.5 to 11 mA with a VGS of 1 V. Thus, if it is essential that ID be maintained within narrow limits to get a given amplifier characteristic, then both fixed bias and self-bias must be used.

Combined Fixed Bias and Self-Bias.   Assume that it is desired to limit ID to a range between 3 and 7 mA (points A and B of Fig. 1–30). Although this cannot be done with self-bias alone (Fig. 1–29), Fig. 1–30 shows two circuits that restrict ID to the desired range of values.

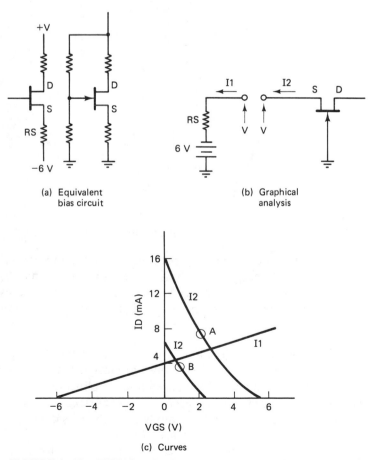

FIGURE 1-30   FET circuits incorporating both fixed and self-bias.

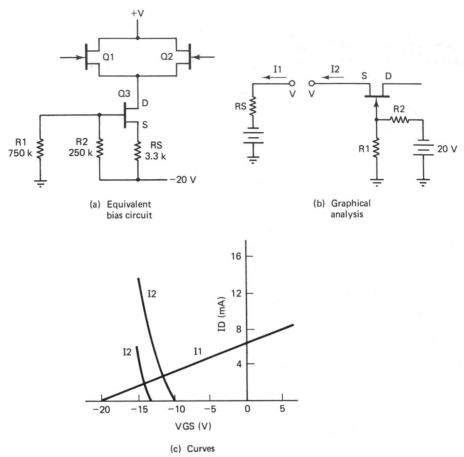

(a) Equivalent bias circuit

(b) Graphical analysis

(c) Curves

FIGURE 1-31   FET bias circuit for differential amplifier.

A representation for the input portion of the two combination circuits is given in Fig. 1-30b. The graphical analysis of Fig. 1-30c shows that, by proper selection of power-supply and resistance values, ID can be bounded by points A and B. The step-by-step procedures for finding the values are described in Chapter 3.

Constant-Current Bias.   Figure 1-31 shows a bias network for a differential amplifier. Transistor Q3 is biased as a constant-current generator to improve the common-mode rejection ratio of the differential amplifier Q1/Q2. (Differential amplifiers are discussed in Chapter 7.) Figure 1-31b shows the bias network, and Fig. 1-31c shows a graphical analysis of the bias circuit. Note that the two transfer curves of Fig. 1-30 are shifted to the left in Fig. 1-31 by the amount of negative voltage that appears at the gate terminal.

2

MOUNTING TECHNIQUES
AND THERMAL
CONSIDERATIONS

This chapter is devoted to mounting techniques and thermal considerations for amplifier components, particularly transistors and ICs. The components used in amplifiers are the same as for all other electronic equipment. The main concern is with *power amplifiers*, both audio and RF, where thermal problems can arise in design and service. This chapter summarizes such problems and their solutions.

2-1 TEMPERATURE-RELATED DESIGN PROBLEMS

There are two basic temperature-related design problems in amplifier design. First, datasheets specify component (transistor, diodes, rectifiers, ICs, and the like) parameters at a given temperature. Many of these parameters change with temperature. Since components rarely operate at the exact temperature shown on the datasheet, it is important to know the parameters at the actual operating temperature.

For example, in the case of transistors, the three most critical parameters (from an amplifier design standpoint) are current gain, collector leakage, and power dissipation. To compound this problem, changes in parameters can affect transistor temperature (an increase in current gain or power dissipation results in a temperature increase).

Methods for determining transistor parameters at frequencies other than specified on datasheets are discussed in Chapters 3 and 4.

Second, in addition to knowing the effects of temperature on parameters, it is important to know how *heat sinks* or component mounting can be used to offset the effects of temperature. For example, if a transistor (or diode, rectifier, or IC) is used with a heat sink or is mounted on a metal chassis that acts as a heat sink, an increase in temperature (from any cause) can be dissipated into the surrounding air.

The following paragraphs describe methods for approximating important parameters (mostly transistor) at temperatures other than those shown on datasheets. Methods for determining the proper dissipation characteristics of heat sinks are also discussed.

Temperature-related design techniques for ICs are discussed in Sec. 2–3.

2-1.1 Effects of Temperature on Transistor Collector Leakage

Collector leakage (Icbo) increases with temperature. As a guideline, collector leakage doubles with every 10 °C increase in temperature for *germanium* transistors and doubles with every 15 °C increase for *silicon* transistors.

Collector leakage also increases with voltage applied at the collector. For example, a typical datasheet leakage figure could be 2 μA at 25 °C and 50 μA at 150 °C. However, the 25 °C figure is with a collector–base voltage of 30 V, while the 150 °C figure is with 5 V. If the temperature is raised from 25 °C to 150 °C with 30 V at the collector, the collector leakage is about 500 μA.

As a result, always consider the possible effects of a different collector voltage when approximating collector leakage at temperatures other than those on the datasheet (as discussed in Chapters 3 and 4).

2-1.2 Effects of Temperature on Transistor Current Gain

Current gain (hfe) increases with temperature. As a guideline, current gain doubles when the temperature is raised from 25 °C to 100 °C for *germanium* transistors and doubles when the temperature is raised from 25 °C to 175 °C for *silicon* transistors.

It is obvious that silicon transistors are less temperature sensitive than germanium transistors. Datasheets usually specify a maximum operating temperature, or ambient temperature. If this temperature is not given or is unknown, the guideline is as follows: do not exceed 100 °C for germanium transistors or 200 °C for silicon transistors.

2-1.3 Effects of Temperature on Power Dissipation

The power-dissipation capabilities of a transistor (and diode or rectifier) must be carefully considered when designing any circuit. Of course, in small-signal circuits (not power amplifiers) the power dissipation is usually less than 1 W, and heat sinks are not needed. In such amplifier circuits, the only concern is that the rated power dissipation (as shown on the datasheet) not be exceeded. As a guideline, do not exceed 90% of the maximum power dissipation, as shown on the datasheet, for any small-signal amplifier circuit.

As with other characteristics, transistor (and diode, rectifier, or IC) manu- facturers specify maximum power dissipation in a variety of ways on datasheets. Some manufacturers provide *safe-operating-area curves* for temperature and/or power dissipation. Other manufacturers specify *maximum power dissipation* in relation to a given ambient temperature or a given case temperature. Still others specify a *maximum junction temperature* or a *maximum case temperature.*

2-1.4 Thermal Resistance

Transistors, diodes, and rectifiers designed for amplifier applications often have some form of *thermal resistance* specified to show the power-dissipation capabilities. Thermal resistance can be defined as the *increase in temperature of the component junction* (with respect to some reference), divided by the power dissipated, or °C/W.

Power-transistor datasheets usually specify thermal resistance at a given temperature. This is also the case for many diodes and rectifiers. For each in- crease in temperature from the specified value, there is a change in the tem- perature-dependent characteristics of the component.

Since there is a change in temperature with changes in power dissipation of the component, the junction-to-ambient temperature also changes, result- ing in a characteristic change. As a result, the component characteristics can change with ambient-temperature changes and with changes produced by varia- tion in power dissipation.

In power transistors, diodes, and rectifiers, thermal resistance is normal- ly measured from the component *junction to the case.* This results in the term θ JC. (A lowercase Greek letter theta is used by engineers to indicate thermal resistance, just to show that they are engineers!)

On those components where the case is bolted directly to the mounting surface with a built-in threaded bolt or stud, the term θ MB (thermal resistance to mounting base) or θ MF (thermal resistance to mounting flange) is used. These terms take into consideration only the thermal paths from junction to case (or mount). For power components in which the junction is mounted directly on a header or pedestal, the total internal thermal resistance from junction to case (or mount) varies from about 50°C/W to less than 1°C/W.

2-1.5 Thermal Runaway

The main problem in operating a transistor near the maximum power limits is the thermal-runaway condition described in Sec. 1-7. (Diodes and rectifiers are generally not as affected by runaway since such components do not have current gain. However, thermal runaway can apply to diodes and rectifiers.)

2-1.6 *Operating Transistors without Heat Sinks*

If a transistor is not mounted on a heat sink, the thermal resistance from case to ambient air, θ CA, is so large in relation to that from junction to case (or mount) that the total thermal resistance from junction to ambient air, θ JA, is primarily the result of the θ CA term.

The following is a list of case-to-ambient (air) thermal resistances for a number of common transistor cases (both old and new). As shown, heavy-duty cases such as TO-3 have a small temperature increase (for a given wattage) in comparison to such cases as the TO-5 (because heavy-duty cases dissipate the heat into the ambient air).

Case	θ CA ($°C/W$)
TO-3	30
TO-5	150
TO-8	75
TO-18	300
TO-36	25
TO-39	150
TO-46	300
TO-60	70
TO-66	60

This information can be used to approximate the maximum power dissipation of transistors (without heat sinks) when such information is not shown on the datasheet. Assume that a germanium transistor with a TO-5 case is used, and the absolute maximum power dissipation (without a heat sink) must be found.

The case-to-ambient thermal resistance factor for a TO-5 case is 150. As discussed in Sec. 2-1.2, germanium transistors should not be operated above 100 °C. Assuming a 25 °C ambient temperature, the transistor temperature should not be allowed to increase more than 75 °C maximum. With a factor of 150 for the TO-5 case and a 75 °C increase, the case should dissipate 0.5 W ($75/150 = 0.5$).

However, as discussed in Sec. 2-1.2, current gain doubles when the temperature is raised from 25° to 100 °C. Assuming that the voltage remains constant, the maximum dissipation allowable is 0.25 W. This is an absolute maximum figure, assuming a germanium transistor, TO-5 case, and an ambient temperature of 25 °C. (For practical design purposes, the 0.25-W figure is safe if the case is mounted on a metal chassis, which will act as a heat sink.)

2-1.7 *Operating Transistors with Heat Sinks*

After about 1 W (or less), it becomes impractical to increase the size of the case to make the case-to-ambient thermal-resistance factor comparable to the junction-to-case factor. For this reason, most power transistors (and some power diode/rectifier components) are designed for use with an external heat sink. In some amplifier circuits, the chassis or mounting area serves as the heat sink. In other amplifiers, a heat sink is attached to the transistor case. Either way, the primary purpose of the heat sink is to increase the effective heat-dissipation area of the case and to provide a low heat-resistance path from case to ambient.

Section 2–2 discusses the practical aspects of heat-sink design and selection. The following paragraphs describe the basic calculations involved.

To properly design (or select) a heat sink for a given amplifier circuit, the thermal resistance of both the transistor and heat sink must be known. For this reason, power-transistor datasheets specify the θ JA that must be combined with the heat-sink thermal resistance to find the total power-dissipation capability.

Note that some power-transistor datasheets specify a *maximum case temperature* rather than θ JA. As discussed in Sec. 2–1.8, maximum case temperature can be combined with heat-sink thermal resistance to find maximum power dissipation.

Heat-Sink Ratings.   Commercial fin-type heat sinks are available for various transistor case sizes and shapes. (Refer to Sec. 2–2.) Such heat sinks are especially useful when the transistors are mounted in Teflon sockets, which provide no thermal conduction to the chassis or PC board.

Commercial heat sinks are rated by the manufacturer in terms of thermal resistance, usually in terms of °C/W. When heat sinks involve the use of washers, the °C/W factor usually includes the thermal resistance between the case and sink, θ CA. With a washer, only the sink-to-ambient, θ SA, thermal-resistance factor is given. Either way, the thermal-resistance factor represents temperature increase (in °C) divided by wattage dissipated.

For example, if the heat-sink temperature rises from 25° to 100°C (a 75°C increase) when 25 W is dissipated, the thermal resistance is 75/25 = 3. This can be listed on the datasheet as θ SA, or simply as 3°C/W.

All other factors being equal, the heat sink with the *lowest thermal resistance* (°C/W) is best. For example, a heat sink with 1°C/W is better than a 3°C/W heat sink. Of course, the heat sink must fit the transistor case and the space around the transistor. Except for these factors, selection of a suitable heat sink should be no particular problem.

Calculating Heat-Sink Capabilities.   The thermal resistance of a heat sink can be calculated if the following factors are known: material, mounting provisions, exact dimension, shape, thickness, surface finish, and color. Even if all these factors are known, the thermal resistance calculations are approximate. As a *very approximate guideline,*

$$\text{heat-sink thermal resistance (in °C/W)} = \sqrt{\frac{1500}{\text{area}}}$$

where the area (total area exposed to the air) is in square inches, material is $\frac{1}{8}$ in. thick, and the shape is a flat disc.

From a practical design standpoint, *it is better to accept the manufacturer's specifications for a heat sink.* The heat-sink thermal resistance actually consists of *two series elements:* (1) the thermal resistance from the case to the heat sink caused by conduction (θ CS), and (2) the thermal resistance from the heat sink to the ambient air caused by convection and radiation (θ SA).

Practical Heat-Sink Considerations.   To operate a transistor at full-power capabilities, there should be no temperature difference between the case and ambient air. This occurs only when the thermal resistance of the heat sink is zero, and the only thermal resistance is that between the junction and case. It is not practical to manufacture a heat sink with zero resistance. However, the greater the ratio θ JC/θ CA is, the nearer the maximum power limit (set by θ JC) can be approached.

When transistors are to be mounted on heat sinks, some form of electrical insulation must be provided between the case and heat sink (unless a grounded-collector circuit is used). Because good electrical insulators are (usually) good thermal insulators, it is difficult to provide electrical insulation without introducing some thermal resistance between case and heat sink.

The best materials for electrical insulation of heat sinks are mica, beryllium oxide (Beryllia), and anodized aluminum. The properties of these three materials for case-to-heat-sink insulation of a TO–3 case are compared as follows:

Material	Thickness (in.)	°C/W	Capacitance (pF)
Beryllia	0.063	0.25	15
Anodized aluminum	0.016	0.35	110
Mica	0.002	0.4	90

For small, general-purpose transistors with a TO–5 case, a beryllium oxide washer can be used to provide insulation between the case and a metal chassis or PC board. The use of a zinc oxide-filled silicon compound (such as Dow Corning 340 or Wakefield 1201) between the washer and chassis, together with a moderate amount of pressure from the top of the transistor, helps to decrease thermal resistance.

If the transistor is mounted within a heat sink, a beryllium cup should also be used between the transistor and heat sink. Figure 2–1 shows both types of mounting. Section 2–2 describes some mounting techniques for power transistors used in amplifiers.

Any insulation between collector and the chassis (as is produced by the washer between the case and heat sink) also results in capacitance between the two metals. This capacitance can be a problem and must be considered in RF amplifier design (Chapter 4). However, the problem of collector-to-chassis capacitance is usually noticed at frequencies above about 100 MHz.

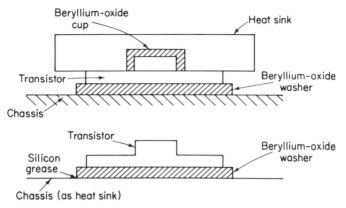

FIGURE 2-1 Typical mounting arrangements for transistor heat sinks.

2-1.8 *Calculating Power Dissipation under Steady-State Operating Conditions*

For practical amplifier design, the no-signal d-c collector voltage and current can be used to calculate power dissipation when a transistor is operated under steady-state conditions (such as in an audio amplifier). Actually, there are other currents that result in power dissipation (collector–base leakage current, emitter–base current). However, these can be ignored, and the power dissipation (in watts) can be considered as the d-c collector voltage times the collector current.

Once the power dissipation has been calculated, the *maximum power dissipation capability* must be found. Under steady-state conditions, the maximum dissipation capability depends on three factors: (1) the sum of the series thermal resistances from the transistor junction to ambient air, (2) the maximum junction temperature, and (3) the ambient temperature.

Following are some examples of how power dissipation can be calculated. Assume that it is desired to find the *maximum power dissipation of a transistor* (in watts) under the following conditions: a maximum junction temperature of 200 °C (typical for a silicon power transistor), a junction-to-case thermal resistance of 2 °C/W, a heat sink with a thermal resistance of 3 °C/W, and an ambient temperature of 25 °C.

First, find the total junction-to-ambient thermal resistance:

$$2 \,°C/W \,+\, 3 \,°C/W \,=\, 5 \,°C/W.$$

Next, find the maximum permitted power dissipation:

$$\frac{200 \,°C \,-\, 25 \,°C}{5 \,°C/W} \,=\, 35 \text{ W (maximum)}$$

If the same transistor is used without a heat sink, but under the same conditions and with a TO–3 case, the maximum power can be calculated as follows: First find the total junction-to-ambient thermal resistance:

$$2\,°C/W + 30\,°C/W = 32\,°C/W$$

Next find the maximum permitted power dissipation:

$$\frac{200\,°C - 25\,°C}{32\,°C/W} = 5 \text{ W (approximate)}$$

Some power-transistor datasheets specify a *maximum case temperature* rather than a maximum junction temperature. Assume that a maximum case temperature of 130 °C is specified instead of a maximum junction temperature of 200 °C. In that event, subtract the ambient temperature from the maximum permitted case temperature:

$$130° - 25\,°C = 105\,°C.$$

Then divide the case temperature by the heat-sink thermal resistance:

$$\frac{105\,°C}{3\,°C} = 35 \text{ W maximum power}$$

2-1.9 Calculating Power Dissipation under Pulse Operation Conditions

When transistors are operated by pulses (such as pulse amplifiers), the maximum permitted power dissipation is much greater than with steady-state operation. The following are some examples of power dissipation calculations for both repetitive and nonrepetitive pulses.

For a single, nonrepetitive pulse, the *transient thermal resistance* must be calculated. Usually, the transient thermal resistance is shown on the safe-operating-area curves in the form of a *power multiplier* for a specific case temperature and a given pulse width.

For example, a 2N3055 transistor has power multipliers of 2.1 for 100-ms pulses, 3.0 for 1-ms pulses, and on up to 7.7 for 30-ms pulses. The steady-state (or direct-current) maximum power is multiplied by these factors to find the maximum permitted single-pulse power. Assuming that the maximum permitted steady-state power is 100 W for a given set of conditions, the single-pulse maximum power is 300 W if a 1-ms pulse is used.

Usually, the datasheets specify the power multipliers for a given case temperature. If the case temperature is increased, the factor must be *derated*. The temperature derating factor is found by

$$\frac{\text{case temperature—ambient temperature}}{\text{maximum junction temperature—ambient temperature}}$$

For example, assume that a transistor has a maximum junction temperature of 200 °C, that the case temperature (under these conditions) is 130 °C, that the ambient temperature is 25 °C, and that the transistor datasheet specifies a power multiplier of 3.0 for 1-ms pulses, with a case temperature of 25 °C. The *derating factor* is

$$\frac{130\,°C\text{-}25\,°C}{200\,°C\text{-}25\,°C} = \frac{3}{5}$$

The maximum single-pulse power is found by

multiplier × (1-derating factor) × steady-state power

In our example, the *maximum single pulse power* is

$$3.0 \times (1\text{-}\frac{3}{5}) \times 35 = 42 \text{ W}$$

These calculations for single-pulse operation are based on the assumption that the heat-sink capacity is large enough to prevent the heat-sink temperature from rising between pulses.

For repetitive pulses, both the case and heat-sink temperatures rise. This increase must be taken into account when determining the maximum power dissipation.

The maximum permitted power dissipation for a transistor operated with repetitive pulses is calculated by

$$\frac{\text{power multiplier} \times (\text{maximum junction temperature—ambient temperature})}{\text{JC resistance} + (\text{power multiplier} \times \text{duty cycle} \times \text{JA resistance})}$$

Assume that it is desired to find the maximum permitted power of the same transistor described for steady-state operation, but now operated by 1-ms pulses repeated at 100-Hz intervals. The conditions are power multiplier = 2.0, JC = 2 °C/W, JA = 5 °C/W (including heat sink), maximum junction temperature = 200 °C, ambient temperature = 25 °C, and a duty cycle of 0.1 (10% of 1 ms on and 9 ms off). The *maximum permitted power* is

$$\frac{3.0(200\,°C - 25\,°C)}{2\,°C} + (3.0 \times 0.1 \times 5\,°C) = 150 \text{ W}$$

Peak pulse power is obtained by multiplying the collector voltage by the collector current (assuming that the transistor is operated in a grounded-emitter or grounded-base configuration and that the transistor is switched full on and full off by the pulses).

For example, assume an ambient temperature of 25 °C, a duty cycle of 10%, a total JA of 5 °C/W, and peak pulses of 120 W (say, a collector voltage of 60, and a collector current of 2 A). The case temperature is

$$\text{case temperature} = (120 \times 0.1 \times 5) + 25 = 85 \,°C$$

2-2 PRACTICAL HEAT-SINK/COMPONENT MOUNTING TECHNIQUES

As discussed in Sec. 2–1, proper mounting procedures must be followed if the *interface thermal resistance* between the component (transistor, diode, rectifier, and the like) package (or case) and heat sink is to be minimized. Proper mounting provides cooler and more reliable operation of the components.

This section summarizes mounting techniques for metal-packaged power transistors (and other semiconductors such as diodes and rectifiers). Included in this section are discussions concerning preparation of the mounting surface, use of thermal compounds, and fastening techniques. Typical interface thermal resistance is given for a number of diode and transistor packages.

As discussed, the junction-to-ambient *total thermal resistance* is the sum of the individual thermal resistances (junction to case, case to heat sink, heat sink to ambient). The junction-to-case and sink-to-ambient resistances are characteristics of the component and heat sink, respectively. So only the case-to-sink resistance can be changed, depending on mounting techniques.

The case-to-sink resistance is often referred to as the interface thermal resistance and is sometimes listed as RθCS. No matter what it is called, the *lowest value of interface thermal resistance is best* since this produces the lowest junction-to-ambient resistance. The following sections summarize practical methods to get lowest junction temperature under all mounting conditions.

2-2.1 Typical Values of Interface Thermal Resistance

Figure 2–2 shows the interface thermal resistance (in °C/W) for a small number of diode and transistor cases. Always check these values against datasheet listings. The values shown are approximate, are subject to change, and are given here for reference only. Figure 2–2 also shows recommended hole and drill sizes, as well as torque for the mounting nuts or screws.

Note that the interface thermal resistance changes quite drastically for different mounting conditions. For example, assume that a TO–3 case is involved. If the case is mounted (on the heat sink or chassis) with an insulator and no thermal compound or lubrication is used, the interface thermal resistance is 1.45 °C/W. If a thermal compound is used, the resistance drops to 0.8 °C/W. If circuit conditions make it possible to eliminate the insulator, the thermal resistance drops to 0.1 °C (with thermal compound) or 0.2 °C (without compound).

Package type and data				Interface thermal resistance (°C/W)				
JEDEO outline number	Description	Recommended hole and drill size	Torque in-lb	Metal-to-metal		With insulator		
				Dry	Lubed	Dry	Lubed	Type
DO-4	10-32 stud 7/16 hex	0.118, No. 12	15	0.41	0.22	1.24	1.06	3-mil mica
DO-5	1/4-28 stud 11/16 hex	0.25, No. 1	30	0.38	0.20	0.89	0.70	5-mil mica
DO-21	Pressfit, 1/2	See Fig. 7-5	*	0.15	0.10	–	–	–
TO-3	Diamond	0.14, No. 28	*	0.20	0.10	1.45	0.8	3-mil mica
						0.8	0.4	2-mil mica
						0.4	0.35	Anodized aluminum
TO-66	Diamond	0.14, No. 28	*	–	0.50			
TO-83	1/2-20 stud	0.50, 0.50	130	–	0.10			

*Can be tapped for 10-24 machine screw

FIGURE 2-2 Interface thermal resistance for a number of diode and transistor cases.

2-2.2 Fastening Techniques

The various types of transistor/diode packages shown in Fig. 2–2 require different fastening techniques. Mounting details for *stud, flat-base, press-fit* and *disc-type* transistors are shown in Figs. 2–3, 2–4, 2–5, and 2–6, respectively. Again, there are many other types of fastening techniques for transistors used in amplifiers. The following notes supplement the few examples shown here.

With any of the mounting schemes, *the screw threads should be free of grease* to prevent inconsistent torque readings when tightening nuts. Maximum allowable torque should always be used to reduce thermal resistance. However, care must be exercised not to exceed the torque rating of parts. Excessive torque applied to disc- or stud-mounted parts (Figs. 2–3 through 2–6) could cause damage to the semiconductor die.

To prevent galvanic action from occurring when components are used with aluminum heat sinks in a corrosive atmosphere, many devices are nickel or gold plated. Take precautions *not to mar the surface*.

With press-fit components (Fig. 2–5), the hole edge must be chamfered as shown to prevent shearing off the knurled edge of the component during press-in. The pressing force should be applied evenly on the shoulder ring to avoid tilting or canting of the device case in the hole during the pressing opera-

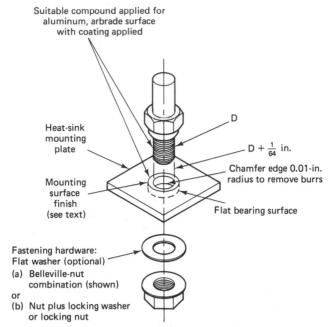

FIGURE 2-3 Mounting details for stud-mounted semiconductors.

tion. Also, thermal compound (Sec. 2–2.4) should be used to ease the component into the hold.

Typically, the pressing force varies from 250 to 1000 lb, depending on
the heat-sink material. Recommended hardnesses for typical heat-sink materials
are copper, less than 50 on the Rockwell F scale; aluminum, less than 65 on
the Brinell scale. A heat sink as thin as $\frac{1}{8}$ in. may be used, but the thermal
resistance increases in proportion to the reduction in contact area. A thin chassis
requires the addition of a backup plate.

With the disc-type mounting (Fig. 2–6), a self-leveling type of mounting
clamp is recommended to assure that the contacts are parallel and that there
is an even distribution of pressure on each contact area. A swivel-type clamp
or a narrow leaf spring in contact with the heat sink is usually acceptable.

The clamping force should be applied smoothly, evenly, and perpendicular
to the disc-type package to prevent deformation of the device or the sink-
mounting surfaces during installation. The spring used should provide a mounting force within the range recommended by the component manufacturer.
Typical clamping forces for disc-type transistors are 800 to 2000 lb.

Installation of a disc-type device between the heat sinks should be done
so as to permit *one heat sink to move with respect to the other*. Such movement avoids stresses being developed due to thermal expansion, which could
damage the component.

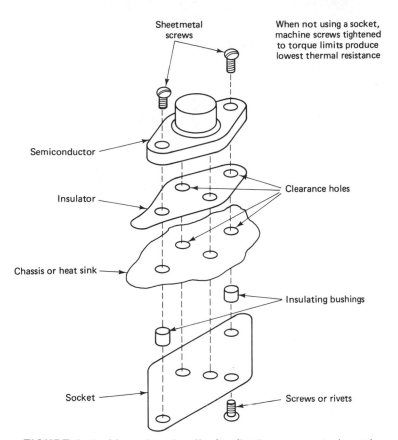

FIGURE 2-4 Mounting details for flat-base-mounted semi-conductors.

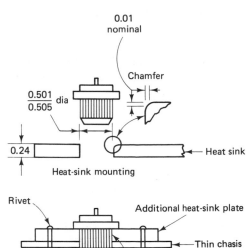

FIGURE 2-5 Mounting details for press-fit semiconductors.

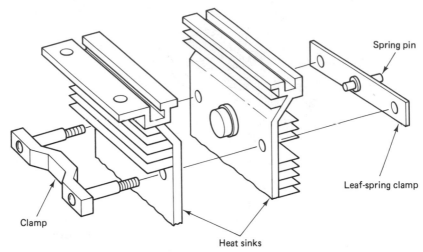

FIGURE 2-6   Mounting details for disc-type semiconductors.

When two or more components are to be operated electrically in parallel, *one of the heat sinks can be common to both (or all) components.* Individual heat sinks must be provided against the other mounting surfaces of the components so that the mounting force applied in each case is independently adjustable.

2-2.3 *Preparation of Mounting Surface*

In general, the heat sink should have a flatness and finish comparable to that of the component. For the typical experimenter or hobbyist, the heat-sink surface is satisfactory if the surface *appears flat* against a straight-edge and is free of any deep scratches. Of course, in commercial amplifiers it may be necessary to measure the actual flatness with special tools and indicators.

Most commercially available case or extruded heat sinks require *spot-facing*. In general, milled or machined surfaces are satisfactory if prepared with tools in good working condition.

The surface must be free from all foreign material, film, and oxide (freshly bared aluminum forms an oxide layer in a few seconds). Unless used immediately after machining, it is good practice to polish the mounting area with No. 000 steel wool, followed by an acetone or alcohol rinse. Thermal grease should be applied *immediately*.

Many aluminum heat sinks are *black anodized* for appearance, durability, performance, and economy. Anodizing is an electrical and thermal *insulator* that offers resistance to heat flow. As a result, anodizing should be removed from the mounting area.

Another aluminum finish is *iridite* (chromate acid dip), which offers low resistance because of the thin surface. For best performance, the iridite finish *must be cleaned of oils and films* that collect in the manufacturing and storage of the sinks.

Some heat sinks are *painted* after manufacture. Paint of any kind has a high thermal resistance (compared to metal). For that reason, it is essential that *paint be removed* from the heat-sink surface where the component is attached.

2-2.4 Thermal Compounds

Thermal compounds (also called *joint compounds* or *silicon greases*) are used to fill air voids between the mating surfaces. This improves contact between the component and heat sink.

A typical compound has a resistivity of about 60 °C-in./W, compared to about 1200 °C-in./W for air. As a result, the thermal resistance of voids, scratches, and imperfections filled with a compound or grease is about one-twentieth of the original resistance.

Thermal compounds are a formulation of *fine zinc particles in a silicon oil* that maintain a greaselike consistency with time and temperature. There are two commonly used methods for applying the compounds. With one technique, the compound is applied in a very thin layer with a spatula or lintless brush, wiping lightly to remove excess material. The other technique involves applying a small amount of pressure to spread the compound. Any excess compound is then removed after the mounting is completed. The excess compound is wiped away using a cloth moistened with acetone or alcohol.

Some recommended thermal compounds for amplifier heat sinks are the following:

Astrodyne, Conductive Compound 829

Dow Corning, Silicon Heat Sink Compound 340

Emerson & Cuming, Inc., Eccotherm, TC-4

General Electric, Insulgrease

George Risk Industries, Thermal Transfer Compound XL500

IERC, Thermate

Thermalloy, Thermacote

Wakefield, Thermal Compound Type 1201

2-3 POWER DISSIPATION PROBLEMS IN INTEGRATED CIRCUITS

The basic rules for amplifier ICs regarding power dissipation and thermal considerations are essentially the same as those for discrete transistors and diodes, as discussed in Sec. 2-1.

The maximum allowable power dissipation (usually specified as PD or "maximum device dissipation" on IC datasheets) is a function of the maximum storage temperature TS, the maximum ambient temperature TA, and the thermal resistance from the semiconductor chip to case. The basic relationship is

$$PD = \frac{TS - TA}{\text{thermal resistance}}$$

All IC datasheets do not necessarily list all these parameters. It is quite common to list only the maximum power dissipation for a given ambient temperature and then show a "derating" factor in terms of maximum power decrease for a given increase in temperature.

For example, a typical IC might show a maximum power dissipation of 110 mW at 25 °C, with a derating factor of 1 mW °/C. If such an IC is operated at 100 °C, the maximum power dissipation is 100 − 25, or 75 °C increase; 110 − 75 = 35 mW.

In the absence of specific datasheet information, the following typical temperature characteristics can be applied to the basic IC package types. *No IC should have a temperature in excess of 200 °C.*

Ceramic flat pack:
   Thermal resistance = 140 °C/W
   Maximum storage temperature = 175 °C
   Maximum ambient temperature 125 °C
TO–5 style (metal can) package:
   Thermal resistance 140 °C/W
   Maximum storage temperature = 200 °C
   Maximum ambient temperature = 125 °C
Dual-in-line (ceramic) package:
   Thermal resistance = 70 °C/W
   Maximum storage temperature = 175 °C
   Maximum ambient temperature = 125 °C
Dual-in-line (plastic) package:
   Thermal resistance = 150 °C/W
   Maximum storage temperature = 85 °C
   Maximum ambient temperature = 75 °C

2–3.1 Working with Power ICs

Most present-day ICs require lower power (typically less than 1 W) and can be operated without heat sinks. However, some power ICs must be used with a heat sink (either an external heat sink or by direct contact with a metal chassis). Power ICs (typically those found in audio power amplifiers) generally use some form of metal package.

The datasheets for these power ICs usually list sufficient information to select the proper heat sink. Also, the datasheets or other literature often make recommendations as to mounting for the power IC. Always follow the IC manufacturer's recommendations. In the absence of such data, and to make the reader more familiar with the terms used, the following sections summarize considerations for power ICs.

2-3.2 *Maximum Power Dissipation*

From an amplifier design standpoint, an IC is a complete, predesigned, functioning circuit that cannot be altered in regard to power dissipation. That is, if the power-supply voltages, input signals, output leads, and ambient temperature are at their recommended levels, the power dissipation will be well within the capabilities of the IC.

With the possible exception of the data required to select or design heat sinks, the amplifier designer or experimenter need only follow the datasheet recommendations. Of course, if the IC must be operated at a temperature higher than the rated ambient, the power dissipation must be derated as described previously.

2-3.3 *Thermal Resistance for Power ICs*

ICs designed for power applications usually have some form of thermal resistance specified to indicate the power-dissipation capability, rather than a simple maximum device dissipation. Thermal resistance can be defined as the increase in temperature of a semiconductor material, with regard to some reference, divided by the power dissipated, or °C/W.

Power IC datasheets often specify thermal resistance at a given temperature. For each increase in temperature from this specified value, there is a change in the temperature-dependent characteristics of the IC. Since there is a change in temperature with changes in power dissipation, the semiconductor chip temperature also changes, resulting in a characteristic change.

IC characteristics can change with ambient-temperature changes and with changes produced by variations in power dissipation. As discussed in Sec. 2-3.4, most ICs have circuits to offset the effects of temperature.

In power ICs, thermal resistance is normally measured from the semiconductor chip to the case. On those ICs where the cases are bolted directly to the mounting surfaces with a threaded bolt or stud (similar to that shown in Fig. 2-3), the thermal resistance is measured from the chip to the mounting stud or flange.

2-3.4 *Thermal Runaway in Power ICs*

Power ICs (and other ICs) are subject to the problems of thermal runaway (as discussed for transistors in Sec. 1-7 and 2-1.5). However, unlike discrete transistors and diodes, most ICs have built-in circuits to prevent thermal runaway. The usual arrangement is to place a diode in the reverse-bias circuit for one or more of the transistors in the IC. The diode is fabricated on the same semiconductor chip as the transistors and thus has the same temperature characteristics.

The thermal protection circuit is arranged so that the reverse bias is increased (by a decrease in diode resistance) when there is an increase in temperature. When temperature increases because of an increase in internal transistor current (or vice versa), the diode resistance changes and increases the reverse bias. This offsets the initial change in current caused by temperature changes.

Many different temperature-compensation circuits have been developed by IC amplifier manufacturers. However, the IC user need not be concerned so long as the datasheet limits are observed, since the circuit is already designed in the IC.

2–3.5 IC Heat-Sink and Power-Dissipation Calculations

The calculations for IC heat sinks and power dissipation are essentially the same as for transistors and diodes as described in Secs. 2–1 and 2–2 and are not repeated here. However, where a large number of ICs (as well as transistors and diodes) are operated in a confined area, it may be necessary to use *forced air* (from a fan, blower, or the like) to keep the ambient air within an acceptable tolerance (particularly when large operating currents are involved).

3

AF AMPLIFIERS

The average human ear is able to hear sounds ranging up to about 15,000 vibrations per second. The corresponding electrical signals (ranging in frequency up to about 15 kHz) are known as AF or audio-frequency signals. However, the frequency range from 15 to 20 kHz is also considered as part of the audio range, since some humans can hear sounds at these frequencies. In this book, AF signals are defined as any frequency up to about 20 kHz.

3-1 FREQUENCY LIMITATIONS OF AMPLIFIER COMPONENTS

Were it not for reactance, a transistor (by itself) should be capable of operating at any frequency from zero (direct current) on up. The top frequency limit should be set only by the transit time of electrons across the transistor junctions. However, limitations are placed on the operating frequency of any transistor by the transistor characteristics. Likewise, the other components (capacitors, resistors, inductors, and so on) used in the amplifier circuit also limit the operating frequency. In this section, we shall see how each of the components affects amplifier operation.

Every electronic component has some impedance and is thus *frequency sensitive*. The component does not attenuate (or pass) signals of all frequencies equally. Even a simple length of wire has impedance. Wire, being a conductor, has some resistance. If alternating current is passed through the wire, there is some inductive reactance. If the wire is near another conductor (or metal chassis), there is some capacitance between the two conductors, and thus some capacitive reactance. The reactance and resistance combine to produce impedance, which, in turn, varies with frequency.

Of course, many of the impedances presented by components are of little practical concern. On the other hand, certain impedances have a very pronounced effect on amplifier design and operation. Four major components are used in AF amplifier designs: transistors, resistors, capacitors, and inductances (coils and transformers). Let us examine how the impedances and reactances of these components affect AF amplifier operation.

3-1.1 *Transistor Frequency Limitations*

As shown in Fig. 3-1, all transistors have some capacitance between the junctions (emitter–base and collector–base). If any of the elements is common or ground, the remaining elements have some capacitance to ground. For example, in a common-emitter amplifier, there is some capacitance from base to ground (CBE across the input) and collector to ground (CCE across the output). Likewise, there is capacitance from collector to base (which forms a feedback path from output to input).

Capacitive reactance decreases with an increase in frequency, and vice versa. A capacitance in series with a conductor presents less attenuation to the signal as frequency increases. A capacitance across a conductor (for example, in parallel from the conductor to ground) acts as a short to signals of increasing frequency.

Consider a common-emitter amplifier where transistor capacitances are across the input and output. As frequency increases, the capacitive reactance drops, producing a short across the input and output, and increases attenuation of the signal. At some frequency, the attenuation equals the transistor amplification, so there is no gain. At higher frequencies, the attenuation exceeds amplification, and there is a loss, even though the transistor may still operate.

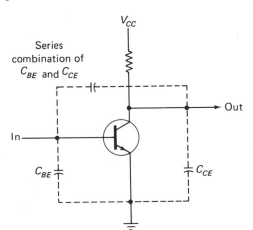

FIGURE 3-1 Capacitances associated with transistor elements.

From a practical standpoint, the input and output capacitances of transistors have little effect at audio frequencies. Most transistors operate well beyond the AF range and generally produce equal (or flat) frequency response. All signals up to about 20 kHz (or higher) are amplified by the same amount. (Refer to Chapter 9 for a further discussion of frequency response.) However, with most transistors, amplification begins to drop as signal frequencies increase into the RF range (Chapter 4).

All transistors have some inductance in their leads. This produces inductive reactance in series with the transistor elements. Inductive reactance increases with frequency. In the AF range, this inductive reactance is of little concern. However, as discussed in Chapter 4, the inductive reactance can produce considerable attenuation in the RF range.

3-1.2 Resistor Frequency Limitations

In the AF range, resistors offer relatively few problems, since resistors attenuate signals equally. Only at very high frequencies, where the resistor leads and body could produce some kind of resistance, is there any particular concern about frequency limits imposed by resistors. However, resistors do produce voltage drops that can be a problem when considering interstage coupling methods (Sec. 3-2) and when used with coupling capacitors.

3-1.3 Capacitance Frequency Limitations

Capacitors have three major uses in AF amplifiers: bypass, decoupling, and coupling.

Bypass Capacitors.   As shown in Fig. 3-2, bypass capacitors are used to provide a signal path around high resistances. For example, if the power supply of an audio amplifier does not have a filter capacitor or a battery is used, the collector-emitter current must pass through a high resistance. This can impede the a-c component of the signal. A bypass capacitor provides a signal path, as shown in Fig. 3-2.

When several stages of amplification are connected, the stages all join at one point, the *common power supply*. In multistage amplifiers, there is the possibility of one stage feeding back through the power supply to a previous stage, thus causing interference with the signal. To avoid this feedback, one or more of these stages may be *decoupled* from the power supply.

Figure 3-2 shows a typical decoupling network. Resistor R is placed in series between the load resistors of the stages and the power supply. This produces a high-resistance path for the a-c signal to the power supply. Capacitor C offers a low-resistance path for the signal, and thus decouples (or bypasses) the signal to ground.

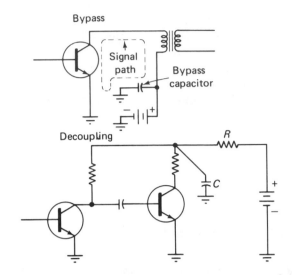

FIGURE 3-2 Examples of bypass and decoupling capacitors.

In practical terms, the functions of bypass and decoupling capacitors are the same, and the terms are interchanged. In either case, the main concern is that the *reactance be low at the lowest frequency involved*. This requires a capacitor of a given value, which increases as frequency decreases. For example, assume that the lowest frequency involved is 100 Hz, and the minimum required reactance is 100 Ω. This requires a capacitance value of about 10 μF. If the frequency is decreased to 10 Hz, the capacitance value must be raised to 160 μF to keep the reactance below 100 Ω.

Frequency Limitations of Coupling Capacitors.   As discussed in Sec. 3–2, coupling capacitors are often used at the input and output of amplifier stages to block direct current. The values of coupling capacitors depend on the low-frequency limit at which the amplifier operates and on the resistances with which the capacitors operate. This is shown in Fig. 3–3.

As frequency increases, capacitive reactance decreases, and the coupling capacitors become (in effect) a short to the signal. As a result, the high-frequency limit need not be considered in audio circuits. In Fig. 3–3, C1 forms a high-pass RC filter with RB. Capacitor C2 forms another high-pass filter with the input resistance of the following stage (or the load).

Input voltage to the filter is applied across the capacitor and resistor in series. The filter output voltage is taken across the resistance. The relation of input voltage to output voltage is

$$\text{output voltage} = \text{input voltage} \times R/Z$$

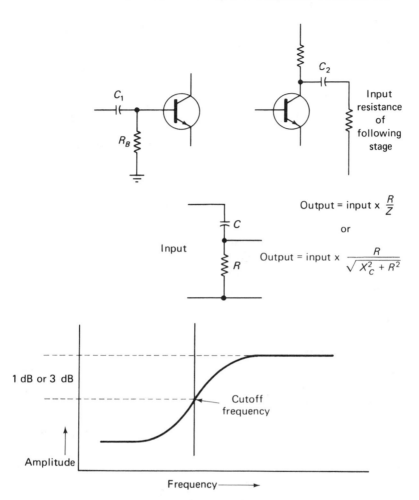

FIGURE 3-3 Formation of high-pass (low-cut) RC filter by coupling capacitors and related resistances.

where R is the d-c resistance value and Z is tne impedance (obtained by the vector combination of series capacitive reactance and d-c resistance).

When the reactance drops to about one-half the resistance, the output drops to about 90% of the input (or about a 1-dB loss). Using the 1-dB loss as the low-frequency cutoff point, the value of C1 or C2 can be found by

$$\text{capacitance} = \frac{1}{3.2 \ FR}$$

where capacitance is in farads, F is the low-frequency limit in hertz, and R is resistance in ohms.

If a 3-dB loss at the low-frequency cutoff point can be tolerated, the value of C1 or C2 can be found by

$$\text{capacitance} = \frac{1}{6.2 \text{ FR}}$$

3-1.4 Inductance Frequency Limits

Both coils and transformers are used in audio amplifiers. As discussed in Sec. 3-2, coils are sometimes used in place of the collector resistor as a load. This permits the collector to be operated at a higher voltage. Likewise, transformers are used for coupling between stages (mostly in older audio equipment). This provides impedance matching, as discussed in Sec. 3-2.

The inductive reactance of coils and transformers increases with frequency. At the high end of the AF range, the reactance of a typical transformer drops to a few ohms. This low impedance acts as a short across the line and attenuates the signal. Thus, coils and transformers tend to attenuate signals at both the high and low ends of the AF range.

3-1.5 Stray Impedances

As discussed, any conductor (wiring, terminals, and the like) can have resistance, reactance, and impedance. Care must be used in the routing of wires and placement of terminals to minimize the effects of this stray impedance. The effects of stray impedances can alter the characteristics of components. A classic example of this is *stray capacitance*, which is added to the input and output capacitances of transistors. The effects of stray impedances are usually not critical at audio frequencies. However, as discussed in Chapter 4, the effects of stray impedance on amplifiers operating in the RF range can be of considerable importance.

3-2 COUPLING METHODS

All amplifiers require some form of coupling. Even a single-stage audio amplifier must be coupled to the input and output devices. If more than one stage is involved, there must be *interstage coupling*. Amplifiers are often classified as to coupling method. For example, the four basic coupling methods are *capacitor (or capacitance) coupling, inductive coupling, direct coupling, and transformer coupling*.

All four coupling methods require resistance and could be called *resistance coupled* (amplifiers). However, the term resistance coupled is generally used to indicate that the amplifier does not have inductances or transformers between stages and that the input and/or output impedance is formed by a resistance. Capacitor coupling is often called *resistance–capacitance* (or RC) coupling.

In this section, we discuss how different methods affect the operation of practical audio or amplifiers. Figure 3–4 shows the four coupling methods.

3-2.1 Direct Coupling

With direct coupling (Fig. 3–4a), the collector of one transistor is connected to the base of the following transistor. The outstanding characteristic of a direct-coupled amplifier is the ability to amplify direct current and low-frequency signals. Because of the special nature of direct-coupled amplifiers, all of Chapter 5 is devoted to direct coupling.

3-2.2 Capacitor Coupling or RC Coupling

As shown in Fig. 3–4b, capacitor or RC coupling is formed by load resistor RL1 of stage 1, the base resistor RB2 of stage 2, and the coupling capacitor C2. The input signal is acted on by stage 1 and appears in amplified form as

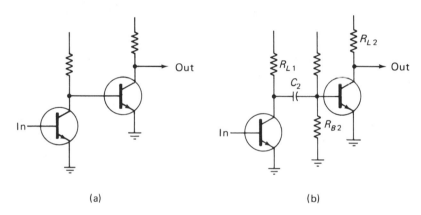

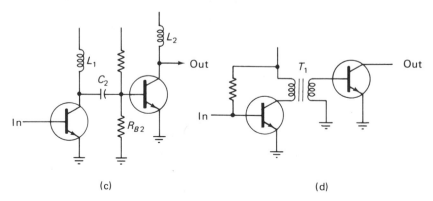

FIGURE 3-4 Four basic types of coupling used in audio amplifiers.

the voltage drop across RL1. The d-c component of the amplified signal is blocked by C2, which passes the a-c component to the input section of stage 2 for further amplification. If necessary, more stages may be coupled to the output of stage 2 for further amplification of the signal.

The main advantage of capacitor or RC coupling is that amplification is uniform over nearly the entire audio range, since resistor values are independent of frequency changes. However, as discussed in Sec. 3–1.3, RC-coupled amplifiers have a low-frequency limit imposed by reactance of the capacitor (which increases as frequency decreases). RC coupling is also small, light, and inexpensive and produces no magnetic field to interfere with the signal. One disadvantage of RC coupling is that the supply voltage is dropped (usually one-half) by the load resistance, so the collectors operate at a reduced voltage.

3–2.3 *Inductive or Impedance Coupling*

As shown in Fig. 3–4c, the load resistors are replaced by inductors L1 and L2. The advantage of impedance coupling over resistance coupling is that the ohmic resistance of the load inductor is less than that of the load resistor. For a power supply of a given voltage, there is a higher collector voltage.

Impedance coupling also suffers from a number of disadvantages. First, impedance coupling is larger, heavier, and costlier than RC coupling. To prevent the magnetic field of the inductor from affecting the signal, the inductor turns are wound on a closed, iron core and usually shielded.

The main disadvantage of impedance coupling is frequency discrimination. At very low frequencies, the gain is low because of the coupling capacitor reactance, as in the RC-coupled amplifier. The gain increases with frequency leveling off at the middle frequencies of the AF range. (However, the frequency spread of this level portion is not as great as for the RC amplifier.) At very high frequencies, the gain drops off because of the increased reactance. Impedance coupling is rarely, if ever, used at frequencies above the audio range. In present-day equipment, inductive coupling is used only where there is a special need for the inductor or coil.

3–2.4 *Transformer Coupling*

As shown in Fig. 3–4d, transformer T1 serves several purposes. As the fluctuating collector current of the first stage flows through the primary winding of T1, the current induces an alternating voltage with similar waveform in the secondary of T1. This voltage is the input signal to the second stage, and there is no need for a coupling capacitor. Also, since the T1 secondary provides a return path for the transistor base, there is no need for a base resistance.

Compared to the RC-coupled amplifier, the transformer-coupled amplifier has essentially the same advantages and disadvantages as the impedance-coupled amplifier. The transistor collectors can be operated at higher voltages. The impedances are set by the transformer primary and secondary windings. However, transformers are frequency sensitive (impedances can change with frequency), so the frequency range of transformer-coupled amplifiers is limited.

The inductances and transformers used in AF work are generally of the iron-core type. If air-core transformers are used at audio frequencies, the inductive reactance (and the impedance) is so small as to be ineffective. At frequencies above the audio range (or at the high end), the reactance of iron-core inductances and transformers is so large that signals cannot pass (or are greatly attenuated), so air-core transformers and inductances are used for higher-frequency amplifiers (Chapter 4).

Coupling transformers also provide for impedance matching between stages. Because the transistor is a current-operated device, impedance matching between the output of one stage to the input of the next is desirable for maximum transfer of power. This is done by making the primary and secondary transformer windings of different impedance.

Typically, the input impedance of a transistor stage is less than the output impedance, so the secondary impedance of an interstage transformer is (usually) lower than the primary impedance. When two common-emitter stages are impedance matched, the overall gain is greater than when identical stages are resistance coupled.

Transformer coupling is also effective when the final amplifier output must be fed to a low-impedance load. For example, the impedance of a typical loudspeaker is on the order of 4 to 16 Ω, whereas the output impedance of a transistor stage is several hundred (or thousand) ohms. A transformer at the output of an audio amplifier can offset the obviously undesired effects of such a mismatch.

3-2.5 Effects of Coupling on AF Amplifier Frequency Response

Figure 3–5 is a simplified frequency-response graph or curve. (A more comprehensive graph, as well as the procedures for producing such graphs, is discussed in Chapter 9.) The graph of Fig. 3–5 is provided here to illustrate the effects of coupling methods on amplifier frequency response. The response is measured by amplifier gain at various frequencies in the AF range.

Note that the gain falls off at very low frequencies. In an RC-coupled amplifier, this drop in gain (generally referred to as *rolloff*) at the low end is due to the capacitive reactance of the coupling capacitor. Since the coupling capacitor is between the output of the first stage and the input to the second stage, the signal is attenuated by the voltage drop across the capacitor. The lower the frequencys the larger the capacitive reactance, and the smaller is the signal input to the second stage. With impedance- or transformer-coupled amplifiers, the low-frequency rolloff is caused by the very low inductive reactance, which

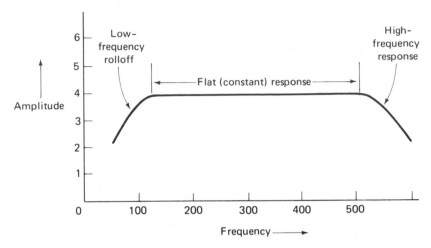

FIGURE 3-5   Simplified frequency-response graph.

acts a a short across the signal path. In effect, the low reactance bypasses some of the signal to ground.

As shown in Fig. 3–5, the gain also falls off at the higher frequencies. In RC-coupled amplifiers, this high-frequency rolloff is due to the output capacitance of the first stage, the input capacitance of the second stage, and the stray capacitance of the coupling network. These capacitances bypass some of the signal to ground.

The higher the frequency is the smaller the capacitive reactance becomes, and the greater the amount of signal so bypassed. As a result, the overall gain drops as frequency increases. With impedance- or transformer-coupled amplifiers, the high-frequency rolloff is produced by the large inductive reactance that attenuates the signal.

To sum up, RC coupling produces the lowest gain and transformer-coupling the highest. As a guideline, three stages of RC-coupled amplification produce about the same gain as two stages of comparable transformer-coupled amplification. RC coupling produces the least frequency distortion. Transformer coupling has the added advantage of immpedance matching.

Note that transformer coupling is used primarily where power is required. In present-day equipment, transformer coupling is replaced with power ICs, as discussed in Secs. 3–10 and 3–14.

3-3   HOW TWO-JUNCTION TRANSISTOR RATINGS AFFECT AMPLIFIER CHARACTERISTICS

The characteristics of discrete-component amplifier circuits are directly related to the ratings or capabilities of the transistors involved. For example, the current gain of an individual amplifier stage can be no greater than the maximum possible current gain of the transistor in that stage. Whether the problem is one of discrete-component amplifier design or amplifier test and troubleshooting, it is always helpful (and often necessary) to know the characteristics of the transistors involved. These characteristics are given in transistor manufacturer's *datasheets*.

Most of the amplifier-related characteristics for a particular transistor can be obtained from the datasheet. There are some exceptions to this rule. For high-frequency work (Chapter 4) and in digital work where switching characteristics are of particular importance, it may be necessary to test the transistor under simulated operating conditions.

In any event, it is always necessary to interpret datasheets. Each manufacturer has its own system of datasheets. It is impractical to discuss all datasheet formats here. Instead, we cover typical information found on datasheets and see how this information affects amplifier circuits.

Figure 3–6 is the datasheet for a 2N332 transistor. The 2N332 is an old-line industrial transistor, suitable for low-power audio or RF (up to about 10 MHz), as well as for switching.

3-3.1 Maximum Voltage

In Fig. 3–6, the maximum collector voltage is listed as VCBO of 45 V. Actually, this is a test voltage rather than an operating or design voltage. (VCBO usually indicates collector–base breakdown voltage, with the emitter circuit open. Transistors do not operate this way in circuits.) However, for design purposes, the 45-V figure can be considered as the *absolute maximum* collector voltage.

Keep the following in mind when considering maximum collector voltage. Except for RF circuits used in transmitters, most transistors are operated with the collector at some voltage value less than the source voltage. For example, in a typical class A circuit, the collector voltage is half the source voltage at the normal operating point. However, the collector voltage rises to or near the source voltage when the transistor is at or near cutoff. Therefore:

Never design any circuit in which the collector is connected to a source higher than the maximum voltage rating, even through a resistance.

The next maximum-voltage design problem to be considered is the type of source voltage. A battery cannot deliver more than the rated voltage. However, any electronic power supply is subject to some voltage variation. Therefore:

Always allow for some variation in source voltage when an electronic power supply is used.

Another factor that affects maximum voltage is temperature. Note that in Fig. 3–6 the maximum voltage is specified at 25 °C. Usually, breakdown occurs at a lower voltage when temperature is increased. (The topic of how operating temperatures affect transistor design is discussed in Chapter 2 and is not repeated here.)

TYPICAL TRANSISTOR SPECIFICATION 2N332

ABSOLUTE MAXIMUM RATINGS (25°C.)

Voltages:

Collector to base (emitter open)	V_{CBO}	45 volts
Emitter to base (collector open)	V_{EBO}	1 volt

Collector current

	I_C	25 ma

Power*

Collector dissipation (25°C.)	P_C	150 mw
Collector dissipation (125°C.)	P_C	50 mw

Temperature range:

Storage	T_{STG}	−65°C. to 200°C.
Operating	T_A	−55°C. to 175°C.

ELECTRICAL CHARACTERISTICS (25°C.)

(Unless otherwise specified, $V_{CB} = 5$ v; $I_E = -1$ ma; $f = 1$ kc)

Small signal characteristics:		min.	nom.	max.	
Current transfer ratio	h_{fe}	9	15	20	
Input impedance	h_{ib}	30	53	80	ohms
Reverse voltage transfer ratio	h_{rb}	.25	1.0	5.0	$\times 10^{-4}$
Output admittance	h_{ob}	0.0	.25	1.2	μmhos
Power gain					
($V_{CE} = 20$ v; $I_E = -2$ ma; $f = 1$ kc;					
$R_G = 1$ K ohms; $R_L = 20$ K ohms)	G_e		35		db
Noise figure	NF		28		db

High frequency characteristics:					
Frequency cutoff					
($V_{CB} = 5$ v; $I_E = -1$ ma)	f_{ab}		15		mc
Collector to base capacity					
($V_{CB} = 5$ v; $I_E = -1$ ma; $f = 1$ mc)	C_{ob}		7		$\mu\mu$f
Power gain (common emitter)					
($V_{CB} = 20$ v; $I_E = -2$ ma; $f = 5$ mc)	G_e		17		db

D-c characteristics:					
Collector breakdown voltage					
($I_{CBO} = 50$ μa; $I_E = 0$; $T_A = 25$°C.)	BV_{CBO}	45			volts
Collector cutoff current					
($V_{CB} = 30$ v; $I_E = 0$; $T_A = 25$°C.)	I_{CBO}		.02	2	μa
($V_{CB} = 5$ v; $I_E = 0$; $T_A = 150$°C.)	I_{CBO}			50	μa
Collector saturation resistance					
($I_B = 1$ ma; $I_C = 5$ ma)	R_{SC}		80	200	ohms

Switching characteristics:					
($I_{B_1} = 0.4$ ma; $I_{B_2} = -0.4$ ma;					
$I_C = 2.8$ ma)					
Delay time	t_d		.75		μsec
Rise time	t_r		.5		μsec
Storage time	t_s		.05		μsec
Fall time	t_f		.15		μsec

*Derate 1mw/°C increase in ambient temperature.

FIGURE 3-6 Typical transistor datasheet (2N332).

In Fig. 3–6, the maximum base–emitter voltage is listed as VEBO of 1 V. Again, this is a test voltage rather than an operating design voltage. Usually, the base–emitter junction has some current flowing at all times. The voltage drop across the junction is about 0.2 to 0.4 V for germanium and 0.5 to 0.7 for silicon transistors.

The lower voltage drops (0.2 to 0.5 V) produce some current flow, while the higher drops (0.4 to 0.7 V) produce heavy current flow. Bias circuits designed to produce the desired drop are discussed in Chapter 1. Either the higher or lower drops can be used, depending on the desired results.

In general, the lower drops produce less current drain and lower no-signal power dissipation. The higher drops may result in operation on a more linear portion of the transistor characteristics. For the purpose of standardization, the lower voltage drops are used in this book. However, if desired, the higher drops can be used as an alternative by slight changes of the bias-circuit values given in the design examples.

In practical design, it is often necessary to select a bias (base–emitter voltage) on the basis of input signal, rather than on some arbitrary point of the transistor's characteristic curve. Keep in mind that the input signal to a transistor can come from an external source, or a previous stage, or in the case of an op-amp (Chapter 7) it can be feedback. Therefore:

Always consider any input signal that may be applied to the base–emitter junction, in addition to the normal operating bias.

3-3.2 Collector Current

In Fig. 3–6, the collector current is listed as IC of 20 mA at 25 °C. As discussed in Chapter 2, collector current increases with temperature (and temperature increases as current increases). Therefore:

Do not operate any transistor at or near the maximum current rating.

Of course, if you can be absolutely certain that the transistor will dissipate any and all temperature increases (a practical impossibility); the circuit could be designed to operate near the maximum current.

In practical design applications, it is the *power dissipated in the collector circuit* (rather than a given current) that is of major concern. For example, assume that the collector operates at 45 V and 25 mA. This results in a power dissipation of over 1 W, far above the 150 mW specified for 25 °C.

3-3.3 Power and Temperature Range

The power-dissipation capabilities of a transistor in any circuit are closely associated with the temperature range. As shown in Fig. 3–6, power dissipation is 150 mW at 25 °C and 50 mW at 125 °C, and must be derated 1 mW for each degree (°C) increase in ambient temperature (as discussed in Chapter 2).

3-3.4   Small-Signal Characteristics

Small-signal characteristics can be defined as those where the a-c signal is small compared to the d-c bias. For example, *hfe* or *forward current transfer ratio* (also known as *a-c beta* or *dynamic beta*) is properly measured by noting the change in collector alternating current for a given change in base alternating current, without regard to static base and collector currents.

Small-signal characteristics do not provide a truly sound basis for practical design. As discussed in relavent chapters, the performance of a transistor in a working circuit can be controlled by the circuit component values (within obvious limits, of course). There are two basic reasons for this approach.

First, not all manufacturers list the same small-signal characteristics on datasheets. To further complicate matters, manufacturers call the same characteristic by different names (or even use the same name to identify different characteristics).

Second, the small-signal characteristics listed in datasheets are based on a set of *fixed operating conditions*. If the conditions change (as they must in any practical circuit), the characteristics change. For example, beta changes drastically with temperature, frequency and operating point. Therefore:

Use small-signal characteristics as a starting point for simplified design, not as hard-and-fast design rules.

3-3.5   High-Frequency Characteristics

High-frequency characteristics are especially important in the design of RF, IF, and VF amplifiers (but not too important for most audio work). As discussed in Chapter 4, networks (such as RF stages in a transmitter) provide the dual function of frequency selection and impedance matching between transistor and load. Unfortunately, the high-frequency information provided in many datasheets (such as Fig. 3–6) is not adequate for amplifier design. Fortunately, manufacturers who are trying to sell their transistors for high-frequency use generally provide a set of curves showing the characteristics over the anticipated frequency range. This is discussed further in Chapter 4 and in Sec. 3–3.8.

3-3.6   Direct-Current Characteristics

Direct-current characteristics, while important in the design of basic bias circuits (Chapter 1), do not have too critical an effect on operation of the final amplifier circuit. The d-c characteristics shown in Fig. 3–6 are primarily test values rather than design parameters. The important point to remember regarding such d-c characteristics is that they serve as a starting point for bias design and that they change with temperature.

3-3.7 Switching Characteristics

Switching characteristics are important in the design of pulse amplifiers. The switching times shown in Fig. 3–6 are defined in Fig. 3–7. These time factors (delay, storage, rise, fall) determine the operating limits for switching circuits.

For example, if an amplifier uses a transistor with a 20-ns rise time to pass a 15-ns pulse, the pulse is hopelessly distorted. Likewise, if there is a $1.5\text{-}\mu s$ delay added to a $1\text{-}\mu s$ pulse, an absolute minimum of $2.5\text{-}\mu s$ is required before the next pulse can occur. This means the amplifier can pass pulses with a maximum repetition rate of 400 kHz ($1/2.5^{-6} = 400{,}000$ Hz). Actually, the maximum repetition rate is lower since there is some "off" time between pulses.

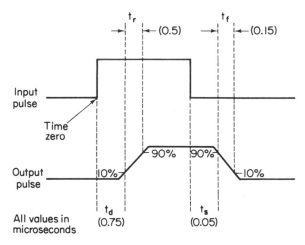

FIGURE 3-7   Definition of switching-time characteristics.

3-3.8 Determining Transistor Parameters at Different Frequencies

Datasheets specify most transistor parameters at some given frequency. Unless the amplifier is to operate at that exact frequency, it is essential to know the parameter at other frequencies (in the frequency range where the amplifier is used). The following paragraphs describe methods for converting from one transistor parameter to another at different frequencies.

Transistor Parameter Terms.   Before we get into the parameters, let us define some typical terms found on transistor datasheets.

Term	Definition
hfb	Common-base a-c forward current gain (alpha)
hfbo	Value of hfb at 1 kHz
hfe	Common-emitter a-c forward current gain (beta)
hfeo	Value of hfe at 1 kHz
fab	Common-base current-gain cutoff frequency. Frequency at which hfb has decreased to a value 3 dB below hfbo (where hfb = 0.707 hfbo)
fae	Common-emitter current-gain cutoff frequency. Frequency at which hfe has decreased to a value 3 dB below hfeo (where hfe = 0.707 hfeo)
fT	Gain–bandwidth product. Frequency at which hfe = 1 (0 dB)
Gpe	Common-emitter power gain
fmax	Maximum frequency of oscillation. Frequency at which Gpe = 1 (O dB)
K	Phase-shift factor (phase shift of current in transistor base)

Common-Base Parameters.   The quantity hfbo (the value of hfb at 1 kHz) remains constant as frequency is increased until a top limit is reached. After the top limit, hfb begins to drop rapidly. The frequency at which a significant drop in hfb occurs provides a basis for comparison of the expected frequency performance of different transistors. This frequency is known as fab and is defined as that frequency at which hfb is 3 dB below hfbo.

Figure 3–8 shows a curve of hfb versus frequency for a transistor with an fab of 1 MHz. This curve has the following significant characteristics:

1. At frequencies below fab, hfb is nearly constant and approximately equal to hfbo.

2. hbf begins to decrease significantly in the region of fab.

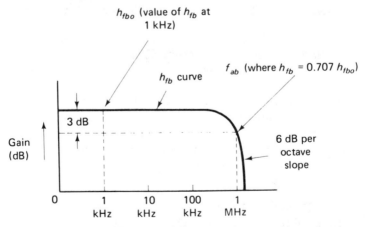

FIGURE 3-8 Curve of hfb versus frequency for a transistor with an fab of 1 MHz.

3. Above fab, the rate of decrease for hfb (with increasing frequency) approaches 6 dB/octave.

The curve of common-base current gain versus frequency for any transistor has the same characteristics and the same general appearance as the curve of Fig. 3–8.

Common-Emitter Parameters. The common-emitter parameter that corresponds to fab is fae, the common-emitter current-gain cutoff frequency. This fae is the frequency at which hfe (beta) has decreased 3 dB below hfeo. Figure 3–9 shows a curve of hfe versus frequency for a transistor with an fae of 100 kHz. The curve of Fig. 3–9 has the same significant characteristics as those described for Fig. 3–8. That is, hfe is considered to be decreasing at a rate of 6 dB/octave at fae.

These characteristics allow such a curve to be constructed for a particular transistor by knowing only hfeo and fae. With the curve constructed, hfe can be determined at any frequency. Also, if fae is not known, a curve can also be constructed if hfeo and *hfe at any frequency above fae* are known.

Gain–Bandwith Product. Gain–bandwith product or fT is sometimes specified on datasheets instead of fae (or instead of hfe at some frequency greater than fae). This fT is the frequency at which gain drops to unity (O dB). fT can be approximated when fae is multiplied by hfeo (fT = fae × hfeo).

On those datasheets where hfe is specified at some frequency greater than fae, fT can be approximated when the specified frequency is multiplied by the specified hfe.

It should be noted that fT is a common-emitter parameter and should not be used with common-base calculations. It should also be noted that common-emitter fT is approximately equal to the common-base parameter of fab. Usually, fT is slightly less than fab.

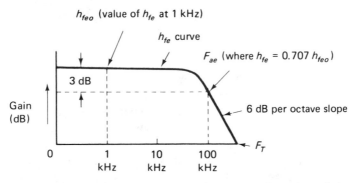

FIGURE 3-9 Curve of hfe versus frequency for a transistor with an fae of 100 kHz.

Maximum Operating Frequency.   Although common-emitter current gain is equal to 1 at fT, there may still be considerable *power gain* at fT due to different input and output impedance levels. fT is not necessarily the highest useful frequency of operation for a transistor connected as a common emitter amplifier. An additional parameter, the maximum frequency of oscillation, or fmax, may be of value.

The term fmax is the frequency at which common-emitter power gain is equal to 1. A plot of common-emitter power gain versus frequency has the same characteristics as the voltage-gain plot and appears similar to Fig. 3–9.

Fmax is found by measuring power gain at some frequency on the 6 dB/octave (slope) portion of the power-gain curve and then multiplying the square root of the power-gain (in magnitude) by the frequency of measurement.

The problem here is that datasheets do not always specify if the power-gain figure is on the slope of the power-gain versus frequency curve. However, some clues can be used to estimate the location of the power-gain figure on the curve.

If two power-gain figures are given, the high-frequency figure can be considered to be on the slope and can be used to find fmax. For example, the datasheet of Fig. 3–6 shows a power gain of 35 dB (magnitude 3000) at 1 kHz and 17 dB (magnitude 50) at 5 MHz. The 17 dB (magnitude 50) at 5 MHz figure can be used to find fmax by

$$\text{fmax} = \text{frequency of measurement} \times \sqrt{\text{power gain in magnitude}}$$

$$= 5 \times \sqrt{50}$$

$$= 35 \text{ MHz (approximately)}$$

If a power-gain figure is given for a frequency higher than fae, that figure can be considered as being on the slope and can be used to find fmax.

Conversion between Parameters.   One problem with datasheets is the mixing of parameters. For example, the datasheet for a 2N332 shows a nominal common-emitter (beta) current transfer ratio hfe of 15 and a nominal common-base (alpha) cutoff frequency fab of 15 MHz. Likewise, the datasheet of a 2N337 shows a "typical" current transfer ratio hfe of 55 (magnitude) and a "typical" alpha cutoff frequency of 30 MHz. (Actually, the datasheet lists alpha cutoff frequency as fhfb for some obscure reason.)

Many other datasheets and transistor specification books have similar forms of parameter mixing. For design purposes, it is necessary to convert between parameters. Usually, it is necessary to convert from common-base to common-emitter, but the reverse can also be true. The following rules summarize the conversion process:

To find beta when alpha is given, beta = alpha/(1 − alpha).

To find alpha when beta is given, alpha = beta/(1 + beta).

To find hfeo when hfbo is given, hfeo = hfbo/(1 − hfbo).

To find hfbo when hfeo is given, hfbo = hfeo/(1 + hfeo).

To find fae when fab is given, fae = K(1 − hfbo) fab.

To find fab when fae is given, fab = fae/K(1 − hfbo).

Note that the constant K refers to a phase-shift factor. The K constant usually ranges between 0.8 and 0.9, but can be as low as 0.6 for some transistors.

Determining Common-Emitter Parameters at Different Frequencies. The common-emitter configuration is used in 90% of present-day transistor amplifiers. Three frequency-related common-emitter parameters are of particular importance to design: (1) hfe at some frequency other than specified on the datasheet, (2) fT, and (3) fmax.

The following three rules summarize the procedures for finding these three parameters when the parameters are not listed on the datasheet.

1. *To find hfe at a particular frequency*:
   a. hfe is approximately equal to hfeo when the frequency of interest is less than fae.
   b. hfe is approximately equal to 0.7 hfeo when the frequency of interest is near fae.
   c. hfe decreases at a rate of 6 dB/octave and is approximately equal to fT/frequency when the frequency of interest is above fae.
   d. hfe is equal to 1 (unity gain) at fT.

2. *To find fT*:
   a. When hfeo and fae are given, fT = hfeo × fae.
   b. When hfbo and fab are given, fT = hfbo × fab × K.

3. *To find fmax*: As discussed, fmax is a common-emitter parameter and requires that the common-emitter power gain be established at some frequency on the sloping portion of the gain versus frequency curve. With such a gain established:

$$\text{fmax} = \text{frequency of measurement} \times \sqrt{\text{power gain in magnitude}}$$

Examples of Relating Datasheet Frequency Parameters to Design Problems. Assume that a 2N337 transistor is to be used in amplifier circuits. One amplifier must provide a voltage gain of at least 10 (magnitude) at 2.5 MHz. The other amplifier must operate at 25 MHz with "some gain." Both amplifiers must operate in the common-emitter configuration. The K factor (phase shift) is 0.9. Each problem requires that fT be found. There are two approaches to finding fT.

The first approach involves finding fT when fae and hfeo are known. Since the 2N337 datasheet gives a common-emitter hfeo of 55 (magnitude), but a common-base cutoff of 30 MHz, it is necessary to find fae.

$$fae = K(1 - hfbo)fab$$

$$0.9 \times 0.018 \times 30 = 0.486 \text{ MHz}$$

With fae found, then

$$fT = hfeo \times fae$$

$$55 \times 0.486 = 26.73 \text{ MHz}$$

The second approach involves finding fT when fab and hfbo are known. Since the datasheet gives a common-emitter hfeo of 55 (magnitude) and a common-base cutoff of 30 MHz, it is necessary to find hfbo:

$$hfbo = \frac{hfeo}{1 - hfeo} = \frac{55}{56} = 0.982$$

With hfbo found, then

$$fT = K \times hfeo \times fab$$

$$0.9 \times 0.982 \times 30 = 26.514 \text{ MHz}$$

With either approach, hfe = fT/frquency of interest, or 26.73/2.5; so hfe is greater than 10 at 2.5 MHz and the 2N337 can be used in the first circuit.

Also, the fT is greater than 25 MHz, so the 2N337 provides "some gain" at 25 MHz and can be used for the second circuit.

No matter what system is used to find transistor parameters at different frequencies, remember that the parameters depend on both voltage and current, as well as operating point. For example, the high-frequency hfe measured at one collector voltage and current must not be used to calculate fT at another voltage or current without considering the possible effects of the different operating points.

3-3.9 *Interpreting Transistor Characteristic Curves with Load Lines*

Many datasheets show two-junction transistor characteristics by means of curves that are reproductions of displays obtained with a *curve tracer*. The collector voltage–current curves shown in Fig. 3–10 are typical.

Such curves are obtained by applying a series of stepped base currents and then sweeping the collector voltage over a given range. Several curves are made in rapid succession at different base currents. Current gain (beta) is found

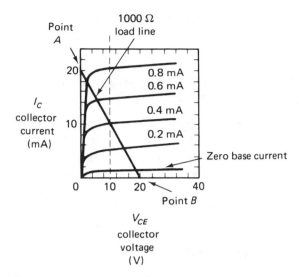

FIGURE 3-10   Typical load line drawn on characteristic curves of a common-emitter transistor.

by noting the difference in collector current for a given change in base current, while maintaining a fixed collector voltage.

Load lines can be superimposed on the datasheet curves to show how the transistor will perform in an amplifier circuit. For example, assume that a collector load of 1 k is used with the transistor shown in Fig. 3–10. When the collector current reaches 20 mA (base current approximately 0.8 mA), the collector voltage drops to zero. Likewise, when the collector current drops to zero, collector voltage rises to 20 V (base current zero).

If a load line is connected between these two extreme points (marked A and B on Fig. 3–10), the instantaneous collector voltage and collector current can be obtained for any base current along the line. *Approximate* transistor gain can be estimated from the load line. For example, with a base current *change* from 0.2 to 0.4 mA (a 0.2-mA change), the collector current changes from about 4 to 9 mA (a 5-mA change). This represents an approximate current gain of 25 (magnitude).

Using datasheet curves and load lines to find transistor characteristics has some drawbacks. The datasheet curves are "typical" for a transistor of a given type and represent an average gain (or, in some cases, the *minimum gain*). Also, as discussed, transistor gain depends on temperature and frequency. The selection of load values, bias values, and operating point on the basis of static gain curves is subject to error. If the transistor beta shifts, the operating point must shift, requiring different bias and load values.

As is discussed throughout this book, the problem of variable-gain transistor amplifiers can be overcome by means of *feedback to stabilize the gain*.

With feedback, amplifier characteristics depend primarily on the relationship of circuit values, rather than on transistor gain characteristics.

3-4   HOW FET TRANSISTOR RATINGS AFFECT AMPLIFIER CHARACTERISTICS

Characteristics of the FET are unique when compared to any other type of transistor. For example, as discussed in Chapter 1, a FET is often biased at the zero-temperature-coefficient, or 0TC, point (where FET drain–source currents remain constant with temperature changes). Likewise, the characteristics shown on FET datasheets do not correspond to those of two-junction transistors (Sec. 3–3). The following paragraphs summarize FET characteristics and datasheet ratings.

3-4.1   FET Datasheet Format

Typically, the first page of the datasheet gives the maximum rating of the FET, mechanical dimensions, and pin layout, similar to a two-junction transistor datasheet. Obviously, if the maximum ratings are exceeded, abnormal circuit operation occurs, and the FET may be destroyed. For example, assume that 70 V is applied to the FET, which has a maximum drain–source rating of 60 V. Damage, if not total destruction, results. All the precautions concerning maximum ratings for two-junction transistors also apply to FETs.

   The second page of the datasheet presents the electrical characteristics and operating conditions under which the characteristics are measured. Both on/off and small-signal characteristics are given. Remember that all characteristics listed on any datasheet are based on a set of *fixed operating conditions*. If the conditions change (as they must in any practical amplifier), the characteristics change.

3-4.2   FET Characteristics

The following characteristics appear on most FET datasheets. The procedures for testing and measuring these characteristics are given in Chapter 9.

   Forward gate current IG(f) is the maximum recommended forward current through the gate terminal. This is a limiting factor in some amplifier applications and is caused by a large forward bias current on the gate. When this condition occurs, the gate current must be limited or degeneration of the FET occurs. A resistor in series with the gate limits the current, but the resistor value determines the variance of gate bias (as affected by the gate leakage current).

   Total device dissipation PD is the maximum power that can be dissipated within a device at 25 °C without exceeding the maximum allowable internal temperature (typically 200 °C). The power is derated according to the thermal resistance (Chapter 2) value. For example, assume that the FET has a 200-mW rating at 25 °C and a 1.33-mW/°C thermal resistance rating. At 125 °C the power dissipated in the FET must not exceed 77 mW ($125 - 25 = 100\,°C$; $100 \times 1.33 = 133$; $200 - 133 = 77$ mW). Operation at the maximum value could damage the FET.

Gate–source breakdown voltage V(BR)GSS is the breakdown voltage from gate to source, with the drain and source shorted. Under these conditions, the gate–channel junction also meets the breakdown specification, since the drain and source are connected to the channel. This means that the drain and source may be interchanged for most amplifier circuits (for symmetrical FETs) without fear of individual junction breakdown.

Gate–source cutoff voltage VGS(off) is the gate-to-source voltage required to reduce the drain current to 0.01 (or preferably 0.001) of the minimum IDSS value.

Pinchoff voltage VP is essentially the same as VGS(off), only measured in a different manner (Chapter 9). Most equations used to describe the operation of a FET in an amplifier use VP, but the value of VGS(off) can be used instead.

Gate leakage (reverse) current IGSS is the gate–channel leakage with the drain shorted to the source and is a measure of the static short-circuit input impedance. Since gate-to-channel is a reverse-biased diode junction (for a JFET), IGSS doubles (approximately) for every $15\,°C$ increase in temperature and is proportional to the square root of the applied voltage.

Zero gate voltage drain current IDSS is the drain-to-source current, with the gate shorted to the source, at a specified drain–source voltage. IDSS is a basic parameter for JFETs and is considered to be a figure of merit for FETs used in amplifiers.

Gate–source voltage VGS is a range of gate-to-source voltages with 0.1 IDSS drain current flowing. The specified drain-to-source voltage is the same as for IDSS. This characteristic gives the min/max variation in VGS for different FETs for a given ID and VDS.

Forward transadmittance Yfs is the magnitude of the common-source forward transfer admittance. Yfs shows the relationship between input signal voltage and output signal current and is a key dynamic characteristic for all FETs. All other factors being equal, an increase in Yfs produces an increase in gain when a FET is used in an *audio amplifier* circuit. Yfs is specified at 1 kHz, with d-c operating conditions the same as for IDSS.

At 1 kHz, Yfs is almost entirely real. At higher frequencies, Yfs includes the effects of gate-to-drain capacitance and may be misleadingly high. However, for FETs operating in the AF range, Yfs is generally an accurate figure. For higher-frequency operation, the real part of transconductance Re(Yfs) should be used.

Forward transconductance Re(Yfs) is the common-source forward transfer conductance (drain current versus gate voltage). For high-frequency operation, Re(Yfs) is considered a figure of merit. The d-c operating conditions are the same as for Yfs, but the test frequency is usually at or near 100 MHz. All other factors being equal, an increase in Re(Yfs) produces an increase in the voltage gain of a FET amplifier operating in the RF range.

In comparing Re(Yfs) with Yfs, the minimum values of the two are quite close, considering the difference in frequency at which the measurements are made. At frequencies of about 30 MHz and above, Yfs increases because of the gate–drain capacitance Cgd and is misleadingly high.

Output admittance Yos is the magnitude of the common-source output admittance and is measured at the same operating conditions and frequency as Yfs. Since Yos is a complex number at low frequencies, only the magnitude is specified on most FET datasheets.

Common-source output resistance ross is the real part of Yos. Note that the lower IDSS FETs have a higher output resistance for the same drain current.

Input capacitance Ciss is the common-source input capacitance with the output shorted and is used in place of Yiss, the short-circuit input admittance. Yiss is entirely capacitive at low frequencies, since the input is a reverse-biased silicon diode (for JFETs). The real part of Yiss can be calculated from IGSS, but is negligible in the AF range.

Reverse transfer capacitance Crss is the common-source reverse transfer capacitance with the input shorted. Crss is used in place of Yrs, the short-circuit reverse transfer admittance, since Yrs is almost entirely capacitive over the useful frequency range of most FETs.

Common-source noise figure NF represents a ratio between input signal-to-noise and is measured in decibels. Short-circuit input noise voltage *en* is the equivalent short-circuit input noise.

As specified on most FET datasheets, NF includes the effects of *en* and *in*, where *in* is the equivalent open-circuit input noise current. For a FET, the combination of *in* is small compared to *en*. As a result, NF and *en* are specified, while *in* is neglected, on many FET datasheets. Typically, NF is independent of operating current and proportional to voltage. However, the voltage effects are slight over the normal operating range of the FET. There are nomographs for converting NF to equivalent input noise voltage for different generator source impedances, as discussed in Chapter 9.

Although the information given in this section may not be necessary for FET amplifiers used at audio frequencies, the data can be of great value for RF amplifiers, as discussed in Chapter 4.

3-5 BASIC AF AMPLIFIER (TWO-JUNCTION TRANSISTOR)

Figure 3–11 is the working schematic of a basic, single-stage AF amplifier using a two-junction transistor. As discussed in Chapter 4, this same basic circuit can also be used at higher frequencies.

Note that the basic AF circuit is similar to the bias circuits discussed in Chapter 1 (Fig. 1–10), except that input and output coupling capacitors C1 and C2 are added. These capacitors prevent direct-current flow to and from exter-

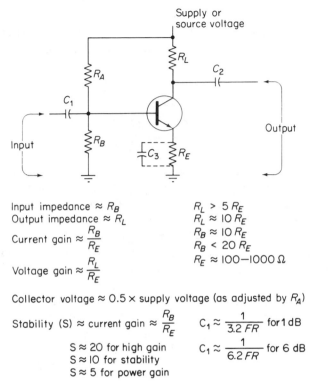

Input impedance $\approx R_B$
Output impedance $\approx R_L$

Current gain $\approx \dfrac{R_B}{R_E}$

Voltage gain $\approx \dfrac{R_L}{R_E}$

$R_L > 5 R_E$
$R_L \approx 10 R_E$
$R_B \approx 10 R_E$
$R_B < 20 R_E$
$R_E \approx 100\text{--}1000 \ \Omega$

Collector voltage $\approx 0.5 \times$ supply voltage (as adjusted by R_A)

Stability (S) $\approx$ current gain $\approx \dfrac{R_B}{R_E}$ $C_1 \approx \dfrac{1}{3.2 \ FR}$ for 1 dB

$S \approx 20$ for high gain $C_1 \approx \dfrac{1}{6.2 \ FR}$ for 6 dB
$S \approx 10$ for stability
$S \approx 5$ for power gain

FIGURE 3-11   Basic audio-amplifier circuit.

nal circuits. Note that a bypass capacitor C3 is shown connected across the emitter resistor RE. Capacitor C3 is required only under certain conditions, as discussed in Sec. 3–6.

Input to the amplifier is applied between base and ground, across RB. Output is taken across the collector and ground. The input signal adds to, or subtracts from, the bias voltage across RB. Variations in bias voltage cause corresponding variations in base current, collector current, and the drop across collector resistor RL. The collector voltage (or circuit output) follows the input signal waveform, except that the output is inverted in phase. (If the input swings positive, the output swings negative, and vice versa.)

Variations in collector current also cause variations in emitter current. This produces a change of voltage drop across RE and a change in the base–emitter bias relationship. As discussed in Chapter 1, the change in bias that results from the voltage drop across RE tends to cancel the initial bias change caused by the input signal and serves as a form of *negative feedback* to increase stability (and limit gain). This form of emitter feedback (current feedback) is known as *stage feedback* or *local feedback* since only one stage is involved. As discussed in the remaining chapters, *overall feedback* or *loop feedback* is sometimes used where several stages are involved.

The outstanding characteristic of the Fig. 3–11 circuit is that circuit characteristics (gain, stability, impedance) are determined (primarily) by circuit values, rather than by transistor characteristics (beta).

3-5.1 Design Considerations for Basic Amplifier

The circuit of Fig. 3–11 is shown with an NPN transistor. Reverse the power-supply polarity if a PNP transistor is used.

If a maximum source voltage is specified in the design problem, the maximum peak-to-peak output voltage is set. For class A operation (Chapter 1), the collector is operated at about one-half the source voltage. This permits the maximum positive and negative swing of output voltage. (The peak-to-peak output voltage cannot exceed the source voltage.) Generally, the absolute maximum peak-to-peak output can be between 90% and 95% of the source. For example, if the source is 20 V, the collector operates at 10 V (Q point) and swings from about 1 to 9 V. However, there is *less distortion* if the output is one-half to one-third of the source.

If a source voltage is not specified, two major factors should determine the value: (1) the maximum collector-voltage rating of the transistor, and (2) the desired output voltage (or the desired collector voltage at the operating point). The maximum collector-voltage rating must not be exceeded. As discussed in Chapter 1, choose a source voltage that does not exceed 90% of the maximum rating. This allows a 10% safety factor. Any desired output voltage (or collector Q-point voltage) can be selected within these limits.

If the circuit is to be battery operated, choose a source voltage that is a multiple of 1.5 V.

If a peak-to-peak output voltage is specified, add 10% (to the peak-to-peak value) to find the absolute minimum source voltage.

If a collector Q-point voltage is specified, double the collector Q-point voltage.

For minimum distortion, use a source that is two to three times the desired output voltage.

If the input and/or output impedances are specified, the values of RB and RL are set, as shown by Fig. 3–11. However, certain limitations for RB and RL are imposed by the trade-off between gain and stability. For example, for a stage current gain of 10 and nominal stability, RB should be 10 times RE. RB should never be greater than 20 times RE (for maximum current gain and minimum stability), nor should RB be less than 5 times RE (for minimum current gain and maximum stability). Since RE is between 100 and 1000 Ω in a typical class A circuit, RB (and the input impedance) should be between 500 (100 × 5) and 20,000 (1000 × 20).

If the input and/or output impedances are not specified, try to match the impedances of the previous stage and following stage, where practical. This provides maximum power transfer.

   The values of coupling capacitors C1 and C2 depend on the low-frequency limit at which the amplifier operates and on the resistances with which the capacitors operate. (This is discussed in Sec. 3–1.3 and shown in Fig. 3–3.) Either the 1- or 3-dB low-frequency cutoff point can be selected.

   The *minimum a-c beta of the transistor* should be higher than the desired gain, even though gain for the circuit is set by circuit values. Since the current of Fig. 3–11 is designed for maximum gains of 20, any transistor with a minimum beta of 20 should be satisfactory.

3-5.2   Design Example

Assume that the circuit of Fig. 3–11 is to be used as a single-stage voltage amplifier. The desired output is 3 V (peak to peak) with a 2000-Ω impedance. The input impedance is to be 1000 Ω. The input signal is 0.3 V (peak to peak). This requires a voltage gain of 10. The low-frequency limit is 30 Hz, with a high frequency limit of 100 kHz. Minimum distortion is desired. (The circuit should not be overdriven.) The source voltage and transistor type are not specified, but the circuit is to be battery operated.

   Supply Voltage and Operating Point.   The 3-V output can be obtained with a 4.5-V battery. However, for minimum distortion, the supply should be two or three times the desired output, or between 6 and 9 V. A 9-V battery provides the maximum insurance against distortion. Therefore, the collector Q-point voltage should be 4.5 V (9/2 = 4.5).

   Load Resistance and Collector Current.   The value of RL should provide the output impedance of 200 Ω. With a 4.5-V drop across RL, the collector current is 2.25 mA (4.5/2000).

   Emitter Resistance, Current, and Voltage.   To provide a voltage gain of 10, the value of RE should be one-tenth of RL, or 200 Ω (2000/10). (If tests provide gain slightly below 10, try reducing the value of RE to the next lower standard value of 180 Ω.) The current through RE is the collector current of 2.25 mA, plus the base current. Assuming a stage current gain of 10, the base current is 0.225 mA (2.25/10). The combined currents through RE are 2.475 mA (2.25 + 0.225). This produces a drop of 0.495 V across RE (2.475/200). For practical design, this can be rounded off to 0.5 V.

   Input Resistance and Current.   The value of RB should provide the input impedance of 1000 Ω. The value of RB should also be at least five times RE, which makes the 1000/200-Ω relationship correct. This relationship provides maximum circuit stability. The base voltage is 0.5 V higher than the emitter voltage, or 1 V (0.5 + 0.5). With a 1-V drop across RB, the current through RB is 1 mA (1/1000).

Base Resistance and Current.   The value of RA should be sufficient to drop the 9-V source to 8 V so that the base is 1 V above ground. The current through RA is the current through RB of 1 mA, plus the base current of 0.225 mA, or 1.225 mA (1.0 + 0.225). The resistance required to produce an 8-V drop with 1.225 mA is 6530 Ω (8/1.225). Use a 6500-Ω standard resistance as the trial value.

Coupling Capacitors.   The value of C1 forms a high-pass filter with RB. The high limit of 100 kHz can be ignored. The low-frequency limit of 30 Hz requires a capacitance value of about 10 μF (if a 1-dB drop at 30 Hz is required). This is found by 1/(3.2 × 1000) or 1/(3.2 × F × R). If a greater drop can be tolerated, the value of C1 can be lowered (about 5μF for 3-dB drop).

The capacitance value of C2 is found in the same manner, except that the resistance value must the load resistance RL, or 2000 Ω. For a 1-dB drop at 30 Hz, the value of C2 should be about 5.2 μF, or 6 μF for practical design. This is found by 1/(3.2 × 30 × 2000). If a greater drop can be tolerated, the value of C2 can be lowered (about 2 or 3 μF for a 3-dB drop).

The voltage values of C1 and C2 should be 1.5 times the maximum voltage involved, or 13.5 V (9 × 1.5). Use 15-V capacitors for practical design.

Transistor Selection.   Some circuits must be designed around a given transistor. In such cases, the source voltage, collector current, power dissipation, and so on, must be adjusted accordingly. In this example a 2N337 transistor is available and can be used, provided that the transistor meets the circuit requirements.

The following table compares 2N337 characteristics and circuit requirements. (Note that any other transistor can be used if the transistor meets the same requirements.)

2N337 Characteristics	*Circuit Requirements*
Collector voltage (max): 45 V	9-V source
Collector current (max): 20 mA	2.25-mA nominal (at saturation, collector current is 4.5 mA; 9/2000)
Power dissipation (max): 125 mW	Approximately 10 mW (4.5 V × 2.25 mA)
(The 2N337 must be derated 1 mW/°C above 25°C ambient temperature.)	
A-c beta (as hfe)(min): 19	10, minimum gain at low frequency
(typical): 55	
Beta (as hfe at 100 kHz): 25	10, minimum gain at 100 kHz

3-6   BASIC TWO-JUNCTION AMPLIFIER
WITH EMITTER BYPASS

Figure 3–11 shows (in phantom) a bypass capacitor C3 across emitter resistor RE. This permits RE to be removed from the circuit as far as the signal is concerned, but leaves RE in the circuit (with regard to direct current). With RE removed from the signal path, the voltage gain is approximately RL/dynamic-resistance (of the transistor), and the current gain is about equal to the a-c beta of the transistor. Use of an emitter-bypass capacitor permits the highly tempera-ture stable d-c circuit to remain intact, while providing a high signal gain.

3-6.1 Design Considerations for Emitter Bypass

An emitter-bypass capacitor creates some problems. Transistor input impedance changes with frequency and from transistor to transistor, as does beta. This means that current and voltage gains can only be approximated. When the emit-ter resistance is bypassed, the circuit input impedance is about equal to beta times the transistor input impedance, so circuit input impedance is even more subject to variation (and is unprectictable). Generally, emitter bypass is used only in those cases where high voltage gain must be obtained from a single stage.

The value of emitter bypass C3 is found by

$$\text{capacitance} = \frac{1}{6.2 \text{ FR}}$$

where capacitance is in farads, F is the low-frequency limit in hertz, and R is input impedance of the transistor in ohms.

3-6.2 Design Example

Assume that C3 is to be used as an emitter bypass for the circuit described in Sec. 3–5 to increase voltage gain. All the circuit values remain the same, as does the low-frequency limit of 30 Hz.

Further assume that a 2N337 transistor is to be used and that the dynamic input resistance is 50 Ω (obtained from the datasheet). This provides a voltage gain of about 40 (2000/50). The desired 3-V output can then be obtained with a 0.075-V input, rather than the 0.3-V input required for the previous example.

The low-frequency limit of 30 Hz requires a C3 capacitance value of 107 μF (110 μF for practical design). This is found by 1/(6.2 × 30 × 50). The voltage value of C3 should be 1.5 times the maximum voltage involved, or 1.5 V (1 × 1.5). Use a 3-V capacitor for practical design.

3-7 BASIC TWO-JUNCTION AMPLIFIER WITH PARTIALLY BYPASSED EMITTER

Figure 3–12 is the working schematic of a basic, single-stage audio amplifier with a partially bypassed emitter. This design is a compromise between the basic design without bypass (Sec. 3–5) and the fully bypassed emitter (Sec. 3–6). The d-c characteristics of both the unbypassed and partially bypassed circuits are essentially the same.

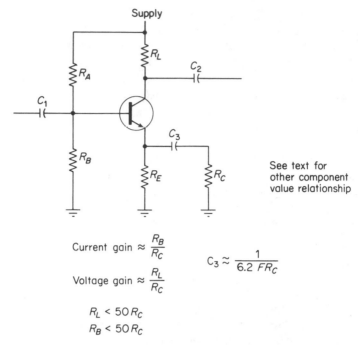

Current gain $\approx \dfrac{R_B}{R_C}$

Voltage gain $\approx \dfrac{R_L}{R_C}$

$C_3 \approx \dfrac{1}{6.2\ FR_C}$

$R_L < 50\,R_C$

$R_B < 50\,R_C$

FIGURE 3–12   Basic audio amplifier with partially bypassed emitter.

All circuit values (except C3 and RC) can be calculated in the same way for both circuits. As shown in Fig. 3–12, the voltage and current gains for a partially bypassed amplifier are greater than for an unbypassed circuit, but less than for the fully bypassed circuit.

3–7.1   Design Considerations for Partially Bypassed Emitter

The design considerations for the circuit of Fig. 3–12 are the same as those for Fig. 3–11, except for the effect of C3 and RC. The value of RC should be chosen on the basis of voltage gain, even though current gain is increased when voltage gain increases. RC should be substantially smaller than RE. Otherwise, there is no advantage to the partially bypassed design. However, a smaller value for RC requires a larger value for C3, since the C3 value depends on the RC value and the desired low-frequency cutoff point.

The value of C3 is found by

$$\text{capacitance} = \frac{1}{6.2\ FR}$$

where capacitance is in farads, F is the low-frequency limit in hertz, and R is input impedance of the transistor in ohms.

3-7.2 Design Example

Assume that the circuit of Fig. 3–12 is to be used in place of the Fig. 3–11 circuit and that the desired voltage gain is 25. Selection of component values, supply voltages, operating point, and the like, is the same for both circuits. The only difference in design is selection of values for C3 and RC. The value of RC should be the value of RL divided by the desired voltage gain, or 80 Ω (2000/25).

The low-frequency limit of 30 Hz requires a C3 capacitance value of 67 μF. This is found by 1/(6.2 × 30 × 80). The voltage value of C3 should be 1.5 times the maximum voltage involved, or 1.5 V (1 × 1.5). Use a 3-V capacitor for practical design.

3-8 BASIC AF AMPLIFIER (FET)

As discussed, the characteristics of FETs are quite different from those of two-junction transistors. This makes for considerable differences in amplifier circuit characteristics. For this reason, we summarize FET amplifier characteristics before going into a basic FET amplifier stage.

Figure 3–13 shows a basic FET bias circuit. This circuit can be converted to a single FET amplifier stage by the addition of input and output coupling capacitors. The basic bias network is modified as necessary to produce a FET stage with the desired characteristics (stage gain, input/output impedance, and so on).

The purpose of the basic bias circuit is to establish a given ID and to maintain that ID (plus or minus some given tolerance) over a given temperature range.

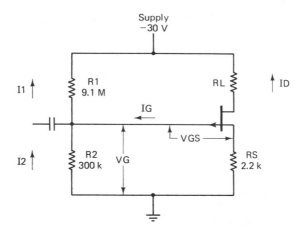

FIGURE 3-13 Basic FET bias circuit.

Often, this is to keep the FET at the 0TC operating point. When the basic FET circuit is used to form a linear amplifier, the output (drain terminal) should be at one-half the supply voltage, if maximum output voltage swing is wanted.

Remember that amplifiers rarely operate under static conditions. The basic bias circuit of Fig. 3–13 is used as a reference or starting point for design or analysis. The actual circuit configuration and (especially) the bias circuit values are selected on the basis of dynamic circuit conditions (desired output voltage swing, expected input signal level, and the like).

3-8.1 Selecting Values for Basic FET Bias Circuit

Assume that the circuit of Fig. 3–13 is to maintain ID at 1 ± 0.25 mA over a temperature range from − 55 to + 125 °C, with a supply voltage of 30 V. The first step is to draw a 1/RS load line on the FET transfer characteristics, as shown in Fig. 3–14. (Note that Fig. 3–14 is similar to Fig. 1–30.)

As shown by the equations, the value of RS is set by the limits of VGS and ID. The value of VGS(min) is the point where the ID(min) of 0.75 mA crosses the high-temperature-limit curve of + 125 °C, or at about 0.8 V. The value of VGS (max) is the point where the ID(max) of 1.25 mA crosses the low-temperature-limit curve of − 55 °C, or at about 1.9 V. Using these values, the first trial value for RS is (1.9 − 0.8)/(1.25 − 0.75) = 2.2 k.

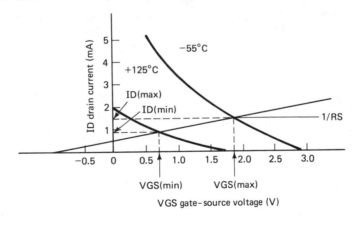

$$RS = \frac{VGS(max) - VGS(min)}{ID(max) - ID(min)}$$

$$R2 = \frac{VGR1}{supply - VG}$$

$$VG = \frac{[ID(min) \times VGS(max)] - [ID(max) \times VGS(min)]}{ID(max) - ID(min)}$$

$$RL(max) = \frac{VRL}{ID(max)}$$

$$VRL = supply - [1.5 \times VGS(off)] - [ID(max) \times RS]$$

FIGURE 3-14 The 1/RS load line for basic FET stage.

The fixed bias voltage VG is determined from the intercept of the 1/RS load with the VGS axis and is computed using the same set of values, as shown in the Fig. 3–14 equations:

$$VG = \frac{(0.75 \times 1.9) - (1.25 \times 0.8)}{0.5} = 0.85V$$

The maximum value of R1 is determined by the maximum gate reverse current, as specified on the datasheet. The variation in VG versus temperature is generally not too great if the R1 value is chosen so that I1 (Fig. 3–13) is at least six times greater than the maximum reverse current.

Assume a maximum gate reverse current of 0.5 μA at the maximum temperature of 125 °C, and the drop across R1 is about 29 V (30 V $-$ a nominal VGS of 1 $=$ 29 V). Six times 0.5 μA is 3 μA. A value of 9.1 M Ω for R1 produces about 3.2 μA for I1.

The value of R2 is found from a simple voltage-divider relationship, ignoring the effect of IG, as shown in the Fig. 3–14 equation:

$$R2 = \frac{0.85 \times 9.1 \times 10^6}{30 - 0.85} = 300 \text{ k (approx)}$$

The maximum value of RL is determined by the voltage drops across RL and ID(max). As shown by the Fig. 3–14 equations, the voltage drop across RL (or VRL) is

$$VRL = 30 - (1.5 \times 3.6) - (1.25 \times 2.2) = 21.85 \text{ V}$$

Note that ID(max) is shown on Fig. 3–14. The approximate value for VGS(off) is found by taking the slope of the curve at VGS $=$ 0 (where the curve intersects the ID axis) and extending the slope to the VGS axis. For example, using the 125 °C curve of Fig. 3–14, the value of VGS(off) is about 3.6 (1.8 $\times$ 2).

With VRL established at 21.85 V and an ID(max) of 1.25 mA, the maximum value of RL is

$$RL(max) = \frac{21.85}{1.25} = 17.48$$

Assuming a standard 1% resistor, a value of 16.9 k can be used for RL.

3-8.2 Small-Signal Analysis of Basic FET Bias Circuit

Figure 3–15 shows a model for analysis of a basic FET bias circuit. This model omits Crss and uses only the real parts of Yis, Yfs, and Yos. Therefore, the model is useful up to about 100 kHz. As discussed in Chapter 4, both the

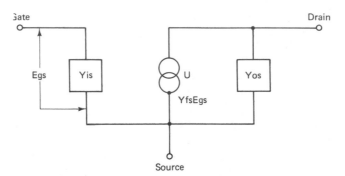

FIGURE 3-15   Model for analysis of a basic FET bias circuit.

capacitance and the imaginary parts of the Y figures must be used in the RF range.

Figure 3–16 shows the schematics for common-source, common-drain (source-follower), and common-gate circuits. The *approximate* equations for voltage gain, input impedance, and output impedance are also included. (The equations for finding the resistance values of R1, R2, RS, and RL are discussed in Sec. 3–8.1.)

For the *common-source circuit* (Fig. 3–16a), the omission of Crss affects the equations for input impedance and voltage gain. However, for frequencies below about 100 kHz, the error is minimal. If only d-c feedback is required, then the source resistor RS is bypassed. With a capacitor across RS, the effects on the design equations are to set RS at zero. Under these conditions, voltage gain is the product of RL and Yfs.

With RS not bypassed, the circuit characteristics are virtually independent of FET parameters (with the exception of Yfs). Instead, the circuit characteristics (impedance, gain, and so on) depend on RL and RS. By using precision resistors with close temperature coefficients, the common-source circuit can be made very stable over a wide temperature range.

The *common-drain (source-follower) configuration* (Fig. 3–16b) is a very useful basic circuit. Some of the properties are a voltage gain always less than unity with no phase inversion, low output impedance (essentially set by the value of RS), high input impedance, large signal swing, and active impedance transformation.

The *common-gate stage* (Fig. 3–16c) offers impedance transformation opposite to that of the source follower. Common gate produces low input impedance and high output impedance. The voltage gain is the same as for common source, except that there is no phase inversion.

The circuits of Fig. 3–16 are for P-channel JFETs. Reverse the polarity for N-channel FETs. MOSFET (or IGFET) devices may not conform exactly to the relationships shown, but are sufficiently close for analysis of FET amplifier circuits.

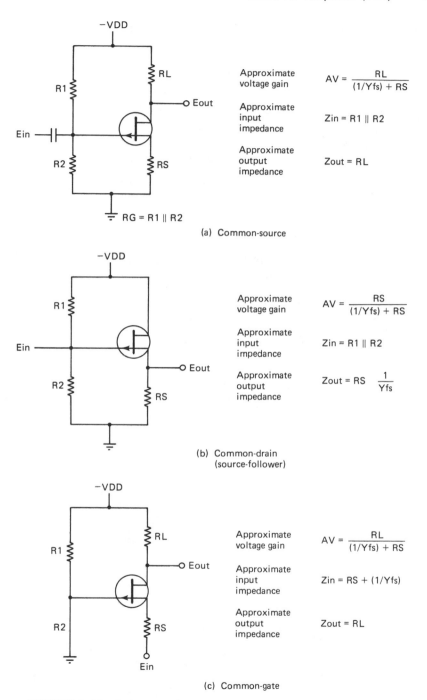

Approximate voltage gain $\quad AV = \dfrac{RL}{(1/Yfs) + RS}$

Approximate input impedance $\quad Zin = R1 \parallel R2$

Approximate output impedance $\quad Zout = RL$

(a) Common-source

Approximate voltage gain $\quad AV = \dfrac{RS}{(1/Yfs) + RS}$

Approximate input impedance $\quad Zin = R1 \parallel R2$

Approximate output impedance $\quad Zout = RS \quad \dfrac{1}{Yfs}$

(b) Common-drain (source-follower)

Approximate voltage gain $\quad AV = \dfrac{RL}{(1/Yfs) + RS}$

Approximate input impedance $\quad Zin = RS + (1/Yfs)$

Approximate output impedance $\quad Zout = RL$

(c) Common-gate

FIGURE 3-16 Schematics for common-source, common-drain (source-follower), and common-gate FET circuits.

3-8.3   *Single-Stage Common-Source FET Amplifier*

Figure 3–17 shows a basic single-stage FET amplifier. Note that the basic amplifier circuit is similar to the basic common-source circuit of Figs. 3–13 and 3–16a, except that input and output coupling capacitors C1 and C2 are added to prevent dc flow to and from external circuits. Bypass capacitor C3, connected across RS, is required only under certain conditions.

Input to the amplifier is applied between gate and ground across R2. Output is taken across the drain and ground. The input signal adds to, or subtracts from, the bias voltage across R2. Variations in bias voltage cause corresponding variations in ID and the voltage drop across RL. The drain voltage (or circuit output) follows the input signal waveform, except that the output is inverted in phase.

Variations in ID also cause variations in voltage drop across RS and a change in the gate–source bias relationship. The change in bias that results from the voltage drop across RS tends to cancel the initial bias change caused by the input signal and serves as a form of negative feedback to increase stability (and limit gain), as described in Sec. 3–5.

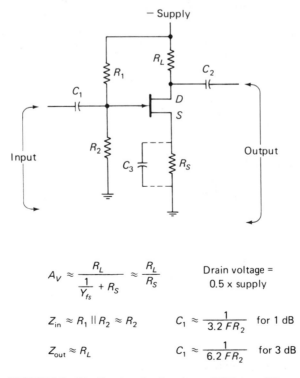

$$A_V \approx \frac{R_L}{\dfrac{1}{Y_{fs}} + R_S} \approx \frac{R_L}{R_S} \qquad \begin{array}{l} \text{Drain voltage =} \\ 0.5 \times \text{supply} \end{array}$$

$$Z_{in} \approx R_1 \| R_2 \approx R_2 \qquad C_1 \approx \frac{1}{3.2\,F R_2} \quad \text{for 1 dB}$$

$$Z_{out} \approx R_L \qquad C_1 \approx \frac{1}{6.2\,F R_2} \quad \text{for 3 dB}$$

FIGURE 3-17   Basic single-stage FET amplifier.

Circuit Analysis.   The outstanding characteristic of the Fig. 3–17 circuit is that circuit gain, stability, impedance, and so on, are determined (primarily) by circuit values, rather than FET characteristics. The circuit shown is a P-channel FET. The power supply polarity must be reversed if an N-channel FET is used.

The *maximum peak-to-peak output voltage* is set by the supply voltage. For class A operation, the drain is operated at about one-half the supply voltage. This permits the maximum positive and negative swing of output voltage. Generally, the absolute maximum peak-to-peak output can be between 90% and 95% of the supply. For example, if the supply is 20 V, the drain operates at 10 V (Q point) and swings from about 1 to 19 V. However, there is less distortion if the output is one-half to one-third of the supply. In any circuit, the maximum drain–source voltage VDS of the FET cannot be exceeded.

The *input and/or output impedances* are set by the resistance values (R1, R2, and RL), as shown in Fig. 3–17. However, certain limitations for R2 and RL are imposed by trade-offs (for gain, impedance match, zero temperature coefficient, operating point, and so on).

For example, the output impedance is set by RL. If RF is increased to match a given impedance, the gain increases (all other factors remaining the same). However, an increase in RL lowers the drain voltage Q point, since the same amount of ID flows through RL, producing a larger voltage drop. This reduces the possible output voltage swing. A reduction in RL produces the opposite effect, increasing the drain-voltage Q point, but still reducing output voltage swing.

When R1 is much larger than R2 (which is generally the case), the input impedance of the circuit is set by the value of R2. If R2 is increased (or decreased) far from the value found in Sec. 3–8.1, the no-signal ID point changes. If the value of ID is chosen for 0TC and R2 is changed drastically, the drain current changes, and the FET is no longer operating at the 0TC point. Generally, this is an undesirable condition.

Typically, the common-source FET circuit is chosen for high input impedance, thus presenting a low current drain to the preceding circuit. If the FET stage must provide a low input impedance, the common-gate circuit of Fig. 3–16c is generally preferred.

The values of C1 and C2 depend on the low-frequency limit at which the amplifier is to operate. As discussed in Sec. 3–1, C1 forms a high-pass (or low-cut) RC filter with R2. Capacitor C2 forms another filter with the input resistance of the following stage (or load).

For a given resistance value, a lower frequency requires a larger capacitor value, as shown by the equations of Fig. 3–17. Of course, if the resistance can be made larger (with the same desired frequency), the capacitor value can be reduced. Since FETs are high-impedance voltage-operated devices, the coupling capacitors in FET amplifiers are generally small (in relation to those of two-junction transistors).

Sufficient Feedback Concept.   The design of a FET amplifier stage can be checked by noting if there is sufficient feedback. Such a condition occurs when the calculated gain is at least 75% of the RL/RS ratio. If so, there is sufficient feedback to be of practical value.

As an example, assume that RL is 16.9 k, RS is 2.2 k, and Yfs is 2000 μmho. The ratio of RL/RS is slightly over 7.6, with the gain slightly over 6:

$$RL/RS = 16.9/2.2 = 7.6+$$

$$\text{gain or AV} = \frac{16.9}{(1/2000 \ \mu mho) + 2.2 \ k} = \frac{16.9 \ k}{2.7 \ k} = 6 +$$

Since 75% of 7.6 is 5.7, the gain of 6 is greater, and there is sufficient feedback. Under these conditions the design should be stable.

3-8.4   Source Bypass for FET Amplifier

Figure 3–17 shows (in phantom) a bypass capacitor C3 across source resistor RS. This arrangement permits RS to be removed from the circuit as far as the signal is concerned, but leaves RS in the circuit (in regards to direct current). With RS removed from the signal path, the voltage gain is about equal to Yfs × RL, so the bypass capacitor permits a temperature-stable d-c circuit to remain intact, while providing a high signal gain.

A source resistance bypass capacitor also creates some problems. The Yfs changes with frequency and from FET to FET, so circuit gain can only be approximated. The source bypass is generally used where maximum gain must be obtained from a single-stage amplifier, and a stable gain is of little concern. The value of C3 is found by

$$\text{capacitance} = \frac{1}{6.2F(RS \times 0.2)}$$

where capacitance is in microfarads, F is low-frequency limit in hertz, and RS is in megohms.

3-8.5   Basic FET Amplifier with Partially Bypassed Source

Figure 3–18 is the working schematic of a basic single-stage FET amplifier with a partially bypassed source resistor. This design is a compromise between the basic design without bypass and the fully bypassed source. The d-c characteristics of both the unbypassed and partially bypassed circuits are essentially the same. All circuit values (except C3 and RC) can be calculated in the same way for both circuits.

As shown in Fig. 3–18, the voltage gain for a partially bypassed FET amplifier is greater than for the unbypassed circuit, but less than for the fully

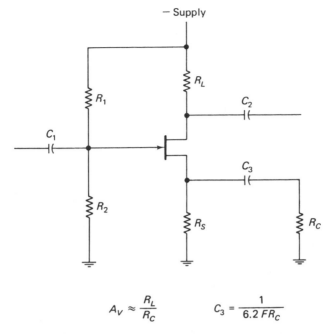

$$A_V \approx \frac{R_L}{R_C} \qquad C_3 = \frac{1}{6.2\,FR_C}$$

FIGURE 3-18   Basic single-stage FET amplifier with partially bypassed source resistor.

bypassed circuit. However, the gain can be set to an approximate value by selection of circuit values, unlike the fully bypassed circuit, where gain is entirely subject to variations in Yfs.

The value of RC is chosen on the basis of desired voltage gain. RC should be substantially smaller than RS. Otherwise, there is no advantage to the partially bypassed design. As shown by the equations, voltage gain is about equal to RL/RC. This holds true unless both Yfs and RC are very low (where 1/Yfs is about equal to RC). In such a case, a more accurate gain approximation is

$$\frac{RL}{(1/Yfs) \, + \, RC}$$

3-8.6   Single-Stage Common-Drain FET Amplifier (Source Follower)

Figure 3–19 is the working schematic of a basic single-stage FET source-follower (common-drain) circuit. Note that this circuit is similar to that of Fig. 3–16b, except that input and output coupling capacitors C1 and C2 are added to prevent d-c flow to and from external circuits.

Input to the source follower is applied between gate and ground across R2. Output is taken across the source and ground. The input signal adds to, or subtracts from, the bias voltage across R2. Variations in bias voltage cause

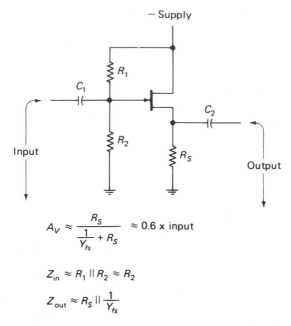

$$A_V \approx \frac{R_s}{\frac{1}{Y_{fs}} + R_s} \approx 0.6 \times \text{input}$$

$$Z_{in} \approx R_1 \parallel R_2 \approx R_2$$

$$Z_{out} \approx R_s \parallel \frac{1}{Y_{fs}}$$

FIGURE 3-19   Basic single-stage FET source-follower (common-drain) circuit.

corresponding variations in ID and in the voltage drop across RS. The source voltage (or circuit output) follows the input signal waveform and remains in phase.

Variations in voltage drop across RS change the gate–source bias relationship and tend to cancel the initial bias change caused by the input signal. This serves as a form of negative feedback to increase stability (and limit gain).

The circuit of Fig. 3–19 is used primarily where high input impedance and low output impedance (with no phase inversion) are required, but no gain is needed. The source follower is the FET equivalent of the two-junction transistor emitter follower described in Sec. 1–6 (and to the vacuum-tube cathode follower, if you can remember that far back!).

Circuit Analysis.   The Q-point voltage at the circuit output (source terminal) is set by ID under no-signal conditions and by the value of RS. Since RS is typically small, the Q-point voltage is quite low (in comparison to the common-source amplifier). In turn, the maximum allowable peak-to-peak output voltage is also low. For example, if the source is at 1 V with no signal, the maximum possible peak-to-peak output is less than 1 V. Of course, the value of RS can be increased as necessary to permit a higher output.

The input and/or output impedances are set by the resistance values (R1, R2, and RS). However, certain limitations for RS are imposed by trade-offs for impedance match and output Q point.

For example, the output impedance is the parallel resistance combination of RS and 1/Yfs. If RS is made very small (less than 10 times) in relation to 1/Yfs, the output impedance is about equal to RS. A low value of RS decreases the source (output) voltage Q point, thus reducing output voltage swing. If RS is made large in relation to 1/Yfs, the output impedance is about equal to 1/Yfs and is subject to variation with frequency and from FET to FET.

There is no voltage gain for a source follower. Typically, the output voltage is about 0.6 times the input voltage, depending on the ratio of 1/Yfs to RS. However, the source follower is capable of current gain and thus *power gain*.

For example, assume that 1 V is applied to the input and 0.6 V is taken from the output. Further assume that the input impedance is 300 k, and the output impedance is 300 Ω. The input power is about 0.0033 mW, while the output power is about 1.2 mW, indicating a power gain of about 350.

3-8.7 Single-Stage Common-Gate FET Amplifier

Figure 3–20 is the working schematic of a basic single-stage FET common-gate amplifier. Note that the basic circuit is similar to that of Fig. 3–16c, except that C1 and C2 are added to prevent d-c flow to and from external circuits.

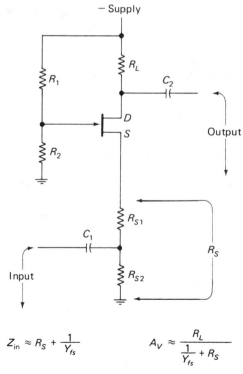

$$Z_{in} \approx R_S + \frac{1}{Y_{fs}} \qquad\qquad A_V \approx \frac{R_L}{\frac{1}{Y_{fs}} + R_S}$$

FIGURE 3-20   Basic single-stage FET common-gate amplifier. $Z_{out} \approx R_L$

Input is applied at the source across a portion of RS. Typically, the value of RS1 is equal to RS2, although some circuits divide the resistance value unequally. The total value of RS (RS1 + RS2) must be considered when calculating the d-c characteristics of the circuit.

Output is taken across the drain and ground. The input signal adds to or subtracts from the bias voltage across RS. Variations in bias voltage cause corresponding variations in ID and in the voltage drop across RL. The drain voltage (or circuit output) follows the input signal in phase.

Variations in voltage drop across RS change the gate–source bias relationship. This change in bias tends to cancel the initial bias change caused by the input signal and serves as a form of negative feedback to increase stability (and to limit gain).

The circuit of Fig. 3–20 is used primarily where low input impedance and high output impedance (with no phase inversion) are required. Gain is determined (primarily) by circuit values, rather than FET characteristics. The common-gate amplifier is the FET equivalent of the two-junction transistor common-base amplifier described in Sec. 1–5 (and to the vacuum-tube common-grid amplifier).

Circuit Analysis.   Although the input and/or output impedances are set by the resistance values, the input impedance of the circuit depends on the reciprocal of Yfs (1/Yfs) factor. This is true unless the value of RS is many times (at least 10) that of 1/Yfs.

Capacitor C1 forms a high-pass RC filter with RS2. Capacitor C2 forms another high-pass filter with the input resistance of the following stage (or the load). Using a 1-dB loss as the low-frequency cutoff point, the value of C1 is approximately 1/3.2FRS2, where capacitance is in microfarads, F is the low-frequency limit in hertz, and RS2 is in megohms.

3–8.8 Basic FET Amplifier without Fixed Bias

Figure 3–21 shows a basic single-stage FET amplifier without fixed bias. Capacitors C1 and C2 prevent d-c flow to and from external circuits. The resistor R1 provides a path for bias and signal voltages between gate and source.

Input is applied between gate and ground across R1. Output is taken across the drain and ground. The input signal adds to, or subtracts from, the bias voltage across R1. Variations in bias voltage cause corresponding variations in ID and the voltage drop across RL. The drain voltage (or circuit output) follows the input signal waveform, except that the output is inverted in phase.

Variations in ID also cause variation in voltage drop across RS and a change in the gate–source bias relationship. The change in bias that results from the voltage drop across RS tends to cancel the initial bias change caused by the input signal and serves as a form of negative feedback to increase stability (and limit gain).

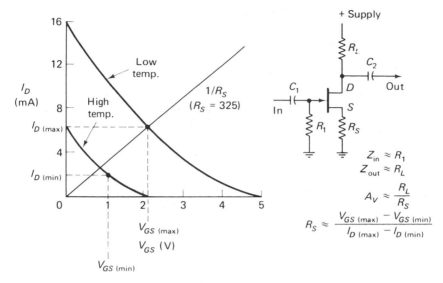

FIGURE 3-21   Basic single-stage FET amplifier without fixed bias.

The major difference in the circuit of Fig. 3-21 and a FET amplifier with fixed bias is that the amount of ID at the Q point is set entirely by the value of RS. It may not be possible to get a desired ID with a practical value of RS, so it may not be possible to operate at the zero-temperature-coefficient point. If this is of less importance than minimizing the number of circuit components (elimination of one resistor), the circuit of Fig. 3-21 can be used in place of the fixed-bias FET amplifier.

Circuit Analysis.   The input and/or output impedances are set by the resistance values of R1, RS, and RL. However, certain limitations for the resistance values are imposed by trade-offs (for gain, impedance match, operating point, and so on).

For example, the value of RS sets the amount of bias, and thus the amount of ID. At the same time, the ratio of RL/RS sets the amount of gain. Going further, the value of RL sets the output impedance. If RS is changed to change the ID, both the gain and Q point are changed. If RL is changed to match a given impedance, both the gain and Q point change.

The input impedance is set by R1. A change in R1 has little effect on gain, operating point, or output impedance. However, R1 forms a high-pass RC filter with C1. A decrease in R1 requires a corresponding increase in C1 to accommodate the same low-frequency cutoff point. As a general rule, the value of R1 is high (in the megohm range). This minimizes current drain on the stage ahead of the FET.

3-8.9 Basic FET Amplifier with Zero Bias

Figure 3–22 is the working schematic of a basic single-stage FET amplifier operating at zero bias and without feedback. Capacitors C1 and C2 prevent d-c flow to and from external circuits. Resistor R1 provides a path for signal voltages between the gate and source.

Input to the amplifier is applied between gate and ground across R1. Output is taken across the drain and ground. Variations in gate voltage cause corresponding variation in ID and the voltage drop across RL, so the drain voltage (or circuit output) follows the input signal waveform (except that the output is inverted in phase).

Any FET has some value of ID at zero VGS. If the FET has characteristics similar to those of Fig. 3–22, the ID varies between about 1.75 and 4 mA, depending on temperature, and from FET to FET. Therefore, with a zero-bias circuit, it is impossible to set the Q-point ID at any particular value. Likewise, drain voltage Q point is subject to considerable variation. Since there is no source resistor, there is no negative feedback, and there is no means to control this variation in ID. For these reasons, the zero-bias circuit is used where circuit stability is of no particular concern.

Circuit Analysis.   The input and/or output impedances are set by the values of R1 and R2. However, as in the case of the other circuit, there are trade-offs for gain, impedance match, operating point, and so forth.

For example, the value of RL sets the amount of gain (with a stable Yfs) and the drain voltage operating point (with a stable ID). A change in Yfs (which is usually accompanied by a change in ID) causes a change in gain (and probably a shift in Q-point voltage). At best, the zero bias voltage is unstable, even though the input and output impedances remain fairly constant. In analyzing

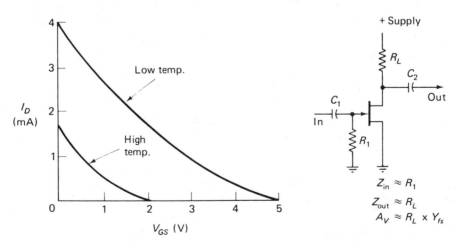

FIGURE 3–22   Basic single-stage FET amplifier with zero bias.

the zero-bias circuit, both the minimum and maximum values of ID must be considered, as well as the minimum and maximum Yfs.

3-9   MULTISTAGE AMPLIFIERS

When stable voltage gains greater than about 20 are required and it is not practical to bypass the emitter (or source) resistor of a single stage, two or more transistor amplifier stages can be used in *cascade* (where the output of one transistor is fed to the input of a second transistor).

In theory, any number of two-junction or FET amplifiers can be connected in cascade to increase voltage gain. In practice, the number of stages is usually limited to three. The overall gain of the amplifier is equal (approximately) to the cumulative gain of each stage, multiplied by the gain of the adjacent stage.

As an example, if each stage of a three-stage amplifier has a gain of 10, the overall gain is (approximately) 1000 ($10 \times 10 \times 10$). Since it is possible to design a very stable single stage with a gain of 10 and adequately stable stages with gains of 15 to 20, a three-stage amplifier can provide gains in the 1000 to 8000 range. Generally, this is more than enough voltage gain for most practical applications. Using the 8000 figure, a 1-μV input signal (say from a low-voltage transducer or delicate electronic device) can be raised to the 8-mV range, while maintaining stability in the presence of temperature and power-supply variations.

IC Amplifiers.   In present design, IC amplifiers are generally used when it is not practical to get sufficient gain with one or two transistors. For this reason, we do not concentrate on the design of multistage amplifiers in this book. However, certain factors must be considered when using more than one transistor as an amplifier. The following paragraphs summarize these factors.

3-9.1   Basic Considerations for Multistage Amplifiers

Any of the single-stage amplifiers described in this chapter can be connected to form a two-stage or three-stage *voltage amplifier*. For example, the basic stage (without emitter or source bypass) can be connected to two like stages in cascade. The result is a highly temperature stable voltage amplifier. Since each stage has its own feedback, the gain is precisely controlled and very stable.

It is also possible to mix stages to get some given design goal. For example, a three-stage amplifier can be designed using a highly stable, unbypassed amplifier for the first stage and two bypassed amplifiers for the remaining stages. Assuming a gain of 10 for the unbypassed stage and gains of 30 for the bypassed stages, this results in an overall gain of 9000. Of course, with bypassed stages, the gain depends on the transistor characteristics (dynamic impedance, hfe, Yfs, and so on) and is therefore unpredictable. However, once the gain is established for a given amplifier, the gain should remain fairly stable.

Since design of a multistage, capacitor-coupled voltage amplifier is essentially the same as for individual stages, no specific circuit example is given. In practical terms, each stage is analyzed and designed as described in this chapter. However, a few precautions must be observed.

Distortion and Clipping.   As with any high-gain amplifier, the possibility of *overdriving* a multistage amplifier is always present. If the maximum input signal is known, check this value against the overall gain and the maximum allowable output signal swing.

As an example, assume an overall gain of 1000 and a supply voltage of 20 V. Typically, this implies a 10-V Q point (for the output collector or drain) and a 20-V (peak-to-peak) output swing from 0 to 20 V. In practice, a swing from about 1 to 19 V is more realistic. Either way, a 20-mV input (peak to peak), multiplied by a gain of 1000, will drive the final output to the limits and possibly into distortion or clipping.

Feedback.   When each stage of a multistage amplifier has feedback (local or stage feedback), the most precise control of gain is obtained. However, such feedback is often unnecessary. Instead, *overall feedback* (or *loop feedback*) can be used, where part of the output from one stage is fed back to the input of the previous stage.

Usually, feedback is through a resistor (to set the amount of feedback), and the feedback is from the final stage to the first stage. However, it is possible to use feedback from one stage to the next (second stage to first stage, third stage to second stage, and so on).

Feedback Phase Inversion.   There is a problem of phase inversion when using loop or overall feedback. In a common-emitter or common-source amplifier, the phase is inverted from input to output. If feedback is between two stages, the phase is inverted twice, resulting in *positive feedback*. This usually produces *oscillation*. In any event, positive feedback does not stabilize gain. The phase-inversion problem can be overcome, when multiple stages are involved, by connecting the output collector (or drain) of the second stage back to the emitter (or source) of the first stage. This produces the desired *negative feedback*.

As an example, if the base (or gate) of the first stage is swinging positive, the collector (or drain) of that stage swings negative, as does the base (or gate) of the second stage. The collector (or drain) of the second stage swings positive, and this positive swing can be fed back to the emitter (or source) of the first stage. A positive input at the emitter (or source) has the same effect as a negative input at the base (or gate). This produces the desired negative feedback to stabilize gain.

Low-Frequency Cutoff.   Unless direct coupling is used (as described in Chapter 5), coupling capacitors must be used between stages, as well as at the input and output. Such capacitors form a low-pass RC filter (Sec. 3–2) with the base-to-ground (or gate-to-ground) resistance. Each stage has its own low-pass filter. In multistage amplifiers, the *effects of these filters are cumulative*.

As an example, if each filter causes a 1-dB drop at some given cutoff frequency and there are three filters (one at the input and two between stages), the result is a 3-dB drop at that frequency in the final output. If this cannot be tolerated, the RC relationship must be redesigned. In practical terms, this means increasing the value of C, since a change in R usually produces some undesired shift in operating point or other circuit characteristic.

3–9.2   Direct-Coupled and Differential Multistage Amplifiers

One method of eliminating the RC-filter problem created by coupling capacitors is to use *direct coupling*. This eliminates the interstage coupling capacitor, as well as some of the interstage resistances. In addition to the direct-coupled amplifier, a number of multistage circuits can be used to provide voltage, current, and even power amplification at audio frequencies.

The circuits for the most important of these are the Darlington-pair configuration (or compound), the phase inverter or splitter, the emitter-coupled amplifier, the transformerless series-output amplifier, the quasi-complementary amplifier, and the full-complementary amplifier. Since all these circuits involve some form of direct coupling, the circuits are discussed in Chapter 5, along with an analysis of FET multistage d-c amplifiers. Due to their highly specialized nature, differential amplifiers used at audio frequencies (as well as any other frequency) are discussed in Chapter 6.

3–10   MULTISTAGE AUDIO AMPLIFIERS WITH TRANSFORMER COUPLING

Figure 3–23 shows the schematic of a classic transformer-coupled audio-amplifier circuit. This basic circuit (or one of the many variations) was used extensively in audio equipment of all types (home entertainment, transmitter modulator, receiver audio amplifier, and the like). However, in present design, the IC power amplifier described in Sec. 3–14 has largely replaced the transformer-coupled amplifier. For that reason, we do not concentrate on the design of transformer-coupled amplifiers in this book. However, the following paragraphs summarize the characteristics of transformer coupling (in case you happen to find such a circuit when troubleshooting older equipment).

The circuit of Fig. 3–23 has a class A input or *driver* stage and a class B push–pull *output stage*. The class A stage provides both voltage and power

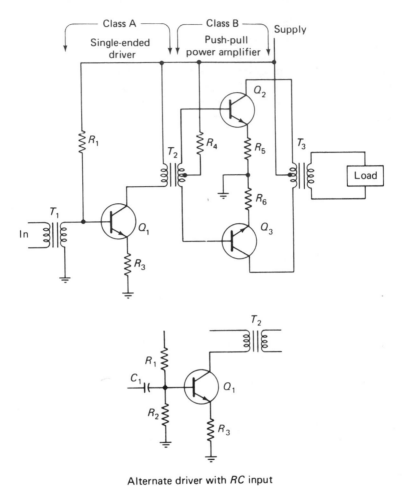

FIGURE 3-23   Classic transformer-coupled audio-amplifier circuit.

amplification as needed to raise the low input signal to a level suitable for the class B power output stage.

The class A stage can be transformer coupled or RC coupled at the input, as needed. Transformer coupling is used at the input where a specific impedance-match problem must be considered in design. The class A input stage can be driven directly by the signal source or can be used with a preamplifier for very low level signals. When required, a high-gain voltage amplifier is used as a *preamplifier*.

The push–pull output stage may be operated as a class B amplifier, where the transistors are cut off at the Q point and draw collector current only in the presence of an input signal. Class B is the most efficient operating mode for audio amplifiers, since the least amount of current is drawn (and no current where there is no signal). However, class B operation can result in cross-over distortion, as discussed in Chapter 1.

The efficiency of an amplifier is determined by the ratio of collector input-to-output power. An amplifier with 70% efficiency produces 7-W output for a 10-W input (with power input being considered as collector source voltage multiplied by total collector current).

Typically, class B amplifiers can be considered as 70% to 80% efficient. Class A amplifiers are typically in the 35% to 40% efficiency range, with class AB amplifiers showing 50% to 60% efficiency.

In all cases, any amplifier circuit that produces an increase in collector current at the Q point produces corresponding lower efficiency. This results in a trade-off between efficiency and distortion.

The efficiency produced by a class of operation also affects the heat-sink requirements. Any design that produces more collector current at the Q point requires a greater heat-sink capability (Chapter 2). As a guideline, a class A amplifier requires double the heat-sink capability of a class B, all other factors being equal.

3-11 AUDIO-AMPLIFIER OPERATING AND ADJUSTMENT CONTROLS

The most common operating controls for audio circuits used with music or voice reproduction equipment (hi-fi, stereo, public address, and the like) are the *volume* or *loudness control*, the *treble control*, and the *bass control*. The other most common audio control is the *gain control* (found on such circuits as operational amplifiers and power-control amplifiers).

The gain and volume controls are often confused, since both controls affect output of the amplifier circuit. A true gain control sets the *gain of one stage* in the amplifier. A true volume control sets the *level of the signal* passing through the amplifier, without affecting the gain of any or all stages. A gain control is usually part of a stage, whereas a volume control is usually found between stages or at the input to the first stage.

In addition to volume, bass, and treble controls, most stereo amplifier systems have some form of *balance control* (so that both channels of audio can be balanced). Also, most hi-fi–stereo systems have a form of *playback equalization* (for tape and phonograph playback). In the following paragraphs, we concentrate on analysis of the basic operating and adjustment controls to see how the controls affect the related audio-amplifier equipment.

3-11.1 Volume-Control Circuits

As shown in Fig. 3–24, the basic volume control is a variable resistor or potentiometer connected as a voltage divider. The voltage output (or signal level) depends on the volume-control setting.

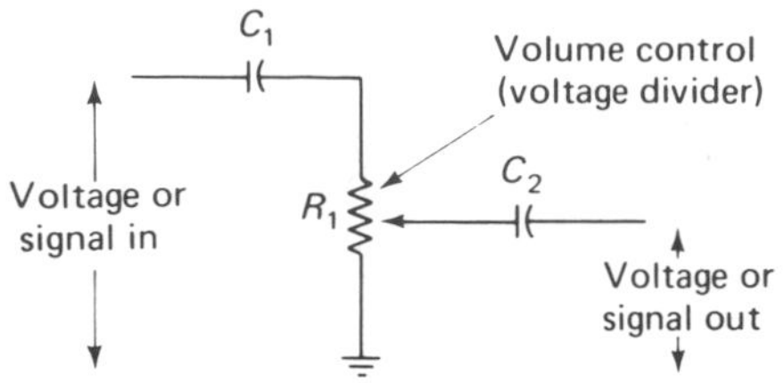

$$C = \frac{1}{3.2FR} \quad \text{for 1 dB}$$

$$C = \frac{1}{6.2FR} \quad \text{for 3 dB}$$

C in farads, F in hertz, R in ohms

FIGURE 3-24 Basic volume-control circuit and equations.

If the audio circuit is to be used with voice or music, the volume control is usually of the *audio taper* type where the voltage output is not linear throughout the setting range. (The resistance element is not uniform.) This produces a nonlinear voltage output to compensate for the human ear's nonlinear response to sound intensity. (The human ear has difficulty in hearing low-frequency sounds at low levels and responds mainly to the high-frequency components.)

If the audio circuit is not used with voice or music, the volume control is usually of the linear type (unless there is some special circuit requirement). With such a control, the actual voltage or signal is directly proportional to the control setting.

No matter what type of volume control is used, the control should be isolated from the circuit elements. If a volume control is part of the circuit (such as a collector or base resistance), any change in volume setting can result in a change of impedance, gain, or bias.

The simplest method for isolating a volume control is to use coupling capacitors (as shown in Fig. 3–24). However, the capacitors create a low-frequency response problem. As in the case of coupling capacitors described in previous sections, capacitor C1 forms a high-pass RC filter with volume potentiometer R1. Coupling capacitor C2 forms another high-pass filter with the input resistance of the following stage.

The volume control should be located at a low-signal-level point in the amplifier circuit. The most common location for a volume control is at the amplifier input stage, or between the first and second stages.

When a volume control is located at the amplifier input, the control resistance forms the input impedance (approximately). Volume controls are available in standard resistance values. Select the standard resistance value nearest the desired impedance.

When a volume control is located between stages, the resistance value should be selected to match the output impedance of the previous stage. Use the nearest standard value to produce the least signal loss.

Very little current is required for a volume control that is isolated as shown in Fig. 3–24, so the power rating (in watts) required is quite low. Usually, a 1- or 2-W rating is more than enough for any volume control used in transistor audio-amplifier circuits. Wirewound potentiometers should not be used for any audio application. The inductance produced by a wirewound potentiometer can reduce the frequency response of the circuit.

Figure 3–24 shows the equations for low-frequency cutoff versus RC-value relationships of typical audio volume controls. Note that these are the same as for the high-pass filters discussed in Sec. 3–1.

Volume Control Using Attenuators and Pads.   In some applications, the volume of audio signals is controlled by various attenuators and pads (such as T, L, O, and H pads). These attenuators and pads are made up of several inter-related resistances, all mechanically coupled to a common control shaft. Such attenuators and pads are commercially available, and no detailed analysis is given here.

Electronic (IC) Volume Control.   In present design, audio amplifiers are sometimes provided with electronic volume controls. Such controls are usually in IC form and are (in turn) under the control of a microprocessor.

Figure 3–25 shows some typical electronic volume-control circuits. Note that these circuits are found in the amplifier module of a modular home-enter-tainment system and that only the left stereo channel is shown.

The amplifier volume up, volume down, muting, and loudness functions are controlled by electronic volume control (or attenuator) IC604. Volume is adjusted in 40 steps, including full muting. (Note that most electronic volume controls provide for adjustment of the volume in steps, rather than continuous adjustment.) Volume control IC604 is under the control of system-control micro-processor IC901.

Commands from front-panel switches or a system-control bus are applied to IC901, which generates a *volume-control code* that is applied to IC604. The code includes clock, data, and strobe information at pins 10, 11, and 12 of IC604, respectively. The code is applied to decoder circuits within IC604 and causes IC604 switches and attenuators to be selected.

Figure 3–26 shows a typical volume-control code. IC604 receives 20 bits of serial data from IC901. The 20 bits are transmitted at the clock rate (4 MHz), or 1 bit per clock pulse. Once the 20 bits are transmitted, the strobe signal is transmitted and instructs IC604 to produce the correct amount of attenuation.

Data bits 1 and 2 are for the left and right channel selection. Bit 3 is for loudness on/off. Bits 4 through 8 are for 2-dB attenuation steps, with bit 4 for 0 dB and bit 8 for −8 dB. Bits 9 through 15 are for 10-dB attenuation steps, with bit 9 for 0 dB and bit 16 for −70 dB. Bits 17 through 20 are for an IC select code. (To select IC604, bits 17, 18, and 19 are low, with bit 20 high.)

As an example, assume that the volume is already at −10 dB, without the loudness function, and it is desired to turn the loudness function on, with

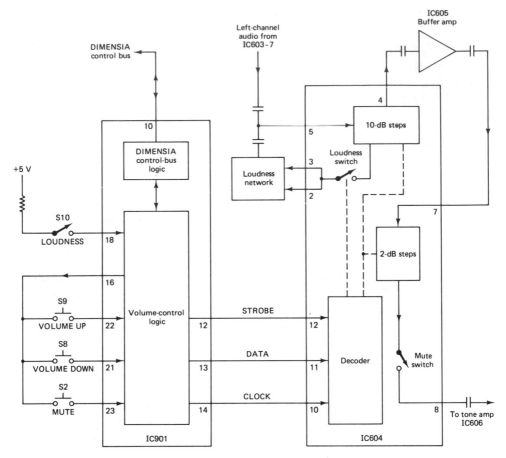

FIGURE 3-25 Typical electronic-volume-control circuits.

a volume of − 8 dB. The amplifier front-panel LOUDNESS button is pressed (once), and the VOLUME UP button is held until the desired − 8 dB is obtained. Under these conditions, bits 1, 2, 3, 8, 9, and 20 are selected (data line high).

Audio from circuits ahead of IC604 is applied to 10-dB attenuators in IC604 through pin 5. The output of the 10-dB attenuators is applied to 2-dB attenuators through buffer amplifier IC605. The output of the 2-dB attenuators is applied to the tone and power output circuits of the amplifier (Sec. 3–14) through a mute switch in IC604.

The 10-dB attenuator in IC604 has eight steps, 0 to − 70 dB, while the 2-dB attenuator has six steps, 0 to − 8 dB. With the volume-control setting at minimum volume (− 78 dB attenuation), continuously pressing (or holding) the VOLUME UP button causes the 2-dB attenuator to step up from − 8 to 0 dB, in 2-dB steps.

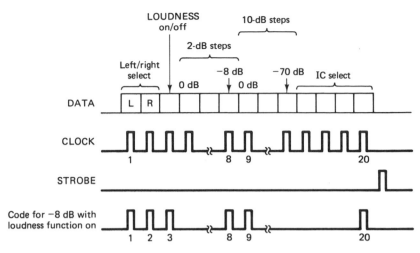

FIGURE 3-26 Typical electronic-volume-control code.

Once the 2-dB attenuator reaches 0 dB, the attenuator resets to −8 dB. At the same time, the 10-dB attenuator switches to −68 dB. This results in a 2-dB step, from −70 to −68 dB.

At low-volume settings, pressing the LOUDNESS button causes the loudness switch in IC604 to close, connecting a loudness network between pins 2, 3, and 5 of IC604. The loudness circuit is a Fletcher–Munson network that attenuates the midrange audio frequencies and passes the bass/treble frequencies. This has the effect of supplying (or reinforcing) positive feedback to the audio at pin 5 of IC604 (at high and low frequencies).

Pressing the MUTE button opens the mute switch in IC604 to interrupt the audio. However, this does not affect the attenuators. When MUTE is pressed again, the audio is restored at the same level of attenuation (unless the attenuation is changed during mute).

Note that troubleshooting for the electronic volume-control circuits shown in Fig. 3-25 is described in Chapter 10.

3-11.2 Gain-Control Circuits

As shown in Fig. 3-27, the basic gain control is a variable resistance or potentiometer, serving as one resistance element in the amplifier circuit. Any of the three resistors (base, emitter, or collector) could be used as the gain control, since stage gain is related to each resistance value (all other factors remaining constant).

The emitter resistance is the most logical choice for a gain control. If the collector resistance is variable, the output impedance of the stage changes as the gain setting is changed. A variable base resistance produces a variable input impedance. A variable emitter resistance (or source resistance in the case

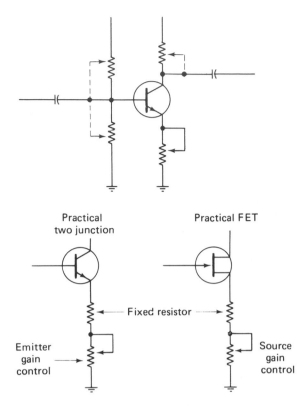

Practical
two junction

Practical FET

Fixed resistor

Emitter
gain
control

Source
gain
control

FIGURE 3-27   Basic gain control for two-junction and FET
stages.

of a FET) has minimum effect on the input or output impedance of the stage,
but directly affects both current and voltage gain.

With all other factors remaining constant, a decrease in emitter resistance
raises both current gain and voltage gain. An increase in emitter resistance lowers
stage gain.

The resistance value of an emitter (or source) gain control should be chosen
on the same basis as the emitter (or source) resistor, except that the desired value
should be the approximate midpoint of the control range. For example, if a
500-Ω fixed resistor is normally used (or if 500 Ω is the calculated value for
proper stage gain, bias stability, and so on), the variable gain control should
be 1000 Ω.

In practical applications, it is usually desirable to connect an emitter gain
control in series with a fixed resistance. If the gain control is set to the minimum
resistance value (zero ohms), there is still some emitter resistance to provide
gain stabilization and prevent thermal runaway. As a guideline, the series
resistance should be no less than one-twentieth of the collector resistor value.
This provides a maximum stage gain of 20.

If the gain control must provide for reduction of the stage voltage gain from some nominal point down to unity, the maximum value of the control should equal the collector resistance.

If reduction to unity current gain is desired, the maximum value of the control should equal the input (base) resistance.

An audio tape potentiometer should not be used as a gain control, unless there is some special circuit requirement. The potentiometer used should be of the noninductive composition type. The wattage rating of an emitter (or source) gain control should be the same as for an emitter (or source) resistor.

Use of a gain control in a power amplifier should be avoided. If a gain control must be used in power amplifiers, the control should be at the input stage where emitter current is minimum.

3–11.3   Tone- and Balance-Control Circuits

Tone (treble and bass) controls are found in most hi-fi amplifier systems. Balance controls are used in stereo amplifiers to balance the gain of both channels.

A *treble control* provides a means of adjusting the high-frequency response of an audio amplifier. Such adjustment may be necessary because of variation in response of the human ear or to correct the frequency response of a particular recording.

A *base control* provides a means of adjusting the low-frequency response of an audio amplifier. Such adjustment may be necessary because of variation in response of the human ear. As discussed, the human ear does not respond as well to low-frequency sounds as low levels as to high-frequency sounds at the same level. Also, coupling capacitors present high reactance to low-frequency signals. Both of these conditions require that the low-frequency signals be boosted (in relation to high-frequency signals).

There are many circuit arrangements for tone controls. Some involve the use of adjustable feedback (mainly in treble controls). Other circuits involve bypassing the coupling capacitors with adjustable reactances (mainly in bass controls). However, the most common tone controls are RC filters using audio-taper potentiometers as the adjustable R portion of the filter.

Typical Tone-Control Network.   Figure 3–28 shows a typical tone-control network for one channel of a stereo preamplifier. The tone-control network of the second channel (not shown) is connected by means of the balance control R19. In theory, the arm of R19 should be set to the exact midpoint. In a practical amplifier, the gain of each channel is not exactly the same, so R19 must be offset from the midpoint to balance both channels. Note that the volume control for the network of Fig. 3–28 is of the type shown in Fig. 3–24, rather than the electronic control of Fig. 3–25.

The bass and treble controls in Fig. 3–28 (R12 and R18) and standard audio-taper potentiometers. At 50% rotation, the resistance is split, 90% on

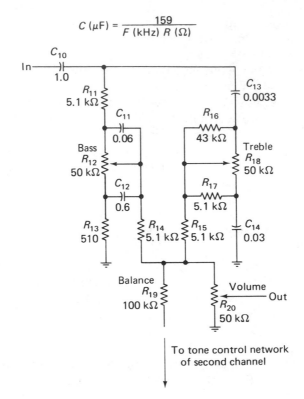

$$C \, (\mu F) = \frac{159}{F \, (kHz) \, R \, (\Omega)}$$

FIGURE 3–28   Typical tone-control network for one channel of a stereo preamplifier.

one side of the wiper and 10% on the other side. Figure 3–29 shows the relationship between wiper position and resistance.

As shown by the bass-response curve of Fig. 3–30a, the frequency response is flat from about 50 Hz to 20 kHz when bass-control R12 is in the center position. The reactance of C11 is made equal to the 45-kΩ portion of R12 at 50 to 60 Hz, and the reactance of C12 is made equal to the 5-kΩ portion of R12 at 50 to 60 Hz. As frequency increases from 50 Hz, C12 couples more signal to the output, while C12 shunts more signal to ground through R13. The net effect is a flat response from about 50 Hz to 20 kHz with a 20-dB insertion loss.

When R12 is in the boost position, C12 with a reactance one-tenth the resistance of R12 at 50 to 60 Hz effectively shunts R11 out of the circuit. This makes R11 and C12 the dominant frequency-response shaping components. Ideally, the full bass boost position supplies an output voltage (at about 50 Hz) that is 20 dB greater than the center position (flat response).

The full-boost position represents zero attenuation of bass frequencies. The amplitude of the output decreases at a 6-dB octave rate to the frequency

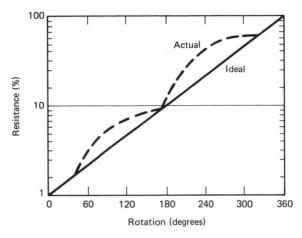

FIGURE 3-29 Relationship of position (rotation) and resistance in tone controls.

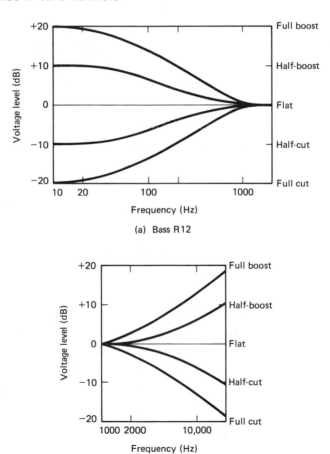

(a) Bass R12

(b) Treble R18

FIGURE 3-30 Normalized tone-control frequency-response curves.

where the reactance of C11 is negligible. The output amplitude is then determined by the ratio of R11 to R13.

When R12 is in the full-cut position, the output amplitude at about 50 Hz is determined by the ratio of C11 reactance to R13 resistance and is about 40 dB below the input voltage. As frequency is increased, the reactance of C11 decreases and the reactance is equal to the resistance of R13, again making the output amplitude dependent on the ratio of R11 to R13.

When R12 is in an intermediate position, the frequency at which rolloff begins (± 3 dB from the flat response curve) varies, but the slope of the rolloff changes only slightly. The boost-cut axis uses the flat response position as the reference point or 0 dB (although, in fact, the point is 20 dB below the input signal, because of the approximate 20-dB insertion loss of the tone-control network).

As shown by the treble-response curve of Fig. 3–30b, the reactances of C13 and C14 are small when compared to the parallel-divider combination of control R18 and fixed resistance R16/R17 (at frequencies below about 2.1 kHz). The resistive divider then provides the 10-to-1 voltage division to maintain the 20-dB insertion loss for the high frequencies. The net result is a 20-dB loss that is flat from about 20 Hz to 20 kHz.

The reactance of C13 should be about one-half the resistance values of R18 at a frequency of 2.1 kHz (or about 25 k). As shown by the equation of Fig. 3–28, the value of C13 should be $159/(2.1 \times 25 \text{ k}) = 0.003 \mu\text{F}$. The value of C14 should be about 10 times the value of C13, or $10 \times 0.003 = 0.03 \mu\text{F}$, to maintain the 10-to-1 voltage division.

The resistance of R16 is about one-tenth the control R18 resistance, with the R17 resistance about 80% of R18. R14 and R15 are isolation resistors made equal to 10% of the control R12/R18 resistances.

IC Tone-Control Networks.   In present design, tone- balance-control networks are often used with IC amplifiers. One advantage to the IC amplifier tone-control design is that any insertion loss presented by the circuit of Fig. 3–28 can be eliminated or minimized as desired.

Figure 3–31 shows one channel of a tone-control network using an IC amplifier. Note that the Fig. 3–31 circuit does not provide for volume control. This is because an electronic volume-control circuit such as shown in Fig. 3–25 is used (Sec. 3–11.2). The tone-control network of Fig. 3–31 is connected between the electronic volume control and the output power amplifier, as described in Sec. 3–14.

3–11.4 *Playback-Equalization Network Circuits*

Many playback network circuits are found in modern audio amplifiers. Most involve the use of frequency-selective feedback between stages or from the output to the input of an amplifier (typically an IC amplifier). A feedback net-

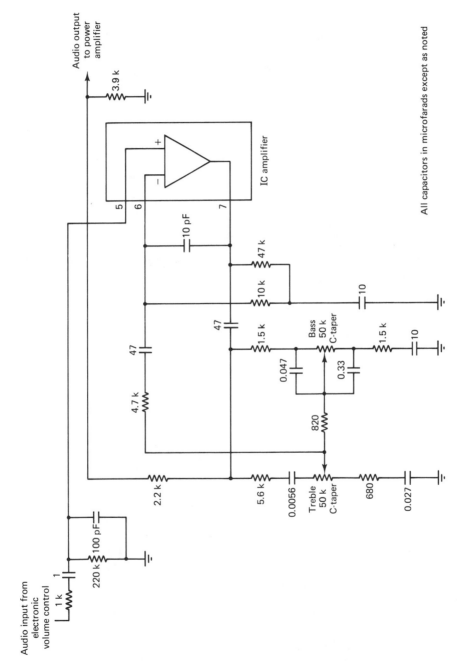

FIGURE 3-31 One channel of a tone-control network using an IC amplifier.

All capacitors in microfarads except as noted

work consists of resistances and capacitances that form a feedback circuit. At any given frequency, the amount of feedback (and thus the frequency response) is set by selection of the appropriate RC combinations. As frequency increases, the capacitor reactance decreases, resulting in a change of feedback (and a corresponding change of frequency response).

Basic Playback-Equalization Network.   Figure 3–32 shows basic play-back-equalization circuit where the voltage gain of two stages (with feedback) is about equal to the feedback-circuit impedance divided by the source impedance. (In this case, the source impedance is the emitter-resistance RE value.) The feedback impedance is the vector sum of the RF resistance value and the CF reactance value. The voltage gain of the two stages can be set to any desired level for any given frequency by means of this simple feedback circuit.

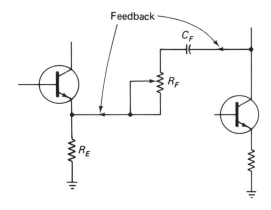

$R_F \approx R_E \times \text{maximum desired voltage gain}$

$C_F \approx \dfrac{1}{125 \times R_E \times \text{high-frequency limit (Hz)}}$

FIGURE 3–32 Basic playback-equalization network.

IC Amplifier Equalization Network.   Figure 3–33 shows an equalization network that uses feedback between the input and output of an IC amplifier (the preamplifier of a stereo system in this case). The closed-loop (with feedback) voltage gain of the preamplifier is set by the ratio of the feedback network to resistor R2. The feedback for playback-equalization) network for phonograph use (RIAA) is composed of C3, C4, R3, and R4, whereas the tape network (NAB) is composed of R5, C6, and C7. Note that CD players normally do not require playback equalization.

RIAA Playback Equalization.   Figure 3–34 shows the standard RIAA equalization curve for phonograph use. The recording curve is the inverse of the playback curve, so addition of the two produces a flat frequency-versus-

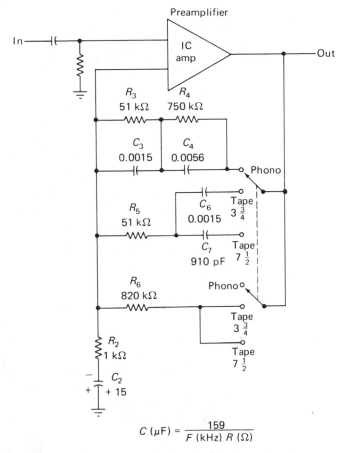

$$C \, (\mu F) = \frac{159}{F \, (kHz) \, R \, (\Omega)}$$

FIGURE 3-33 Playback-equalization network for phonograph (RIAA) and tape (NAB) recordings.

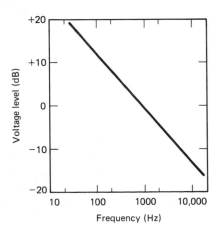

FIGURE 3-34 RIAA playback-equalization curve.

amplitude response. In phonograph recording, the high frequencies are emphasized to reduce effects of noise and the low inertia of the cutting stylus. The low frequencies are attentuated to prevent large excursions of the cutting stylus. It is the job of the frequency-selective feedback network to accomplish the addition of the recording and playback responses.

It is impossible to have the playback network be the exact inverse of the recording compensation, since each recording system is slightly different. However, optional guidelines can be applied. A typical audio range is from 20 Hz to 20 kHz, so there is a rolloff at both the low and high ends.

At the low end, the rolloff should start at some point between 10 and 20 Hz. This is done by making the 10-Hz point about 3 dB down from the 20-Hz point. As frequency increases from 20 Hz, there should be an almost linear rolloff. Ideally, the voltage gain at 20 Hz should be 100 times the gain at 20 kHz and 10 times the gain at 1 kHz. This produces the approximate RIAA curve of Fig. 3–34.

In the circuit of Fig. 3–33, the linear rolloff is produced by dividing the playback network into three sections. The R2C2 section sets the 10-Hz point at 3 dB down from the 20-Hz point, the R4C4 section covers frequencies up to about 1 kHz, and the R3C3 network covers higher frequencies.

The value of R3 should be 1000 times the desired voltage gain at 1 kHz, while the values of R4 should be 15 times that of R3. The value of R2 is also based on the value of R3 and is selected to provide the desired 1-kHz voltage gain (of 50 in this case). That is, the R3/R2 ratio sets the 1-kHz voltage gain.

The preamplifier of Fig. 3–33 produces an arbitrary minimum voltage gain of 5 at the highest frequencies (20 to 24 kHz). (Most of the gain for the complete stereo system is provided by power output amplifiers, such as described in Sec. 3–14.)

Using a minimum gain of 5 at the highest frequency, the gain at 20 Hz must be 100 times that amount, or 500. Likewise, the gain at 1 kHz must be 50. Using these desired gains, the value of R3 is 1000 $\times$ 50 = 50 k (use a 51-k standard). With R3 at 50 k, the value of R4 is 15 $\times$ 50 = 750 k; and the value of R2 is 50 k/50 = 1 k.

At low frequencies, the predominant impedance of the compensation feedback network is that of R4. As frequency increases from about 50 Hz, the reactance of C4 in parallel with R4 begins to decrease the impedance of the R4C4 section. The reactance of C4 is equal to R4 at about 35 to 40 Hz. At about 1 kHz, the net impedance of R4C4 is low compared to R3, and R3 sets the midband gain.

As frequency increases to about 2 kHz, the parallel impedance of capacitor C3 begins to shunt R3, decreasing the impedance of the R3C3 section. The reactance of C3 is equal to R3 at about 2.1 kHz.

Note that the equation shown on Fig. 3–33, based on the frequency breakpoint and corresponding resistor, is used to find the values of C2, C3, and C4.

NAB Playback Equalization. Figure 3–35 shows the standard NAB equalization curves for tape use. Again, the recording curve is the inverse of the playback curve, so addition of the two produces a flat response. Likewise, the high frequencies are emphasized and the lows are attenuated. However, unlike phonograph playback, tape playback tends to flatten out after about 3 to 4 kHz.

A different response is required for different tape speeds. Figure 3–35 shows the playback response curves for both 3¾ and 7½ IPS (inches per second). Up to about 1 kHz, the curves are almost identical. Because there is only one frequency breakpoint (where the curve must start to flatten) for each tape speed, a simple RC compensation network is sufficient (instead of the multisection network used for phono playback).

The breakpoint for 3¾ IPS occurs at about 1.85 kHz. The midband frequency gain is still 50, so the value of R2 remains at 1 k, and R5 is made equal to R3, or 51 k. The reactance of C6 is made equal to 51 k (R5) at 1.85 kHz (the nearest standard value is 0.0015 µF).

The breakpoint for 7½ IPS is at about 3.2 kHz, so C7 must have a reactance of 51 k at this frequency. A C7 value of 910 pF is the nearest standard.

Because C6 and C7 block the direct-current path for the IC preamplifier feedback input, R6 is added when the phono-tape switch is in either tape position. The use of R6 prevents a full 20-dB bass boost because of the shunting action across the tape compensation network. However, the network does provide about 15 dB of boost, which is generally satisfactory.

Note that the accuracy of both the RIAA and NAB compensation is only as good as the components used. In a practical amplifier, it is usually recommended that 5% (or better) tolerance resistance and capacitors be used. Likewise, it may be necessary to trim the values to get an exact (or near exact) performance curve for truly good hi-fi performance.

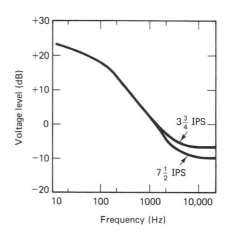

FIGURE 3-35 NAB playback-equalization curve.

3-12 TUNED AUDIO AMPLIFIERS (ACTIVE FILTERS)

In addition to passive audio filters (either LC or RC), it is possible to use amplifiers to form active filters. There are two major advantages for such filters. First, it is possible to get the equivalent of an inductive reactance, without actually using a heavy and bulky inductance required for a typical LC filter. (LC filters are generally not practical in the audio range). Second, an active filter eliminates the signal loss (so-called *insertion loss*) associated with passive filters (either RC or LC).

There are two common approaches to active filters. One approach involves the use of op-amps with feedback networks, such as described in Chapter 7. The other approach is to use low-gain amplifier stages with simple RC feedback networks, as described in this section.

3-12.1 Active Low-Pass (High-Cut) Filter

Figure 3-36 shows the basic circuit of an active low-pass (high-cut) filter, together with the corresponding characteristic curves for several sets of component values. Note that these values are approximate and usually require trimming to get an exact curve. The typical voltage gain is slightly less than 1 (unity) for transistors with a minimum beta of 20.

The amount of gain, as well as the shape of the curves, is set by the amount of feedback in relation to signal (which, in turn, is set by component values). Note that the feedback is positive and thus adds to the signal. However, the feedback amplitude (across the entire frequency range) is just below the point necessary for oscillation. The circuit of Fig. 3-36 is an emitter follower (which typically has no gain).

The circuit of Fig. 3-36 requires a bias of about −10 V (one-half the −20-V supply) at the input. This bias can be taken from a previous stage. If no such stage exists, the bias can be obtained by the addition of the 20-k resistor (shown in phantom as R3) and by changing the value of Rin to 20 k. Such an arrangement introduces a loss of about 6 dB, so it is better to operate the circuit by direct coupling from the output of a previous stage.

3-12.2 Active High-Pass (Low-Cut) Filter

Figure 3-37 shows the basic circuit of an active high-pass (low-cut) filter, together with characteristic curves. The circuit of Fig. 3-37 is the inverse of the Fig. 3-36 circuit. That is, the Fig. 3-37 circuit uses capacitors in series with the base, with feedback obtained through RF rather than CF. The gain and shape of the curves are set by the amount of feedback (determined by circuit values).

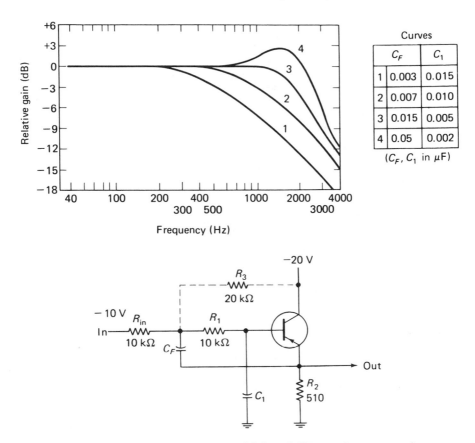

FIGURE 3-36 Active low-pass (high-cut) filter and corresponding response curves.

3-12.3 Active Bandpass Filter

The circuits of Figs. 3–36 and 3–37 can be cascaded to provide a bandpass filter. Any of the curves can be used. However, curve 3 is the most satisfactory because of the sharp break at cutoff. Curves 1 and 2 have considerable slope with no sharp break. Curve 4 produces some peaking at the breakpoints.

If the circuits are cascaded, the low-pass filter (Fig. 3–36) should follow the high-pass filter (Fig. 3–37). This provides the necessary bias at the input of the low-pass filter (− 10 V from the emitter of the high-pass filter).

3-12.4 Active Peaking Filter

The circuits of Figs. 3–36 and 3–27 can be cascaded to provide a peaking filter with the proper selection of components. However, a single-stage tuned amplifier produces the same results. Such a circuit is shown in Fig. 3–38.

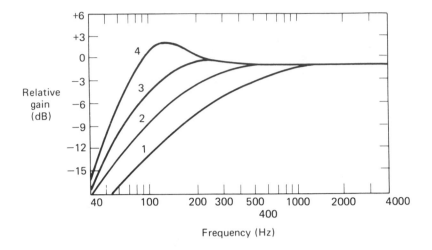

FIGURE 3-37 Active high-pass (low-cut) filter and corresponding response curves.

As shown by the characteristic curve, the center or peak frequency is about 1 kHz. If desired, the center frequency can be changed by as much as 3 decades when *both capacitors* are changed by a common factor. However, in a practical circuit, the input resistance values require some trimming.

Note that the characteristic curve shows a gain of about 24 db. This is a no-load voltage gain at the peak or center frequency. If the circuit is loaded, as it must be in any practical application, the gain is reduced. (Ideally, the circuit should work into an approximate 10-k load.) Although there is a reduction in gain for a load, the shape of the output peak should remain substantially the same.

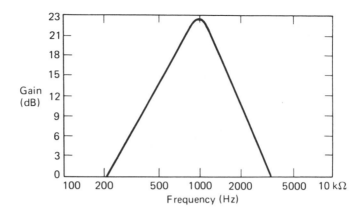

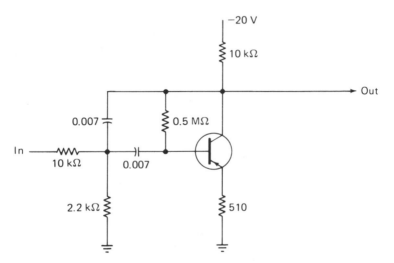

FIGURE 3-38 Active peaking filter and corresponding response curve.

3-13 TRANSFORMERLESS PHASE INVERTER

One advantage of a transformer-coupled amplifier (Sec. 3–10) is that two signal voltages (180° out of phase) can be taken from a center-tapped transformer winding (such as T2/T3 in Fig. 3–23). The same result can be produced with a transformerless phase inverter such as shown in Fig. 3–39.

   In a common-emitter amplifier with a resistive load, the collector and emitter are 180° out of phase with each other. If the input signal voltages for a push-pull stage are obtained from these two points (collector and emitter), the necessary 180° out-of-phase relationship is produced. Also, since approximately the

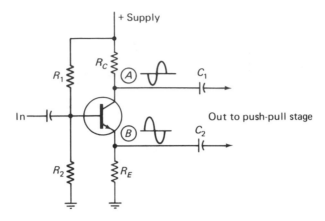

FIGURE 3-39   Transformerless phase-inverter stage.

same collector current flows through RC and RE (if the resistances of RC/RE are equal), the voltage drops are equal. Point A becomes as much negative as point B becomes positive.

In the circuit of Fig. 3–39, resistors R1 and R2 form the voltage divider that forward biases the emitter–base junction of Q1. The collector resistor RC and the emitter resistor RE are equal in value, as are the coupling capacitors C1 and C2. Note that RE is unbypassed. This provides inverse-current feedback that reduces distortion and stabilizes gain.

The main advantage of the phase-inverter stage is to eliminate the need for a transformer. This results in a smaller, lighter, and less expensive amplifier and eliminates the transformer's magnetic field (which can produce distortion of the signal unless the transformer is properly shielded).

3-14   IC AUDIO AMPLIFIERS

A great variety of audio amplifiers is available in IC form. There are also some audio amplifiers in *hybrid* form. Such hybrid circuits consist of resistors, capacitors, diodes, and transistors, all contained in a single, hermetically sealed package. Hybrid circuits are similar to ICs, except that ICs are usually complete functioning circuits. Hybrid audio circuits have generally been replaced by ICs in present design.

Many IC and hybrid audio amplifiers involve some form of *direct coupling* and/or *differential amplifiers*. For that reason, these audio circuits are discussed in the relavent chapters (5 and 6). In this section, we describe modern audio amplifiers (hi-fi/stereo) using IC design. The descriptions include the power-amplifier and output-protection circuits of a modular amplifier, as well as the audio circuits for modular AM/FM receivers, cassette decks, and CD players. Troubleshooting for these specific circuits is discussed in Chapter 10.

3-14.1 Modular Amplifier Output Circuits

Figure 3–40 shows the audio-output circuits of a modular amplifier. (This amplifier is used in the RCA DIMENSIA® system.* The audio-output power amplifiers for both channels are contained within one hybrid integrated circuit IC701. The single IC provides for high reliability and less complex servicing. Because of the larger power rating of 50 W (rms minimum), the IC is mounted on a heat sink (Chapter 2).

The tone amplifiers that provide for treble and bass adjustment are also contained within a single integrated circuit IC606. Both left and right channels use a subsonic filter consisting essentially of capacitor C618 and SUBSONIC FILTER switch S602. The subsonic filter attenuates frequencies below 20 Hz to reduce rumble caused by warped records or defective turntables. The audio output is monitored by the protection circuits described in Sec. 3–14.2.

Audio from volume control IC604 (Sec. 3–11, Fig. 3–25) is applied to the noninverting inputs of IC606 (which is essentially an op-amp functioning as a tone amplifier or control). Op-amps are discussed in Chapter 7.

BASS control R634 and TREBLE control R635 are connected between the output and inverting input of IC606, as shown in Fig. 3–31. Decreasing the bass or treble negative feedback has the effect of boosting the bass or treble, and vice versa.

The output of IC606 is coupled to the balance control and subsonic filter. BALANCE control R637 is connected between the right- and left-channel audio, with the wiper connected to ground. Moving the wiper up reduces the right-channel impedance while increasing the left-channel impedance. R637 is usually set to provide equal audio-signal levels in both channels.

With SUBSONIC FILTER switch S602 in the OFF position, C618 is out of the circuit, and there is no attenuation of low-frequency signals. With S618 in the ON position, C618 acts as a high-pass filter, attenuating all frequencies below 20 Hz. The output of the subsonic filter is applied to the noninverting input of power amplifier IC701.

An RC network is connected between the output of IC701 and the inverting input of IC701. This RC network is used to prevent oscillation. The audio output from IC701 is applied through R716, which acts as the sensing resistor for the protection circuits (Sec. 3–14.2). The audio from R716 is applied to the speaker terminals and headphone jack.

3-14.2 Modular Amplifier Output-Protection Circuits

Figure 3–41 shows the output-protection circuits of the modular amplifier described in Sec. 3–14.1. The DIMENSIA amplifier has three protection circuits.

The *overload-protection* circuit prevents damage to power amplifier IC701 when a low-impedance or shorted speaker is connected. The *midpoint-potential-*

---

*DIMENSIA is a registered trademark of Thomson Consumer Electronics, Inc. (RCA).

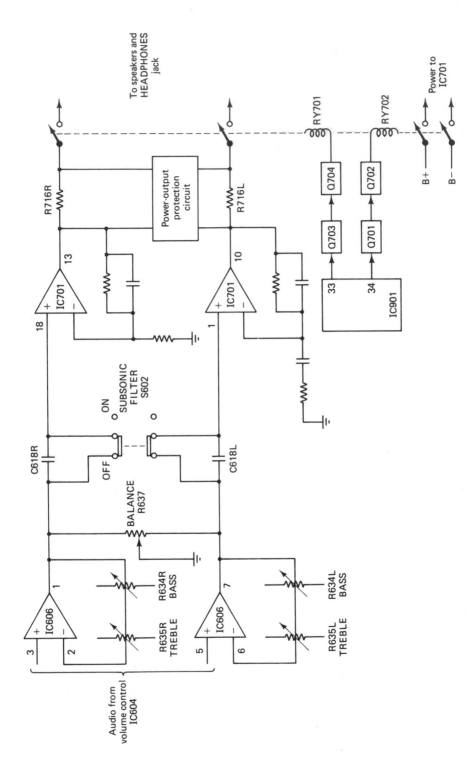

FIGURE 3-40 Audio-output circuits of a modular amplifier.

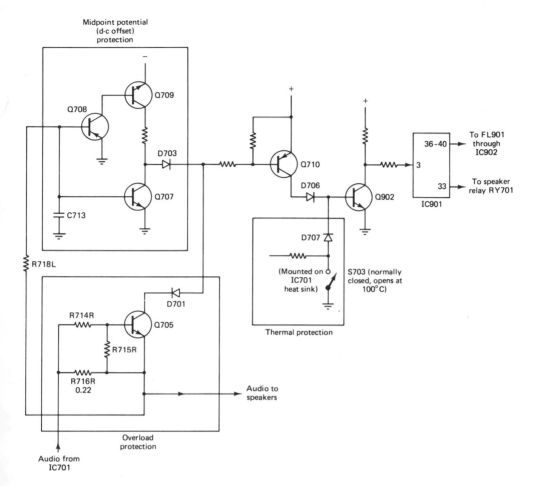

FIGURE 3-41 Output-protection circuits of a modular amplifier.

protection circuit is used to prevent damage to the speakers in case of a defective IC701 and is turned on when a d-c potential (sometimes called *d-c offset*) is present at the output of IC701. (Such outputs are discussed in Chapter 5.) The *thermal-protection* circuit prevents damage of IC701 caused by excessive heat.

These three protection circuits are coupled to pin 3 of system-control microprocessor IC901. When any one of the circuits is turned on, pin 3 of IC901 goes low. IC901 recognizes this as a possible danger condition and produces a high at pin 33. The high is applied to the speaker relay RY701 through Q703/Q704 and causes the RY701 contacts to open, disconnecting the speakers from the IC701 output. IC901 also pulses the function display FL901 through driver IC902. The front-panel function display flashes, indicating to the user that the protection circuits are turned on.

Thermal Protection.   Thermal-protection switch S703 is mounted on the power amplifier IC701 heat sink and is normally closed. This keeps D707 reverse biased and Q902 off. With Q902 not conducting, pin 3 of IC901 is high, and pin 33 of IC901 remains low to keep the speakers connected to the IC701 output.

If the temperature of the IC701 heat sink rises to 100 °C, S703 opens, forward biasing D707. This turns Q902 on and produces a low at pin 3 of IC901. Under these conditions, pin 33 of IC901 goes high to disconnect the speakers from IC701. Simultaneously, pins 36 through 40 of IC901 are pulsed to flash the function-display portion of FL901.

Midpoint-Potential Protection.   The midpoint-potential-protection circuit functions by monitoring the d-c output from IC701. In theory, there should be no d-c output from IC701 to the speakers. (Excessive d-c current can damage the speaker coils.) However, as a practical matter, there may be as much as ± 1.7 V at the IC701 output without damage to the speakers. The midpoint-potential-protection circuit (called the *d-c offset protection circuit* in some literature) is turned on if the 1.7-V value is exceeded.

If there is any d-c output from IC701 to the speakers, this potential causes C713 to charge through R718L/R. C713 charges to the average value of the speaker voltage. During normal operation, with the d-c output from IC701 less than ± 1.7 V, the midpoint-potential-protection circuit is turned off.

If the average charge across C713 increases above $+1.7$ V, Q707 is turned on, forward biasing D703. This applies a low to the base of Q710, turning Q710 on and forward biasing Q706. This applies a high to the base of Q902, turning Q902 on, and causes pin 3 of IC901 to go low. IC901 then produces a high at pin 33 to disconnect the speakers (and to pulse the FL901 function display) as discussed.

Overload Protection.   The overload-protection circuit is the same for both channels, so only the right channel is covered here. Audio output from IC701 to the speakers is applied through R716R, a 0.22-Ω resistor. This resistance is much smaller than the speaker load impedance (typically 4 to 16 Ω).

During normal operation, the voltage across R716R is very small. If a shorted or very low impedance speaker is connected, excessive output current flows through R716R and the voltage across R716R increases sharply.

Resistors R714R and R715R are connected as a voltage divider across R716R to the base of Q705R. As the current through R716R increases (due to a short or low-impedence load), the voltage applied to Q705R increases, turning Q705R on. This forward biases D701, turns on Q710, forward biases D706, turns on Q902, and applies a low to pin 3 of IC901 to disconnect the speakers and pulse the FL901 display.

3-14.3 Modular Tuner Audio Circuits

Figure 3–42 shows the audio-output and muting circuits of a modular AM/FM tuner. (This tuner is used in the RCA DIMENSIA system and is discussed further in Chapter 4.) Note that four signals are used to mute audio output from

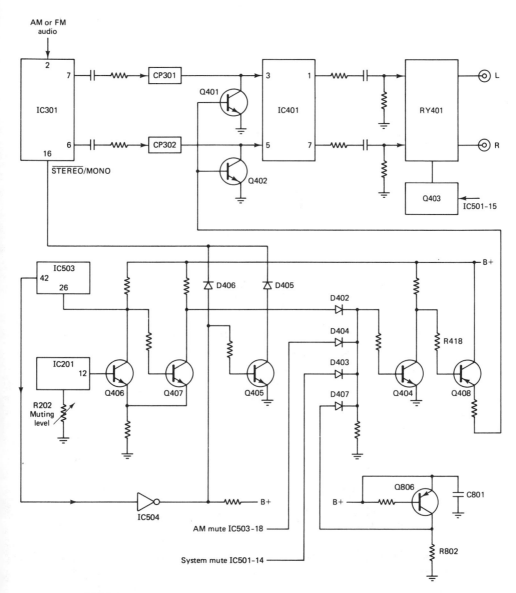

FIGURE 3-42 Audio-output and muting circuits of a modular AM/FM tuner.

the tuner to the amplifier (Secs. 3–14.1 and 3–14.2). Similarly, the affect of these muting signals is changed, depending on the operating mode selected.

For example, when an FM signal of sufficient strength is tuned in during AUTO MODE, pin 12 of IC201 goes low. This low turns Q406 off and Q407 on. With Q407 on, Q404, Q408, Q401, and Q402 are turned off to unmute the audio (which passes from IC301 to the left- and right-channels through amplifier IC401 and relay RY401).

When the FM signal is weak or between FM stations, pin 12 of IC201 goes high. This high is applied to Q406/Q407 (connected as a Schmitt trigger). In AUTO MODE, the collector of Q407 goes high, turning Q404 on (through D402) and turning Q408 on. With Q408 on, B+ is applied to Q401/Q402, turning both Q401 and Q402 on. This grounds both L and R audio lines and mutes the audio.

In the mono moe (AUTO MODE indicator off), pin 42 of IC503 goes low. This low is inverted by IC504, turning Q405 on, and turning Q404, Q408, Q401, and Q402 off. This removes the ground from the left- and right-channel audio lines and unmutes the audio. The high from IC504 also tells IC301 to operate as a mono amplifier (same audio signal to both channels).

In AUTO MODE, pin 42 of IC503 goes high. This high is inverted by IC504 and applied to pin 16 of IC301 through D406, telling IC301 to operate as an FM-stereo multiplexer (L- and R-channel audio). The low from IC504 has no effect on Q405.

The audio can be muted (or unmuted) by signals from pin 18 of IC503 (for AM operation) and pin 14 of IC501 (when the tuner is used as part of a system).

The audio can also be muted by a temporary power interruption. During normal operation, C801 charges to the B+ level of about 13 V. If the B+ drops due to a momentary power interruption, Q806 turns on, and C801 discharges through Q806/R802. The temporary high across R802 turns Q404, Q408, Q401, and Q402 on to mute the audio (temporarily, until C801 discharges).

3–14.4 Cassette Deck Audio Circuits

Figure 3–43 shows the overall record/playback amplifier circuits (for one channel) of a modular cassette deck. (This deck is used in the RCA DIMENSIA system.) The record/playback heads are placed in the playback mode, or record mode, by the record/playback switching circuit. This switching circuit is controlled by an output from pin 24 of the tape-transport mechanism controller IC500.

During the playback mode (IC500 pin 24 low), the head recovers the recorded audio signal from tape and applies the signal to a playback amplifier.

During the record mode (IC500 pin 24 high), the switching circuits place the record/playback heads in the record mode so that both line and microphone audio can be recorded on tape.

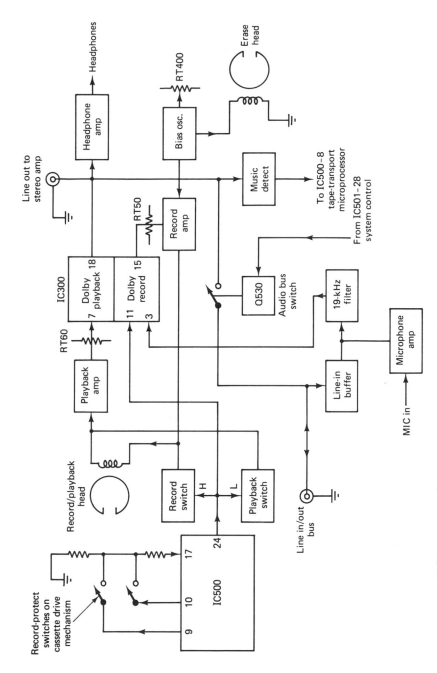

FIGURE 3-43 Overall record/playback amplifier circuits (for one channel) of a modular cassette deck.

139

Record/Playback Switching during Playback.   Figure 3–44 shows the record/playback switching, as well as the playback amplifier used during playback. With pin 24 of IC500 low, both Q62 and Q63 are turned on (grounding the record input to the head), while Q61 is turned off (permitting the recovered audio to pass to IC60). Note that the equalization network connected to IC60 is a standard equalization circuit (Sec. 3–11.4) for many audiocassette tapes, with 3180-μs (R69/C62), 120-μs (R64/C62), and 70-μs (R61/C61) time constants. The voltage gain of IC60 is about 49 dB at 400 Hz.

The output of IC60 is applied to the playback portion of Dolby® processing circuits in IC500 (Sec. 3–14.5) through playback-gain control RT60.* After processing by the Dolby circuits, the audio is applied to the headphone amplifier and line.

Headphones and Line-out.   Figure 3–45 shows the headphone and line-out circuits. Note that OUTPUT level control RV40 controls audio to both the headphones and line (stereo amplifier). The output of the headphone amplifier is typically 80 mV into an 8-Ω impedance when a Dolby calibration tape is used.

Audio-bus switch Q530 permits the line in/out jack to be used as a bidirectional audio bus connection during DIMENSIA operation. Audio-bus switch IC01 is controlled by Q530, which, in turn, is controlled by the signal at pin 28 of IC501 (high for audio bus).

During nonsystem operation (audio-bus switch IC01 open), the line in/out jack acts as a line-in function only. During DIMENSIA operation, the line in/out jack may be used as both an in and out connection (in, switch open; out, switch closed, during playback).

Record/Playback Switching during Record.   Figures 3–46 and 3–47 show the record/playback switching, as well as the buffer, amplifier, and filter circuits used during record. With pin 24 of IC500 high, Q70 is turned off, as are Q62 and Q63 (removing the ground from the record input to the head). Q61 is turned on, completing the ground connection for the record input of the head.

As shown in Fig. 3–46, the audio at the line in/out jack is applied to the line-in buffer circuits. The buffer keeps the line input at an impedance of about 200 Ω. The audio exits the buffer and is applied to a 19-kHz filter that removes a 19-kHz pilot signal from an FM multiplex broadcast, as well as any a-c bias leakage signals. Either of these signals can trigger the Dolby NR circuits, thus upsetting the proper response characteristics of the Dolby NR function. Audio from the filter is applied to the record portion of the Dolby NR IC300, as described in Sec. 3–14.5.

Audio from the microphone input can also be recorded. Operation of these circuits is identical to line-in, except that audio from the microphone is applied to microphone amplifier IC02 as shown in Fig. 3–46.

*Dolby is a trademark of Dolby Laboratories Licensing Corporation.

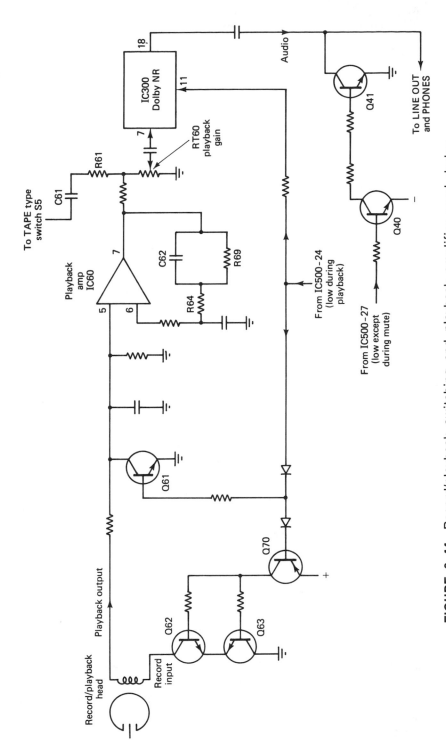

FIGURE 3-44 Record/playback switching and playback amplifier used during playback.

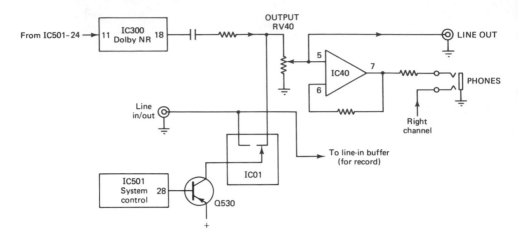

FIGURE 3-45 Headphone and line-out circuits.

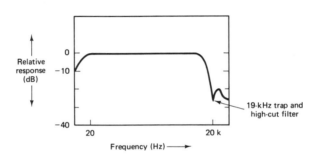

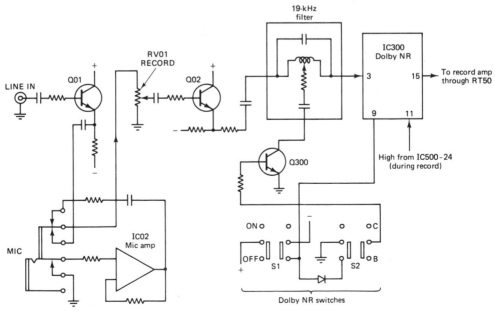

FIGURE 3-46 Record/playback switching and filter circuits.

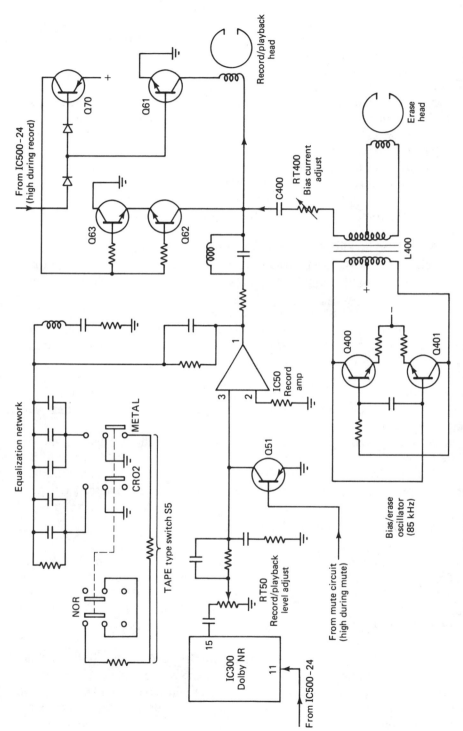

FIGURE 3-47 Record/playback amplifier and bias circuits.

As shown in Fig. 3–47, audio from IC300 is applied to the record amplifier through record/playback level adjustment RT50, together with signals from the bias oscillator. After processing by IC50, the audio is applied to the record input of the record/playback head and recorded onto tape.

Note that a fixed amount of bias current is applied to the erase head. However, the record head receives higher or lower bias current, depending on the position of the TAPE type switch S5. Bias current is also adjusted by RT400.

The record amplifier consists of IC50, with the associated components to compensate for record-current requirements. These components boost both high and low ends of the frequency range (similar to the equalization networks described previously. TAPE type switch S5 cuts in the components as necessary to provide the correct compensation for the three basic types of tape (normal, CrO2, and metal).

As shown in Fig. 3–43, record-protect switches on the cassette drive mechanism prevent IC from placing the deck in the record mode (prevent pin 24 of IC500 from going high) when a prerecorded cassette (with protect tabs removed) is installed.

When a cassette with the tabs intact is installed, the record-protect switches are closed. Scan signals from pins 9 and 10 of IC500 are applied to pin 17, permitting normal control by IC500 (that is, IC500 can go high if so instructed by the front panel controls or the remote unit).

When a cassette with the tabs removed is installed, the record-protect switches are open. This prevents scan signals from being applied to pin 17, thus preventing record operation (pin 24 of IC500 cannot go high), no matter what commands are applied to IC500.

Note that there is a separate record-protect switch for forward and reverse (on this particular deck). Keep this in mind when troubleshooting a "no record in one direction" trouble symptom. Check that *both* tabs are in place on the cassette. Try another cassette or put heavy tape over the missing tab area.

3–14.5 Dolby Processing Circuits

Figure 3–48 shows the Dolby processing circuits for the cassette deck described in Sec. 3–14.4 (Fig. 3–43). Note that the Dolby processing functions are combined in a single module (IC300), commonly called an MD300 Dolby NR (noise reduction) module (and found on many audio devices). The use of a single module for Dolby processing is quite common for decks used in modular home-entertainment systems.

As shown in Fig. 3–48, we do not go into the IC300 circuits in full detail. As is the case with any IC, you must replace the entire module if any of the internal circuits fail to work. However, you should have some knowledge of what is done by the internal circuits to troubleshoot the circuits that are external to the Dolby module.

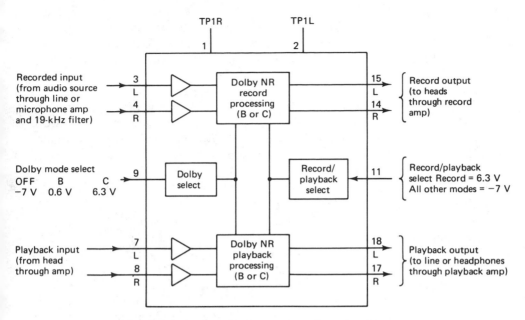

FIGURE 3-48 Dolby processing circuits.

Figure 3–49 shows the basic block diagram of a tape recorder with Dolby noise reduction. Note that Dolby processing occurs both in recording and playback. The input audio to be recorded is processed before the audio is applied to the record amplifier. The audio taken from tape is processed after amplification by the playback amplifier. This is typical for most decks used in home-entertainment systems.

In any tape deck, undesirable "hiss" noise (made up of predominantly high frequencies) is introduced into the audio path at various points (typically at the tape record/playback head, playback amplifier, and bias circuit). This

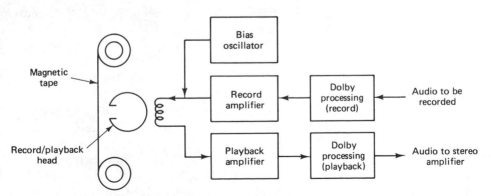

FIGURE 3-49 Basic tape recorder with Dolby noise reduction.

annoying noise can be reduced by passing the signals through processing circuits, as is done in many non-Dolby decks. However, the objectionable noise does not have the same effect on signals of various amplitudes and frequencies. For example, *the noise is most objectionable and noticeable in signals of low amplitude and high frequency.*

In Dolby processing, the high-level and/or low-frequency signals are not affected. Instead, the low-level signals and high-frequency signals are suppressed by about 10 dB with Dolby B or by about 20 dB with Dolby C. This results in an improved dynamic range, as well as an improvement in signal-to-noise ratio. However, any original signal (music, voice, and the like) is also modified by the processing during record. That is, the desired high-frequency and low-level signals are suppressed, along with the hiss or other noise components. So the original signal must be modified to the same extent, but in the opposite direction, during playback.

With Dolby NR, the low-level and high-frequency signals are boosted by the same amount during playback that the signals were suppressed during record. This leaves the output signal an exact reproduction of the input signal, but with the noise component suppressed.

As shown in Fig. 4–48, the Dolby module IC300 is controlled by mode-select signals at pin 9 and record/playback signals at pin 11. *During record*, pin 11 of IC300 is set to about 6 V by IC500. This causes the Dolby record circuits to turn on and the playback circuits to turn off.

Audio signals to be recorded are applied through amplifiers and the 19-kHz filter to pins 3 and 4 of IC300, as described in Sec. 3–14.4. *During playback*, pin 11 of IC300 is set to about − 7 V by IC500, causing the Dolby playback circuits to turn on and the record circuits to turn off.

The audio signals can be processed for either Dolby B or Dolby C or can be recorded without any Dolby processing, depending on the voltage at pin 9 of IC300. In turn, the pin 9 voltage is set by Dolby NR switches S1 (ON/OFF) and S2 (B/C), as shown in Fig. 3–46.

During the record Dolby-off mode (− 7 V at pin 9 of IC300), the audio is routed through IC300 with no processing and exits IC300 at pins 14 and 15 for application to the head through the record amplifier.

During the record Dolby-B mode (0.6 V at pin 9 of IC300), the audio is processed (suppressed by about 10 dB) and is applied to the head through the record amplifier.

During the record Dolby-C mode (6.3 V at pin 9 of IC300), the audio is processed (suppressed by about 20 dB) and is applied to the head through the record amplifier.

Audio signals taken from the tape by the head during playback are amplified and applied to pins 7 and 8 of IC300, as discussed in Sec. 3–14.4. These signals can be processed for either Dolby B or Dolby C or can be played back without any Dolby processing, depending on the voltage at pin 9 of IC300 (as set by Dolby NR switches S1/S2).

During the playback Dolby-off mode (-7 V at pin 9 of IC300), the playback audio is routed through IC300 with no processing and exits IC300 at pins 17 and 18 for application to the line or headphones through the playback amplifier.

During the playback Dolby-B mode (0.6 V at pin 9 of IC300), the audio is processed (boosted by about 10 dB) and is applied to the line or headphones through the playback amplifier.

During the playback Dolby-C mode (6.3 V at pin 9 of IC300), the audio is processed (boosted by about 20 dB) and is applied to the line or headphones through the playback amplifier.

3–14.6 CD Player Audio Circuits

Figure 3–50 shows the audio circuits of a CD player. (Note that the right-channel circuits are stressed.) As discussed in the author's many books on CD, audio is recorded on a compact disc in digital form. The player circuits convert the recorded material to conventional analog audio by means of a D/A (digital to analog) converter. The output of the D/A converter is a single line, containing both the left- and right-channel stereo signals in serial form (left-channel, right-channel, left-channel, and so on). The serial audio is converted to parallel stereo-audio by the circuits shown in Fig. 3–50.

Output from pin 17 of the D/A converter IC403-17 is amplified by IC501 and multiplexed into right- and left-channel audio by sample/hold circuits within IC502, under control of IC402. Note that the audio output from pin 17 of IC403 is passed through front-panel OUTPUT control R542. Also note that the audio signal is still in serial left/right/left/right format and must be converted to parallel stereo-audio. The audio from IC403 also contains a certain amount of digital noise, which must be filtered out to produce a high-quality signal. This is done by the circuits of Fig. 3–50.

Serial-to-Stereo Conversion.   The SHR and SHL signals generated by IC402 are applied to pins 9 and 11 of IC502 and close the proper switch at the correct time to route the left-audio information through the left-channel processing circuits and the right-audio information to the right channel. When one switch in IC502 is closed, the other switch is connected to ground, thus preventing any noise from passing to the processing circuits.

Filtering.   The right-channel audio exits IC502 at pin 3 and is applied to IC503R. The capacitor between the input (pin 2) and output (pin 6) of IC503R removes much of the digital noise present at pin 3 of IC502. The audio is then applied to pin 42 of IC504R, an analog low-pass filter (or LPF). The audio exits IC504R at pin 6 and is applied to pin 3 of IC505R. The first stage of amplification within IC505R occurs between pins 1 and 3. The audio reenters IC505R at pin 6, and is amplified once again and exits at pin 7.

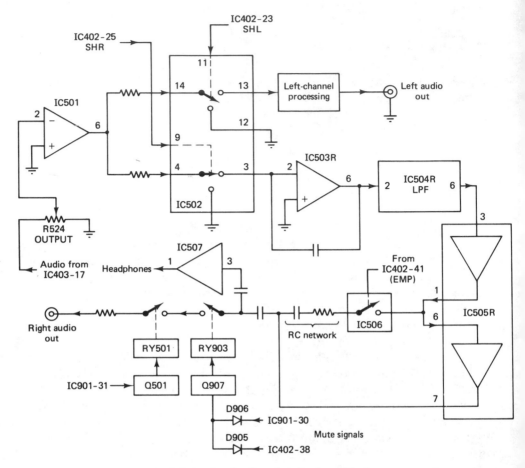

FIGURE 3-50 Audio circuits for CD player.

An RC network is connected across pins 6 and 7 of IC505R for de-emphasis of the high-frequency signals. The network is cut in and out of the circuit by a switch in IC506. The switch is controlled by the EMP (emphasis) signal at pin 41 of IC402. De-emphasis is only required on discs that have pre-emphasis during the recording process. IC402 recognizes a disc with pre-emphasis by means of a "flag" signal recorded on the disc. When the flag is present, IC402 switches in the de-emphasis RC network.

Output Control.   The audio at pin 7 of IC505 is coupled to the rear-panel output jacks through two relays. Relay RY903 is an internal muting relay operated by mute signals from IC901 and IC402. Relay RY501 is the audio-bus relay operated by signals from IC901. The front-panel headphone jacks receive audio ahead of the relays through IC507.

4

RF, IF, AND VF AMPLIFIERS

When electrical signals reach frequencies of about 15 kHz and higher, the signals take on the properties of radio-frequency or RF signals. That is, the signals generate electromagnetic radio waves, which are radiated (transmitted) from the conductor. Amplifiers designed to amplify signals of such frequencies are known as RF amplifiers. Useful radio frequencies may be as high as several thousands of megahertz (MHz) or several gigahertz (GHz).

It is not practical to design any amplifier circuit that covers the entire frequency range or to use all radio frequencies for all purposes. Instead, the RF spectrum is broken down into various *bands*, each used for a specific purpose. In turn, amplifier circuits are generally designed for use in one particular band.

Figure 4-1 shows the most common assignment of radio frequency bands. Both commercial and military bands are shown. Note that the commercial RF bands are from about 3 kHz to 300 GHz, whereas the military (and special-purpose) band assignment run from 225 MHz to 56 GHz.

Radio waves with frequencies greater than about 1 GHz are known as *microwaves*. The amplifier circuits used with microwaves are quite different from those used at lower frequencies. Because of their specialized nature, microwave amplifiers and related circuits are not discussed in this book. Instead, we concentrate on RF amplifiers operating at frequencies up to and including the UHF band.

4-1 TYPES OF RADIO-FREQUENCY AMPLIFIERS

Although there is an infinite variety of amplifier circuits, RF amplifiers may be divided into two general types: *narrowband amplifiers* (with bandwidths up to several hundred kilohertz) and *wideband amplifiers* (with bandwidths on the

Commercial Bands

Very low frequency (VLF) 3-30 kHz
Low frequency (LF) 30-300 kHz
Medium frequency (MF) 300 kHz-3 MHz
High frequency (HF) 3-30 MHz
Very high frequency (VHF) 30-300 MHz
Ultrahigh frequency (UHF) 300 MHz-3 GHz
Superhigh frequency (SHF) 3-30 GHz
Extrahigh frequency (EHF) 30-300 GHz

Military Bands

P-band 225-390 MHz
L-band 390-1550 MHz
S-band 1.55-5.2 GHz
X-band 5.2-10.9 GHz
K-band 10.9-36 GHz
Q-band 36-46 GHz
V-band 46-56 GHz

United States Broadcast Bands

Amplitude modulated (AM) 535-1605 kHz
Frequency modulated (FM) 88-108 MHz
VHF television 54-216 MHz
UHF television 470-890 MHz

FIGURE 4-1 Assignment of radio-frequency bands in the United States.

order of megahertz, or several megahertz). The reason for this division or classification merits some discussion.

As shown in Fig. 4-1, the amplitude-modulated or AM broadcast band for the United States is from 535 to 1605 kHz. The frequencies of transmitting stations within this band are spaced from 10 to 15 kHz apart to prevent interference with each other. In the frequency-modulated or FM broadcast band, the transmitting-station frequencies are spaced 200 kHz apart. In the television broadcast bands, the stations are about 6 MHz apart.

Within a specific band, each transmitting station is assigned a specific frequency at which the station is to operate. However each station transmits not only at this frequency, but at a relatively narrow band of frequencies lying at either side of this assigned frequency. Such a band of frequencies is required if the signal is to convey intelligence. For example, an AM broadcast-band station that is assigned a certain frequency transmits a signal with a band extending 5 to 7.5 kHz on either side of the assigned frequency.

An RF amplifier used in an AM broadcast radio receiver is adjusted to cover a portion of the band about 15 kHz wide, corresponding to the spread of a single station. Under these conditions, the *bandwidth* of the RF amplifier is said to be 15 kHz. The amplifier is adjusted (or tuned) to one station at a time. In the FM broadcast band, where each station is spaced 200 kHz apart, the bandwidth of the RF amplifier is about 150 kHz. Both AM and FM broadcast-band RF amplifiers are essentially *narrowband amplifiers*. In the television bands, where the stations are 6 MHz apart, the RF amplifiers are of the *wideband* type (or *broadband* type), since the transmitted TV signal is about 4.5 MHz wide.

As can be seen, the RF amplifier serves two purposes. One purpose is as a *bandpass filter*, which passes signals from the desired station and rejects all others. The other purpose is to amplify these signals to a suitable voltage (or power) level.

4-2 BASIC NARROWBAND RF AMPLIFIER THEORY

The circuit of Fig. 4-2 is a typical narrowband RF amplifier, such as those found in discrete-component radio receivers. In present design, the circuit is usually part of an IC *tuner package*, as described in Sec. 4-6. For now, let us consider the bandpass filter or tuning function and certain feedback problems.

4-2.1 Tuning Narrowband Amplifiers

The circuit of Fig. 4-2 is a single stage of *tuned radio frequency* (TRF) voltage amplification. Input to the stage is by means of a *tuned RF transformer*, and output is obtained by a similar device. Transformer T1 is the input transformer,

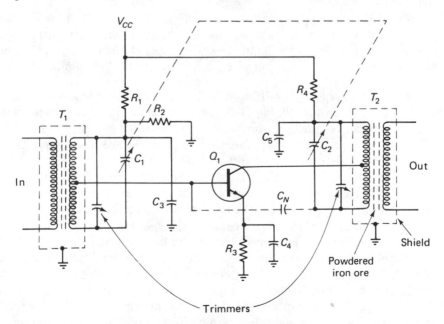

FIGURE 4-2 Typical narrowband RF amplifier in discrete-component form.

and T2 is the output transformer. The secondary of T1 is tuned to resonance at the frequency of the incoming signal by means of variable capacitor C1. T2 is tuned to the same resonant frequency by means of C2.

In many cases, the transformers are tuned by means of adjustable powered-iron cores. In that case, C1 and C2 are fixed capacitors. Since it is impossible to construct two circuits that are exactly alike, *trimmer* capacitors (in parallel with C1/C2) vary the overall capacitance of the tuned circuit slightly to compensate for small differences between the circuits.

Although the use of several tuned circuits increases the overall *selectivity* of the amplifier, the need for manipulating a number of variable capacitors can be a problem. This problem is overcome by means of *ganging*, where the rotors of the variable capacitors are mechanically connected so that all rotors move simultaneously when one dial is manipulated. Any variations in tuning are corrected by the trimmers on each transformer. When the transformers are tuned by means of adjustable cores, ganging is not used, and each tuning circuit is adjusted separately.

4-2.2 Feedback in RF Amplifiers

One difficulty often found in RF amplifiers is the prevention of feedback from the output of a stage to the input or to another stage. There are two types of undesired feedback: *radiated feedback* and *feedback through the transistor* (internal feedback).

Radiated feedback is prevented by shielding. Fortunately, most modern transistors are so constructed that there is little danger of feedback through the transistor at moderately high frequencies. However, at higher frequencies, internal feedback can produce undesired conditions in an RF amplifier.

Miller Effect and Neutralization.   One feedback problem is known as the Miller effect. As shown in Fig. 4–3, there is a capacitance between the base and emitter of a two-junction transistor (or between the gate and source of a FET). This forms the input capacitance of the circuit. There is also a capacitance between the base and collector (or gate and drain). This capacitance feeds back some of the collector signal to the base. The collector signal is amplified and is 180° out of phase with the base signal (in a common-emitter amplified). The collector signal feedback opposes the base signal and tends to distort the input signal. Likewise, the collector–base capacitance is, in effect, in series with the base–emitter capacitance and thus changes the input capacitance.

These conditions make for a constantly changing amplitude-modulated relationship of signals in an amplifier. For example, if the input signal amplitude changes, the amount of feedback changes, changing the input capacitance. In turn, the change in input capacitance changes the match between the transistor and the input tuned circuit, changing the amplitude. Likewise, if the input signal frequency changes, the feedback changes (since the collector–base capacitive reactance changes), and there is a corresponding change in amplification.

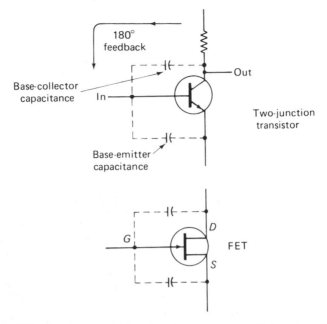

FIGURE 4-3   Input and feedback capacitances in two-junction transistors and FETs.

This Miller effect is not necessarily a problem in all solid-state RF amplifiers. The FET RF amplifier is usually more susceptible to Miller effect than two-junction transistors. However, when the Miller effect becomes severe with any RF amplifier, the effect can be eliminated (or minimized to a realistic level) by neutralization.

Neutralization is a method for reducing the amount of unwanted feedback, either from radiation or internal feedback. With neutralization, a portion of the voltage from the output circuit of the amplification stage is fed back to the input circuit so as to cancel out the base voltage caused by the unwanted feedback. Neutralization is done by impressing a voltage on the base that is equal in magnitude, but opposite in phase, to the undesired feedback. The two voltages "buck" each other out.

The two ends of the output transformer primary winding (such as T2 in Fig. 4–2) are of opposite phase. If this opposite-phase voltage is fed to the base through a neutralizing capacitor (CN of Fig. 4–2), the two voltages cancel out. As a guideline, the neutralizing capacitor should equal the collector–base capacitance (typically a few picofarads).

Common-Base RF Amplifiers.   Another method for reducing the amount of unwanted feedback, without neutralization, is to use the *common-base amplifier* configuration (Chapter 1). A common-base RF amplifier is shown in Fig. 4–4.

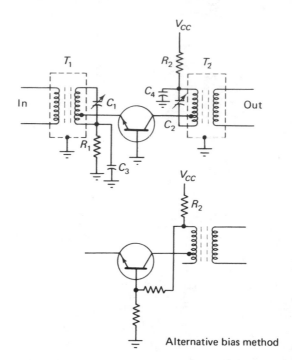

FIGURE 4-4 Common-base (grounded-base) RF amplifier.

Again, the input transformer T1 is tuned to resonance by variable C1, while output transformer T2 is tuned to the same resonant frequency by C2. Resistor R1 is the emitter resistor, and C3 is the emitter-bypass capacitor. Resistor R2 and capacitor C4 form a decoupling network (Sec. 3–1).

The base is grounded, and the input signal is applied to the emitter. The output is taken between the collector and base, which is common to the input and output circuits. The grounded base acts as a shield between the input and output circuits, thus reducing feedback. The circuit of Fig. 4–4 is often found at the antenna input of discrete-component receivers. In addition to minimizing undesired feedback, the ground or common-base circuit provides a low input impedance (to match a 50-, 75-, or 300-Ω antenna).

4-2.3 Intermediate-Frequency Amplifier

Most radio and TV receivers operate on the *superheterodyne* principle where the frequency of the received radio signal is first converted to a lower predetermined frequency called the *intermediate frequency*. The amplifier is fixed to operate at this frequency, rather than being tunable over the entire band.

The circuit of Fig. 4–5 is a typical intermediate frequency or IF amplifier, such as those found in discrete-component radio receivers and TV sets. In present design, the circuit is usually part of an IC package, as described in Sec. 4–6. For now, let us consider the tuning function of the IF amplifier.

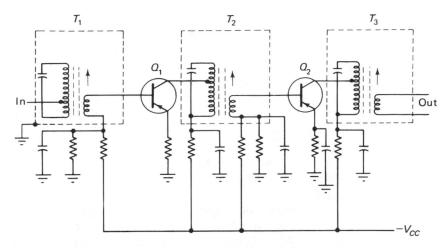

FIGURE 4-5 Typical IF amplifier in discrete-component form.

The two-stage IF amplifier of Fig. 4–5 is similar to the RF amplifier of Fig. 4–2, except that the IF amplifier transformers are tuned to the predetermined frequency by means of small fixed capacitors. Since it is not necessary to tune the IF transformers over the entire band, it is practical to tune both the primary and secondary windings of each transformer to the intermediate frequency. Thus, by adding tuned circuits, the receiver selectivity is increased.

To compensate for small variations between the IF transformers, each transformer has an adjustable powdered-iron core that can be moved in or out, thus varying the inductance slightly. In some IF transformers the inductances are fixed, and the windings are tuned by trimmers.

Note that only the primary windings of the transformers are tuned in the Fig. 4–5 circuit. Since the impedance of the primary is reflected to the secondary, the effect is the same regardless of which winding is tuned. However, because there are fewer tuned circuits, the overall selectivity of the amplifier is somewhat reduced. All other factors being equal, the more tuned circuits there are in any amplifier the greater the selectivity, and vice versa.

The collector of each transistor is connected to a tap on the primary winding of the corresponding output transformer. This is done to match the impedance of the winding to the output impedance of the transistor. Similarly, the secondary of each input transformer has fewer turns than the primary winding so that the winding may match the input impedance of the transistor.

4-2.4 RF Power Amplifier

Most radio transmitters use some form of power amplifier to raise the low-amplitude signal developed by the oscillator to a high-amplitude signal suitable for transmission. For example, most oscillators develop signals of less than 1 W, whereas a solid-state transmitter may require several hundred watts (or more) output.

Figure 4–6 shows two basic RF power-amplifier circuits. In the circuit of Fig. 4–6a, the collector load is a parallel-resonant circuit (called a *tank circuit*), consisting of variable capacitor C1 and coil L1, tuned to resonance at the desired frequency. The output, which is an amplified version of the input voltage, is from L2, which together with L1 forms an output transformer.

The circuit of Fig. 4–6a has certain advantages and disadvantages. The winding of L2 can be made to match the impedance of the load (by selecting the proper number of turns and by positioning L2 in relation to L1). While that may prove an advantage in some cases, it also makes for an *interstage coupling network* that is subject to mismatch and detuning by physical movement or shock. Another disadvantage of the Fig. 4–6a circuit is that all current must pass through the tank circuit coil.

For best transfer of power, the impedance of L1 should match that of the transistor output. Since two-junction transistor output impedances are generally low, the value of L1 must be low, often resulting in an impractical size for L1. The circuit of Fig. 4–6a is a carry-over from vacuum-tube circuits and, as such, is not often found in present-day transmitter circuits. A possible exception is in a few low-power FET amplifier circuits.

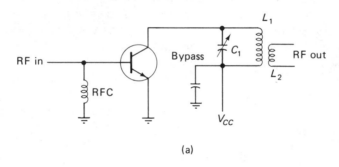

(a)

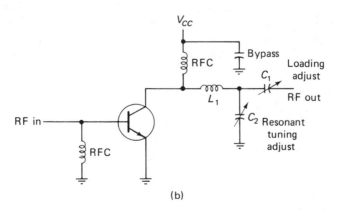

(b)

FIGURE 4-6 Two basic RF power-amplifier circuits.

The circuit of Fig. 4–6a, or one of the many variations, is commonly found in solid-state transmitters using two-junction and high-power FET transistors. The collector load is a resonant circuit formed by the network L1, C1, and C2. Note that C1 is labeled "Loading adjust" whereas C2 is labeled "Resonant tuning adjust." As is discussed in the remaining paragraphs of this chapter, these networks provide the dual function of frequency selection (equivalent to the tank circuit) and impedance matching between transistor and load. To properly match impedances, both the resistive (so-called *real part*) and reactive (so-called *imaginary part*) components of the impedance must be considered (Sec. 4–7).

Both Fig. 4–6 circuits are operated class B, which is typical for RF amplifiers. Class B operation is obtained by connecting the emitter directly to ground and applying no bias to the base–emitter junction. Since any two-junction transistor requires some forward bias to produce current flow, the transistor remains cut off except in the presence of a signal.

4–2.5 RF Multiplier

The circuits of Fig. 4–6 can be used as a frequency multiplier. That is, the collector is tuned to a higher whole-number multiple (harmonic) of the input frequency. Many radio transmitters use some form of multiplier to raise the low-frequency signal developed by the oscillator to a high-frequency signal. For example, most crystals used in oscillators have a fundamental frequency of less than 10 MHz, whereas a solid-state transmitter may produce an output in the UHF range.

Although the circuits of power multipliers and power amplifiers are essentially the same, the efficiency is different. That is, an amplifier operating at the same frequency as the input has a higher efficiency than an identical circuit operating at a multiple of the input frequency.

4–2.6 RF Amplifier–Multiplier Combinations

The circuits of Fig. 4–6 can be cascaded to provide increased power amplification and/or frequency multiplication. Typically, no more than three stages are so cascaded. The stages can be mixed. That is, one or two stages can provide frequency multiplication, with the remaining one or two stages providing power amplification. Such arrangements are discussed in the remaining sections of this chapter.

4–3 BASIC WIDEBAND RF AMPLIFIER THEORY

Except for pure sine waves, all signals are found to contain not only the fundamental frequency, but harmonic and subharmonic frequencies as well. These harmonics are whole-number multiples of the fundamental frequency. Pulse signals have an especially high harmonic content. For example, the pulses used in television contain frequencies ranging from about 30 Hz to 4 or 5 MHz.

An ordinary RF amplifier with a bandwidth of several hundred kilohertz is not able to amplify uniformly signals with such a broad range of frequencies. For this reason, it is necessary to use special *broadband* or *wideband* amplifiers for such application. These amplifiers are usually known as *VF (video frequency) amplifiers* or simply *video amplifiers* in television equipment (TV sets, camcorders, VCRs, and the like) or as *pulse amplifiers* in radar and similar equipment.

The RC amplifiers described in Chapter 3 are, in effect, wideband amplifiers. Such circuits amplify uniformly at all frequencies of the audio range, dropping off only at the low- and high-frequency ends. A wideband amplifier (capable of passing RF signals including pulses) is formed when the uniform amplification is extended to both ends of the frequency range.

A basic RC amplifier circuit is shown in Fig. 4-7. Capacitor Cout represents the output capacitance of Q1. Capacitor CD represents the *distributed capacitance* of the various components and related wiring. Capacitor Cin represents the input capacitance of Q2.

Coupling capacitor CC and base resistor RB form a voltage divider across the input of Q2. At low frequencies, the impedance of CC is large, and relatively little of the signal is applied to the base of Q2. Accordingly, the low-frequency response of the amplifier is lowered.

Capacitances Cout, CD, and Cin, acting in parallel, shunt the load resistor R1 of Q1. This lowers the effective resistance of R1, as well as the high-frequency response of the amplifier. As discussed in Chapters 1 and 3, a lower value of RL lowers the gain, all other factors being equal.

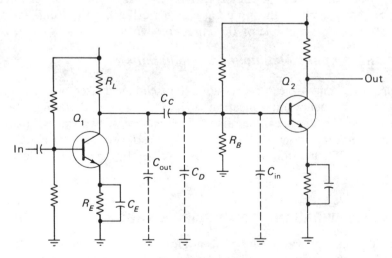

FIGURE 4-7   Basic RC amplifier circuit.

4-3.1 Increasing Wideband Response

There are several methods for improving the low- and high-frequency response of wideband amplifiers (or RC amplifiers designed for wideband use). In all cases, transistors with small input and output capacitances should be used. Likewise, the components must be carefully placed so that their leads and distributed capacitances are kept at a minimum.

Emitter Bypass.   The emitter-bypass capacitor CE of Fig. 4–7 affects the low-frequency gain of the amplifier. As discussed in Chapter 3, the impedance of CE is higher at lower frequencies. Thus, the amplifier gain is lower at lower frequencies. Accordingly, the capacitance of CE must be large enough to offer a low impedance (with respect to RE) at the lowest frequency to be amplified.

Collector Resistance.   The value of collector load resistor RL also affects the frequency response and gain of the amplifier. The graph of Fig. 4–8 shows the effects produced by various values of R.

A large-value collector resistor produces a high gain at the middle frequencies and a steep drop in gain at the high and low frequencies. A small-value collector resistor produces a much smaller overall gain, but the proportional drop in gain at the high and low frequencies is also much less than for the larger-value collector resistors. With the small-value collector resistor, amplifier gain is uniform over a much wider range of frequencies. In effect, the amplifier sacrifices gain for bandwidth. Because of these conditions, wideband amplifiers use low-value collector resistances and transistors with high gain (large betas).

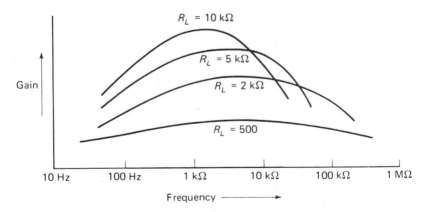

FIGURE 4-8   Effects produced by various values of collector resistance.

Coupling Capacitances.   At low frequencies, the effects of the transistor input/output capacitance and the distributed capacitances are negligible, but the impedance of the coupling capacitor becomes increasingly important. One way to compensate for the effects of the coupling capacitor is shown in Fig. 4-9, which is the video amplifier of a typical discrete-component TV set. (In present design, the circuits of Fig. 4-9 are often found in IC form.)

In the Fig. 4-9 circuit, the load resistance for Q1 is made up of two parts, R5 and R6, connected in series. Capacitor C3 is the bypass capacitor for R6. At the higher frequencies, the collector load is effectively R5, since the small impedance of C3 at these frequencies permits C3 to completely bypass R6 (in effect removing R6 from the circuit).

At low frequencies, the impedance of C3 becomes high, and the bypassing effect is greatly reduced. The collector load resistance then becomes R5 + R6. This greater resistance produces a greater output voltage, thus compensating for the low-frequency drop produced by C3.

Shunt Peaking Coil.   Since the drop in high-frequency response is due to the shunting effect of the transistor capacitances (and distributed capacitances) upon the load resistor, a small coil L1 (called a shunt peaking coil) is inserted in series with the load resistance. At low frequencies, L1 offers very little impedance, and the collector load is, essentially, the resistance of R5 + R6. At high frequencies, the impedance of L1 is high, and the collector load is the sum

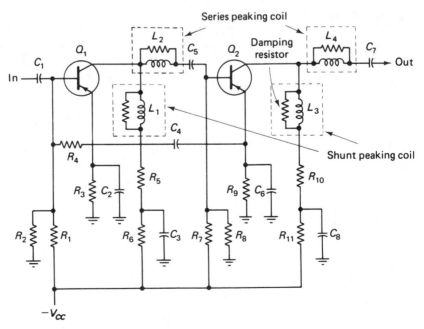

FIGURE 4-9   Video amplifier of a typical discrete-component TV set.

of the R5 + R6 resistances and the resistance of L. Thus, amplifier gain is increased.

Inductor L1 sets up a resonant circuit with the distributed capacitances of the circuit (and the capacitances of the transistor). The value of L1 is selected so that the circuit is resonant at a frequency where the high-frequency response of the amplifier begins to drop. In this way an additional boost is given to the gain, and the high-frequency end of the response curve is flattened.

Series Peaking Coil. Figure 4–9 shows another similar compensation circuit. Series peaking coil L2 is connected in series with the coupling capacitor C5. At high frequencies, L2 forms a low-impedance series-resonant circuit with the capacitances, causing a larger voltage to appear at the base of Q2.

Video and pulse amplifiers often use both shunt and series peaking coils for high-frequency compensation. As a general guideline, the values of these coils are such that resonance is obtained at the *highest desired frequency*. That is, the capacitances are calculated (or measured), and a corresponding value of inductance is chosen for resonance at the high-frequency end.

Damping Resistances. Note that the coils of Fig. 4–9 are shown with resistances connected in parallel. As discussed in Sec. 4–4, when resistances are connected across coils, the resonant point of the coils is flattened or broadened. Such resistances are often known as *damping* resistances.

Inverse Feedback. Another method for overcoming the effects of the drop in gain at the low and high ends of the frequency band involves the use of inverse feedback. As discussed in Chapters 1 and 3, inverse feedback tends to oppose any change in signal level. Thus, when the signal level tends to drop at either end of the frequency range, inverse feedback opposes this change. In the circuit of Fig. 4–9, inverse feedback is provided by C4 and R4.

Camcorder Video Circuit. Figure 4–10 shows some video-amplifier circuits found in a camcorder. Note that the circuits use discrete components, rather than IC, even though the camcorder is of current design. The circuits illustrate several of the points just discussed for wideband video amplifiers.

The circuits of Fig. 4–10 convert weak signal current from the pickup tube of a camcorder to a voltage signal and then amplify the voltage signal into a prevideo signal of suitable level. The circuit has one input from the pickup tube and produces one output at the emitter of Q1001 to the prevideo processor circuits. The circuit also provides for application of the target voltage from the high-voltage power supply to the pickup tube target.

The preamp has a low output impedance and uses negative feedback in an amplifier with a low-noise JFET input (to improve the signal-to-noise ratio). Negative feedback from output to input is applied through C1007/C1006/R1015/R1014/R1016/R1017. (This is similar to the C4/R4 net-

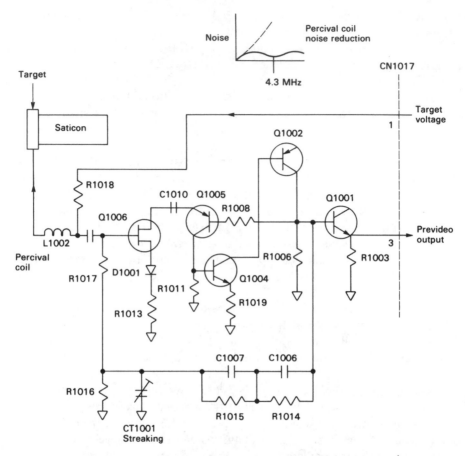

FIGURE 4-10 Video amplifier circuits found in a camcorder.

work of Fig. 4–9.) In the Fig. 4–10 circuit, CT1001 (called the *smear* or *streaking* adjustment) varies the low-frequency (400- to 500-kHz) portion of the negative feedback signal to maintain a flat frequency response through the amplifier.

A *Percival compensation circuit* is used at the input to prevent signal-to-noise deterioration because of impedance mismatch. The Percival coil L1002 resonates at about 4.3 MHz (as shown by the curve on Fig. 4–10) and separates the pickup tube output capacitance from the preamp input capacitance.

4-4 RESONANT CIRCUITS FOR RF AMPLIFIERS

RF amplifier design is based on the use of resonant circuits (or *tank* circuits) consisting of a capacitor and a coil (inductance) connected in series or parallel, as shown in Fig. 4–11. At the resonant frequency, the inductive and capacitive

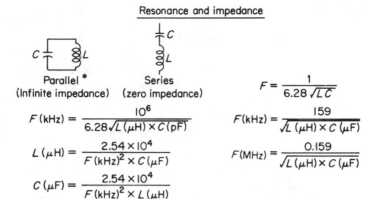

$$F = \frac{1}{6.28\sqrt{LC}}$$

Resonance and impedance

Parallel *
(Infinite impedance)

Series
(zero impedance)

$$F\,(\text{kHz}) = \frac{10^6}{6.28\sqrt{L\,(\mu\text{H}) \times C\,(\text{pF})}}$$

$$L\,(\mu\text{H}) = \frac{2.54 \times 10^4}{F\,(\text{kHz})^2 \times C\,(\mu\text{F})}$$

$$C\,(\mu\text{F}) = \frac{2.54 \times 10^4}{F\,(\text{kHz})^2 \times L\,(\mu\text{H})}$$

$$F\,(\text{kHz}) = \frac{159}{\sqrt{L\,(\mu\text{H}) \times C\,(\mu\text{F})}}$$

$$F\,(\text{MHz}) = \frac{0.159}{\sqrt{L\,(\mu\text{H}) \times C\,(\mu\text{F})}}$$

*Approximate; accurate when circuit Q is 10 or higher

Inductive reactance

$$Z = \sqrt{R^2 + X_L^2} \quad Q = \frac{X_L}{R} \qquad L = \frac{X_L}{6.28\,F}$$

Series

$$Z = \frac{R X_L}{\sqrt{R^2 + X_L^2}} \quad Q = \frac{R}{X_L} \qquad F = \frac{X_L}{6.28\,L}$$

Parallel

$$X_L = 6.28 \times F\,(\text{Hz}) \times L\,(\text{H})$$
$$X_L = 6.28 \times F\,(\text{kHz}) \times L\,(\text{mH})$$
$$X_L = 6.28 \times F\,(\text{MHz}) \times L\,(\mu\text{H})$$

Capacitive reactance

$$Z = \sqrt{R^2 + X_C^2} \quad Q = \frac{X_C}{R} \qquad F = \frac{1}{6.28\,C X_C}$$

Series

$$Z = \frac{R X_C}{\sqrt{R^2 + X_C^2}} \quad Q = \frac{R}{X_C} \qquad C = \frac{1}{6.28\,F X_C}$$

Parallel

$$X_C = \frac{1}{6.28 \times F\,(\text{Hz}) \times C\,(\text{F})}$$
$$X_C = \frac{159}{F\,(\text{kHz}) \times C\,(\mu\text{F})}$$

FIGURE 4-11 Resonant-circuit equations.

reactances are equal, and the circuit acts as a high impedance (in a parallel circuit) or a low impedance (in a series circuit). In either case, any combination of capacitance and inductance has some resonant frequency.

Either (or both) the capacitance or inductance can be variable to permit tuning of the resonant circuit over a given frequency range. When the inductance is variable, tuning is usually done by means of a metal (powdered iron) slug inside the coil. The slug is screwdriver-adjusted to change the inductance (and thus the inductive reactance) as required.

Typical RF amplifier circuits used in receivers (AM, FM, TV, communications, and so on) often include two resonant circuits in the form of a transformer (RF or IF transformer, and the like). Either the capacitance or inductance can be variable. Since such transformers are available commercially, design is not discussed here. However, measurement of RF (and IF) transformer resonant values, as the values affect design, are discussed in Chapter 9.

In the case of RF amplifiers used in transmitters, it is sometimes necessary to design the *coil portion* of the resonant circuit. This is because coils of a given inductance and physical size may not be available from commercial sources. Thus, design information is included for basic RF coils in this section.

4-4.1 Design Considerations for Resonant Circuits

The two basic design considerations for RF resonant circuits are *resonant frequency* and the Q *(or quality) factor*.

Resonant Frequency.   Figure 4–11 contains equations that show the relationships among capacitance, inductance, reactance, and frequency, as these factors relate to resonant circuits. Note that there are two sets of equations. One set includes reactance (inductive and capacitive). The other set omits reactance. The reason for two sets of equations is that some design approaches require the reactance to be calculated for resonant networks. Solid-state RF transmitter–amplifier circuits are a classic example of this.

Q or Selectivity.   A resonant circuit has a Q, or quality, factor. The circuit Q depends on the ratio of reactance to resistance. If a resonant circuit had pure reactance, the Q would be high (actually infinite). However, this is not practical. For example, any coil has some d-c resistance, as do the leads of a capacitor. As frequency increases, the a-c resistance presented by the leads increases as a result of the skin effect. The sum total of these resistances is usually lumped together and is considered as a resistor in series or parallel with the circuit. The total resistance is usually termed the *effective resistance* and is not to be confused with the reactance.

The resonant-circuit Q depends on the individual Q factors of inductance and capacitance used in the circuit. For example, if both the inductance and capacitance have a high Q, the circuit has a high Q, provided that a minimum of resistance is produced when the inductance and capacitance are connected to form a resonant circuit.

From a practical design standpoint, a resonant circuit with a high Q produces a sharp resonance curve (narrow bandwidth), whereas a low Q produces a broad resonance curve (wide bandwidth). For example, a high-Q resonant circuit provides good harmonic rejection and efficiency in comparison with a low-Q circuit, all other factors being equal.

The *selectivity* of a resonant circuit is related directly to Q. A very high Q (or high selectivity) is not always desired. Sometimes it is necessary to add resistance to a resonant circuit to broaden the response (increase the bandwidth, decrease the selectivity). The resistances across the coil in Fig. 4–9 are an example.

Usually, resonant-circuit Q is measured at the point on either side of the resonant frequency where the signal amplitude is down 0.707 of the peak resonant value, as shown in Fig. 4–12. (Resonant-circuit Q measurements are included in Chapter 9.)

Note that Q must be increased for increases in resonant frequency if the same bandwidth is to be maintained. For example, if the resonant frequency is 10 MHz, with a bandwidth of 2 MHz, the required circuit Q is 5. If the resonant frequency is increased to 50 MHz, with the same 2-MHz bandwidth, the required Q is 25.

Circuit Q must be decreased for increases in bandwidth if the same resonant frequency is to be maintained. For example, if the resonant frequency is 30 kHz, with a bandwidth of 2 kHz, the required circuit Q is 15. If the bandwidth is increased to 10 kHz, with the same 30-kHz resonant frequency, the required Q is 3.

4-4.2 Resonant Circuit Design Examples

Assume that it is desired to find the resonant frequency of a 0.002-μF capacitor and a 0.02-mH inductance. Using the equations of Fig. 4–11, first convert the 0.02 mH to 20 μH. Then

$$F = \frac{0.159}{\sqrt{20 \ \mu H \ \times \ 0.002 \ \mu F}}$$

$$F = \frac{0.159}{\sqrt{0.4}}$$

$$= \frac{0.159}{0.2}$$

$$= 0.795 \text{ MHz} \quad \text{or} \quad 795 \text{ kHz}$$

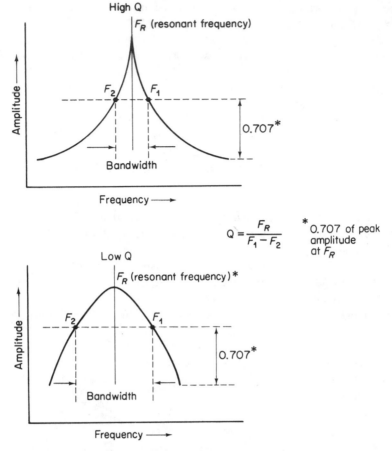

FIGURE 4-12 Relationship of bandpass characteristics to Q of resonant circuit.

Assume that it is desired to design a circuit that resonates at 400 kHz with an inductance of 10 μH. What value of capacitor is necessary? Using the equations of Fig. 4-11,

$$C = \frac{2.54 \times 10^4}{400^2 \times 10}$$

= 0.0158;   use the nearest standard value of 0.016 μF

Assume that it is desired to design a circuit that resonates at 2.65 MHz with a capacitance of 360 pF. What value of inductance is necessary? Using the equations of Fig. 4-11,

$$L = \frac{2.54 \times 10^4}{2650^2 \times (360 \times 10^{-6})}$$

$$= 10 \ \mu H$$

Assume that an RF power amplifier network must operate at 40 MHz with a bandwidth of 8 MHz. What circuit Q is required? Using the equations of Fig. 4–12,

$$FR = 40, \ F1 - F2 = 8$$

$$Q = \frac{40}{8} = 5$$

4-4.3 Design Considerations for Coils

Figure 4–13 shows the equations necessary to calculate the self-inductance of a single-layer, air-core coil. Note that maximum inductance is obtained when the ratio of coil radius to coil length is 1.25, that is, when the length is 0.8 of the radius. RF coils wound for this ratio are the most efficient (maximum inductance for minimum physical size).

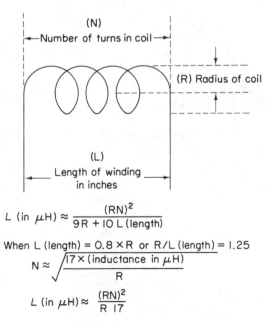

$$L \ (\text{in } \mu H) \approx \frac{(RN)^2}{9R + 10 \ L \ (\text{length})}$$

When L (length) = 0.8 × R or R/L (length) = 1.25

$$N \approx \sqrt{\frac{17 \times (\text{inductance in } \mu H)}{R}}$$

$$L \ (\text{in } \mu H) \approx \frac{(RN)^2}{R \ 17}$$

FIGURE 4-13 Calculations for self-inductance of isngle-layer air-core coil.

4-4.4 Coil Design Examples

Assume that it is desired to design a coil with 0.5-μH inductance on a 0.25-in. radius (air core, single layer). Using the equations of Fig. 4-13, for maximum efficiency, the coil length should be 0.8R, or 0.2 in. Then,

$$N = \frac{17 \times 0.5}{\sqrt{0.25}}$$

$$= \sqrt{34}$$

$$= 5.8 \text{ turns}$$

For practical purposes, use six turns and spread the turns slightly. The additional part of a turn increases inductance, but the spreading decreases the inductance. After the coil is made, the inductance should be checked with an inductance bridge or as described at the end of this chapter.

4-4.5 Tuning Amplifier RF Circuits with Voltage-Variable Capacitors

Voltage-variable capacitors (VVCs) are used in many applications to tune the RF circuits of amplifiers. The capacitance of a VVC is controlled by external voltage. The capacitance is varied when the external voltage is varied. If a VVC is used in an RF tuning circuit of an amplifier, it is possible to vary the circuit capacitance and thus vary the amplifier resonant frequency with a variable external voltage.

A VVC is sometimes called a *voltage-variable diode*, because the device is constructed more like a diode than a capacitor. However, VVC is the more accepted term.

Most VVC resonant circuits are in the form shown in Fig. 4-14 for parallel circuits and in the form in Fig. 4-15 for series circuits. In some cases, for biasing purposes, additional RF chokes are connected in the circuits. The main concern with VVCs in resonant circuits is the *tuning range* of the circuit. All other factors being equal, the tuning range depends on the capacitance range of the VVC.

An example of how VVCs are used to tune IC RF voltage amplifiers is given in Sec. 4-6.

4-5 RF VOLTAGE-AMPLIFIER CIRCUITS (DISCRETE COMPONENT)

RF voltage amplifiers are used primarily in receivers and receiver-type circuits. IF amplifiers, or IF limiter-amplifiers, are examples of RF voltage amplifiers. The input or first stage of a receiver may include a separate RF voltage amplifier

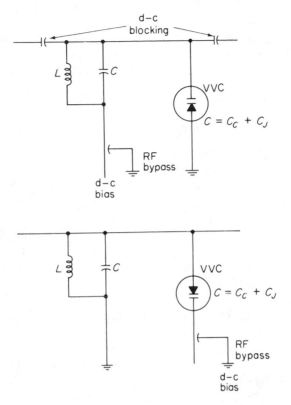

FIGURE 4-14 Typical parallel circuits for VVC control.

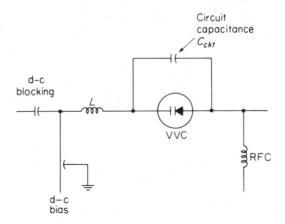

FIGURE 4-15 Typical series circuit for VVC control.

(such as with some discrete-component communications receivers). However, most solid-state receivers combine the RF voltage-amplifier function with that of the local oscillator.

In present design, the trend is to incorporate the RF voltage-amplifier circuits described here into IC form. For that reason, we do not go into full detail on discrete-component voltage amplifiers. However, the circuits covered in this section provide a basis for understanding the IC amplifiers described in Sec. 4–6. Note that the illustrations in this section include equations that show design characteristics for the discrete-component circuits.

4–5.1 RF Voltage Amplifier

Figure 4–16 is the schematic of a typical RF voltage amplifier. Such a circuit could be used as an IF amplifier, IF limiter, or separate RF amplifier with few modifications. Both the input and output are tuned to the desired operating frequency by means of the resonant circuits. In this case, the resonant circuits are composed of transformers with a capacitor across the primary. The capacitors can be variable, but are usually fixed. The resonant circuit is tuned by an adjustable slug between the windings.

4–5.2 Frequency Mixers and Converters

Figure 4–17 shows the working schematic of a typical frequency mixer and converter. Such a circuit is a combination of an RF voltage amplifier and an RF oscillator. The individual outputs of the two sections are combined to produce an IF output. Usually, the RF oscillator operates at a frequency above the RF amplifier, with the difference in frequency being the intermediate frequency.

The resonant circuit of T1 is tuned to the incoming RF signal, T2 is tuned to the oscillator frequency (RF + IF), and T3 is tuned to the intermediate frequency (IF). The resonant circuits of T1 and T2 are usually tuned by means of variable capacitors ganged together so that both the oscillator and RF amplifier remain at the same frequency relationship over the entire tuning range. For example, if T1 tunes from 550 to 1600 kHz, and T3 is at a fixed IF of 455 kHz, T2 must tune from 1005 to 2055 kHz. Usually, trimmer capacitors are connected in parallel with the variable capacitors to permit adjustment over the tuning range.

4–5.3 AVC–AGC Circuits for Amplifiers

Most receivers have some form of AVC–AGC (automatic volume control–automatic gain control) circuit. The terms AVC and AGC are used interchangeably. AGC is a more accurate term since the circuits involved control the gain of an IF or RF stage (or several stages simultaneously), rather than the volume

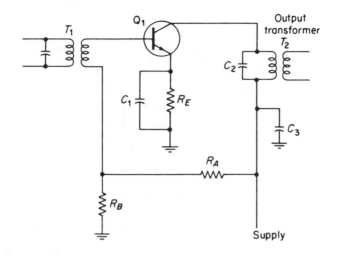

Voltage gain $\approx$ beta $\times(\frac{1}{N})$

where $N = \sqrt{\dfrac{Z_P}{Z_S}}$ $N^2 = Z_P/Z_S$

Voltage drop across	At operating frequency
$R_E \approx$ emitter-base voltage 0.5 for silicon 0.2 for germanium	$X_{C1} \gtrless Z_{in}$ of Q_1 $X_{C2} \gtrless 100\ \Omega$ $X_{C3} \approx X_{primary}$ of T_2

$R_B \approx 10\ R_E$

Voltage drop across R_A = supply – drop across R_B

Supply $\approx$ 3 to 4 times desired output voltage

FIGURE 4-16   Basic RF voltage amplifier.

of an audio signal in an AF stage. However, in a broadcast receiver, the net result is an automatic control of volume. Either way, the circuit provides a constant output despite variations in signal strength. An increased signal reduces stage gain, and vice versa.

Figure 4–18 shows the working schematic of two AGC systems that are common to discrete-component broadcast and communications receivers. Diode CR1 acts as a variable shunt resistance across the input of the IF stages. Diode CR2 functions as both the detector and AGC bias source.

Under no-signal conditions or in the presence of a weak signal, diode CR1 is reverse biased and has no effect on the circuit. In the presence of a very large signal, CR1 is forward biased and acts as a shunt resistance to reduce gain.

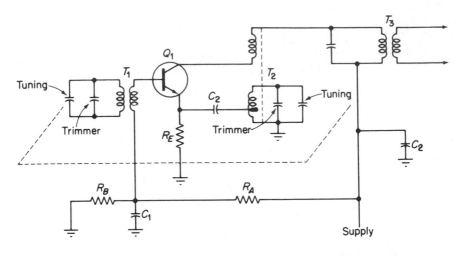

$R_E \approx$ impedance of tap on T_2

Drop across $R_B \approx 0.2 \times$ supply voltage

Drop across $R_A \approx 4 \times R_B$

$R_A + R_B \approx 15{-}20 \times R_E$
at operating frequency X_{C1} and $X_{C2} \approx 50\ \Omega$ (or less)

Power output of $T_2 \approx 0.5 \times \dfrac{\text{collector voltage}^2}{Z \text{ of } T_2 \text{ collector winding}}$

Power output of $T_3 \approx 0.125 \times \dfrac{\text{collector voltage}^2}{Z \text{ of } T_2 \text{ collector winding}}$

FIGURE 4-17 Basic RF mixer and converter (RF amplifier and local oscillator).

The output of CR2 is developed across R1 and applied to the audio stages. Resistor R1 also forms part of the bias network for the IF-stage transistor. The combined fixed bias (from the network) and variable bias (from the detector) is applied to the IF-stage base–emitter circuit. The detector bias varies with signal strength and is of a polarity that opposes variations in signal. That is, if the signal increases, the detector bias is more positive (or less negative) for the base of a PNP transistor, and vice versa for an NPN transistor.

4-5.4 *Television-Amplifier AGC Circuits*

Most discrete-component TV receivers use a *keyed*, saturation-type AGC circuit. The RF-tuner and IF-stage transistors connected to the AGC lines are forward biased at all times. On strong signals, the AGC circuits *increase the*

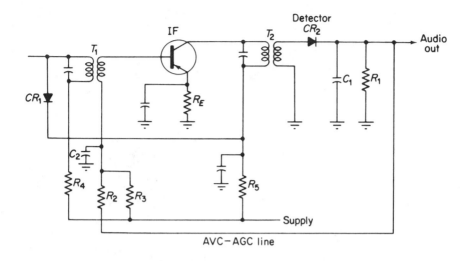

CR₁ is reverse biased with no signal
$C_2 \approx 10 \mu F$
Drop across $R_1 \approx 0.5 - 1.0$ V
Drop across $R_1 + R_2 \approx 1.0 - 2.0$ V
Drop across $R_3 =$ supply $- (R_1 + R_2)$
$R_1 + R_2 \approx 10 \times R_E$

FIGURE 4-18 Two AGC systems common to discrete-component broadcast and communications receivers.

forward bias, driving the transistors into saturation, thus reducing gain. Under no-signal conditions, the forward bias remains fixed.

Although the AGC bias is a d-c voltage, the bias is partially developed (or controlled) by bursts of IF signals. A portion of the IF signal is taken from the IF amplifiers and is pulsed, or keyed, at the horizontal sweep frequency rate (15,750 Hz). The resultant keyed bursts of signal control the amount of d-c voltage produced on the AGC line.

Figure 4-19 is the schematic of a typical AGC circuit for RF and IF amplifiers in a discrete-component TV set. Transistor Q1 is an IF amplifier with the collector tuned to the IF center frequency of 42 MHz by transformer T1. No d-c voltage as such is supplied to Q1. The keying pulses from the horizontal flyback transformer (at 15,750 kHz) are applied to the collector through diode CR1. This produces an average collector voltage of about 1V.

When Q1 is keyed on, the bursts of IF signals pass through T1 and are rectified by CR2. A corresponding d-c voltage is developed across C4 and acts as a bias for AGC amplifier Q2. Transistor Q2 is connected as an emitter follower, with the AGC line being returned to the emitter. Variations in IF signal strength cause corresponding variations in Q2 bias, Q2 emitter voltage, and the AGC line voltage.

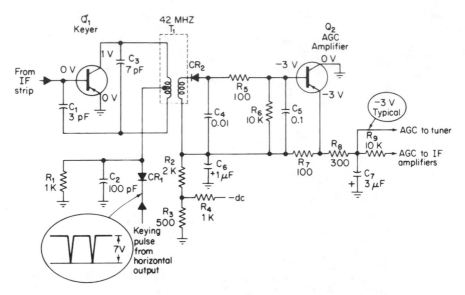

FIGURE 4-19   Typical AGC circuit for RF and IF amplifiers in a discrete-component RV set.

4-6   RF VOLTAGE-AMPLIFIER CIRCUITS (IC)

This section describes the RF amplifier circuits found in a typical AM/FM tuner of IC design. Although most of the amplifier components described here are contained in ICs, there are a number of discrete components between ICs and discrete components used to tune or adjust the amplifier ICs. Compare these circuits to the discrete-component circuits described in Sec. 4–5.

4-6.1   Relationship of Tuner Circuits

Figure 4–20 shows the relationship of the tuner circuits. Note that the audio output and muting circuits are the same as described in Sec. 3–14.3.

Amplifier and Multiplex Decoder.   IC301 is both an amplifier and multiplex decoder and is used in the audio path for AM and FM. Once audio reaches pin 2 of IC301, whether from the AM or FM section, IC301 produces corresponding audio at pins 6 and 7. IC301 has only one adjustment control. This is the multiplex VCO-adjust potentiometer R305 connected at pin 15.

FM Section.   FM broadcast signals are applied to FM tuner package MD101. Note that MD101 contains RF voltage amplifiers, local oscillator, mixer/converter, IF amplifiers, and detector. These are FM versions of the circuits shown in Figs. 4–17 through 4–19.

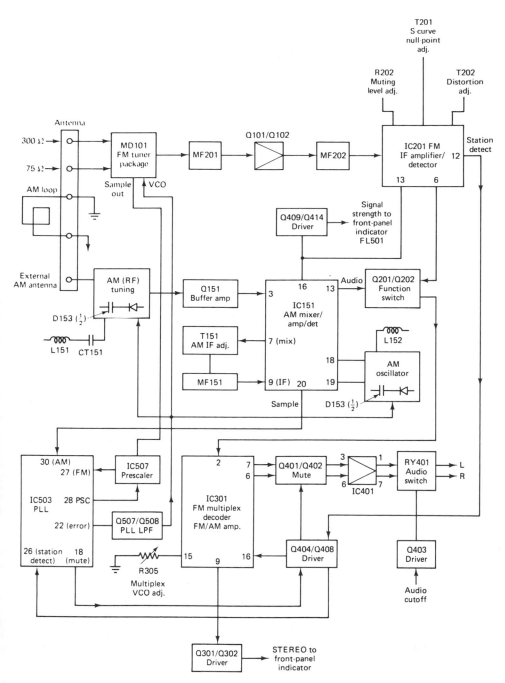

FIGURE 4-20 Relationship of circuits in IC tuner.

The IF output from MD101 is applied to the FM amplifier and detector IC201 through amplifiers Q101/Q102 and ceramic filters MF201/MF202, as shown in Fig. 4–21. This amplifier/filter combination removes any amplitude modulation and passes only signals of the desired frequency. Note that Q101/Q102 are connected as a form of differential amplifier such as discussed in Chapter 6. Note that IC201 has three adjustments: muting-level adjustment R202, S-curve null-point adjustment T201, and FM distortion adjustment T202.

The audio output from the FM section is applied to amplifier/multiplex decoder IC301 through audio-select amplifiers Q201/Q202, as shown in Fig. 4–22. These circuits select the AM or FM audio for the input to IC301.

When FM is selected, FM B + is applied to Q202, turning Q202 on. Since there is no B + applied to Q201, the AM audio does not pass. However, FM audio present at pin 6 of IC201 is amplified and passed to pin 2 of IC301 through Q202, C301, and R301.

When AM is selected, Q201 is turned on, Q202 is turned off (blocking FM audio), and AM audio present at pin 13 of IC151 is passed to pin 2 of IC301 through Q201, C301, and R301.

Note that IC301 has a dual function. When suitable FM is present, IC301 is a stereo amplifier (as well as a decoder). IC301 functions as a mono amplifier when there is no FM stereo present or when the stereo signal is too weak to produce proper FM.

AM Section.   AM broadcast signals are applied to AM IF amplifier/detector IC151 through buffer–amplifier Q151, as shown in Fig. 4–23. The RF circuit (L151/CT151) is tuned by one section of D153 (which is a VVC similar to that described in Sec. 4–4.5). The mixer output of IC151 is tuned by AM IF adjust T151 and applied to the IF portion of IC151 through ceramic filter MF151. Note that IC151 contains RF amplifiers, local oscillator, mixer/converter, IF amplifiers, and detector similar to that shown in Figs. 4–17 through 4–19.

The audio output from the detector of IC151 is applied to IC301 through Q201/Q202 (Fig. 4–22). Note that IC301 functions as a mono amplifier when the AM mode is selected. Since no stereo signal is present in the AM mode, IC301 shifts to mono operation, just as when there is no FM stereo or stereo is weak.

FM Tuning.   The FM section is tuned to the desired frequency (and locked to that frequency) by signals applied to MD101. The VCO signal (a variable d-c voltage, sometimes called the *error voltage*) from PLL microprocessor IC503 is applied to MD101 through Q507/Q508 (which are connected as a direct-coupled amplifier, such as described in Chapter 5). The IC503 error voltage shifts the MD101 oscillator as necessary to tune across the FM broadcast band or to fine-tune MD101 at a selected station.

The frequency produced by MD101 is sampled and applied to the FM input of IC503 through prescaler IC507. (Operation of the prescaler is described

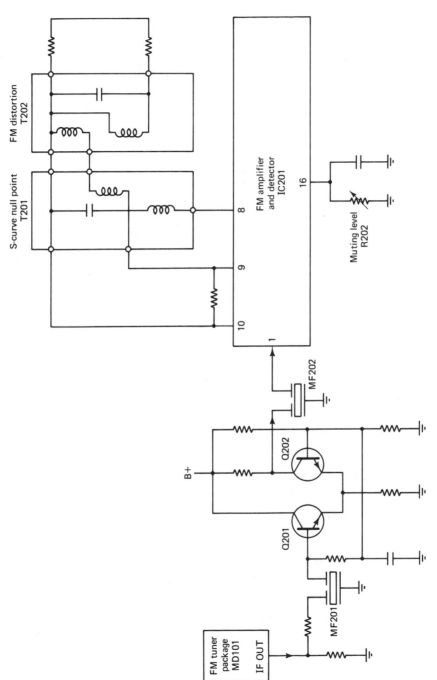

FIGURE 4–21 FM amplifier circuits in IC tuner.

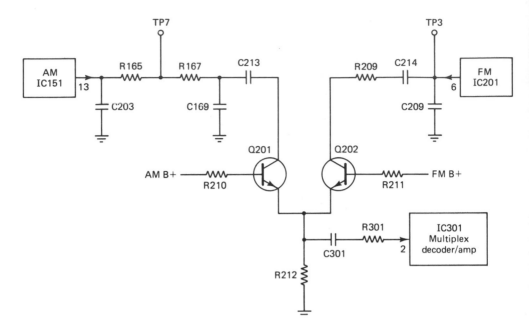

FIGURE 4-22   Audio select circuits in IC tuner.

in Sec. 4–6.2.) The sampled signal serves to complete the FM tuning loop. For example, if MD101 drifts from the frequency commanded by IC503, the error voltage from IC503 changes the MD101 oscillator as necessary to bring MD101 back on frequency.

A station-detect signal is produced at pin 12 of IC201. This signal is used to tell PLL microprocessor IC503 that an FM station has been located and that the station has sufficient strength to produce good FM operation.

In between FM stations or when the FM station is weak, pin 12 of IC201 goes high, turning on Q406. This causes pin 26 of IC503 to go low and causes the audio to be muted (Sec. 3–14.3).

When there is an FM station of sufficient strength, pin 12 of IC201 goes low, turning Q406 off and causing pin 26 of IC503 to go high. Under these conditions, the audio is unmuted, and IC503 fine-tunes the FM section for best reception of the FM station. The muting level is set by R202.

AM Tuning.   The AM section is also tuned to the desired frequency (and locked to that frequency) by signals from IC503 applied to the sections of VVC D153. This is the same error voltage applied to the VCO of FM tuner package MD101. The IC503 error voltage shifts the RF and oscillator circuits as necessary to tune across the AM broadcast band.

The oscillator circuit of IC151 is adjusted by L152, while the RF input circuit is tuned or trimmed by tracking adjustments L151 and CT151.

The frequency produced by IC151 is sampled and applied to the AM input of IC503. The sampled signal serves to complete the AM tuning loop. For

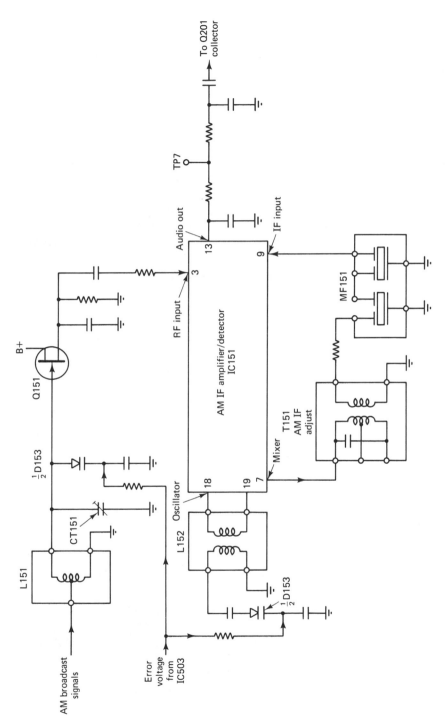

FIGURE 4-23 AM amplifier circuits in IC tuner.

example, if the AM section drifts from the frequency commanded by IC503, the error voltage from IC503 changes both circuits controlled by D153 as necessary to bring the AM section back on frequency.

4-6.2 Frequency Synthesis Tuning of RF Voltage Amplifiers

The RF voltage amplifiers shown in Fig. 4–20 use some form of frequency synthesis tuning (also known as *quartz tuning* or possibly *digital tuning*). Frequency synthesis, or FS, tuning provides for convenient push-button or preset AM and FM station selection, with automatic station search or scan, and automatic fine-tune (AFT) capability.

As shown in Fig. 4–24, the key element in an FS system is the PLL (phase-lock loop) that controls the variable-frequency oscillator (VFO) and/or RF tuning, as required for station selection and fine tuning. Note that the PLL used in the AM/FM tuner of Fig. 4–20 is essentially the same as the PLL used in the FS tuners of TV sets and VCRs (and can be applied to virtually any RF amplifier tuning system).

Basic PLL. As shown in Fig. 4–24a, a PLL is the term used to designate a frequency-comparison circuit in which the output of a VFO (variable-frequency oscillator) is compared in frequency and phase to the output of a very stable (usually quartz-crystal controlled) fixed-frequency reference oscillator. Should a deviation occur between the two compared frequencies or should there be any phase difference between the two oscillator signals, the PLL detects the degree of frequency/phase error and automatically compensates by tuning the VFO up or down in frequency/phase until both oscillators are locked to the same frequency and phase.

The accuracy and frequency stability of a PLL circuit depend on the accuracy and frequency stability of the reference oscillator (and on the crystal that controls the oscillator). No matter what reference oscillator is used, the VFO of most PLL circuits is a VCO where frequency is controlled by an error voltage.

AM FS Tuning. Figure 4–24b shows how the PLL principles are applied to the AM section of our tuner. The 1-kHz reference oscillator of Fig. 4–24a is replaced by a reference signal obtained by dividing down the PLL IC503 clock (4.5 MHz). This reference signal is applied to a phase comparator within IC503. The other input to the phase comparator is a sample of the AM signal at pin 30 of IC503 (taken from pin 20 of IC151).

The phase-comparator output is an *error signal* or *tuning correction voltage* applied through low-pass filter Q507/Q508 to the AM tuning circuits (Fig. 4–23). Q507/Q508 act as a buffer between the comparator and low-pass and tuning circuits.

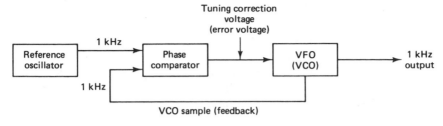

(a) PLL

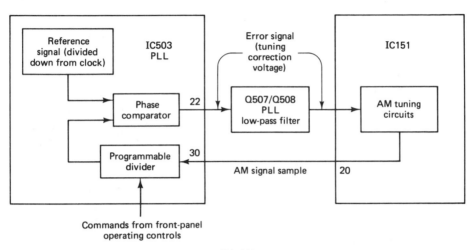

(b) AM

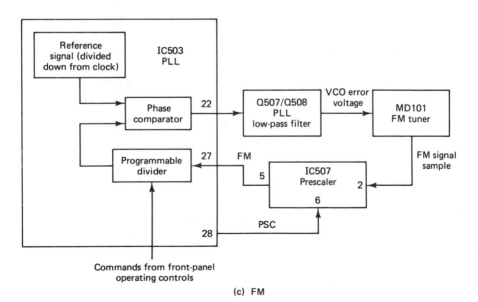

(c) FM

FIGURE 4-24 Elements of an FS system.

Note that the AM sample is applied to the phase comparator through a *programmable divider* or *counter*. The division ratio of the programmable divider is set by commands from the front-panel operating controls (TUNING UP/DOWN, PRESET, SCAN, and so on). In effect, the divider is programmed to divide the AM sample by a specific number.

The variable-divider function makes possible many AM local-oscillator frequencies. An AM frequency change is done by varying the division ratio with front-panel commands. This produces an error signal that shifts the tuning circuits until the AM signal (after division by the programmable divider) equals the reference-signal frequency, and the tuning loop is locked at the desired frequency.

FM FS Tuning.   Figure 4–24c shows the PLL circuits for the FM section of the tuner. This circuit is similar to the PLLs found in TV sets and VCRs in that a *prescaler* is used. The system of Fig. 4–24c is generally called an *extended* PLL and holds the variable-oscillator frequency to some harmonic or subharmonic of the reference oscillator (but with a fixed-phase relationship between the reference and variable signals).

The PLL of Fig. 4–24c uses a form of *pulse-swallow control*, or PSC, that allows the division ratio of the programmable divider to be changed in small steps. As in the case of AM, the division ratio of the programmable divider is set by commands applied to IC503 from the front-panel controls.

The PSC system uses a very high speed prescaler IC507, also with a variable division ratio. The division ratio of the prescaler is determined by the PSC signal at pin 28 of IC503 and can be altered as required to produce subtle changes in the frequency needed for optimum station tuning (fine tuning) in the FM mode.

The PSC signal at pin 28 of IC503 is a series of pulses. As the number of pulses increases, the division ratio of the prescaler also increases. When a given FM station or frequency is selected by the front-panel controls, the number of pulses on the PSC line is set by circuits in IC503 as necessary for each station or frequency.

The overall division ratio for a specific FM station or frequency is the prescaler division ratio, multiplied by the programmable-divider division ratio. The result of division at any FM station or frequency is a fixed output to the IC503 comparator when the FM tuner is set to the desired frequency.

4-7 BASIC TRANSMITTER RF-AMPLIFIER DESIGN APPROACHES

Transmitter RF amplifiers can be designed using two approaches: (1) y-parameters and (2) large-signal parameters.

The *y-parameter approach* involves *two-port networks*. Basically, the method consists of characterizing the transistor (FET or two-junction) as a linear active two-port network (LAN) with admittances (y-parameters) and using the parameters to solve design equations for stability, gain, and input/output admittances. The two-port or y-parameter approach is best suited for voltage amplifiers using small-signal characteristics and is recommended for low-power RF amplifiers.

Design based on large-signal characteristics (transistor input/output resistances and capacitances) is recommended for RF power amplifiers (using high-power transistors). Both design approaches are discussed in the remainder of this chapter.

4-7.1 Design Problems for RF Amplifiers

With either approach, it is difficult at best to provide simple, step-by-step procedures for designing RF amplifiers to meet all possible circuit conditions. In practice, there are several reasons why this procedure often results in considerable trial and error. Here are some typical problems.

First, not all the characteristics are always available in datasheet form. For example, input and output admittances may be given at some low frequency, but not at the desired operating frequency.

Often, manufacturers do not agree on terminology. A good example of this is in y-parameters, where one manufacturer uses letter subscripts and another uses number subscripts (y_{21}). Of course, this type of variation can be eliminated by conversion.

In some cases, manufacturers give the required information on datasheets but not in the required form. For example, some manufacturers may give the input capacitance in farads rather than listing the input admittance in mhos. The input admittance is found when the input capacitance is multiplied by 6.28F (where F is the frequency of interest).

This conversion is based on the assumption that the input admittance is primarily capacitive and thus depends on frequency. The assumption is not always true for the frequency of interest, so it may be necessary to use complex admittance measuring equipment to make actual tests of the transistor.

The input and output tuning circuit of an RF amplifier must perform three functions. First, the circuits (capacitors and coils) must tune the amplifier to the desired frequency. Second, the circuits must match the input and output impedances of the transistor to the impedances of the source and load (or there may be considerable loss of signal). Third, as in the case with any amplifier, there is some feedback between output and input. If the admittance factors are just right, the feedback may be of sufficient amplitude and proper phase to cause *oscillation* in the amplifier. The amplifier is considered as unstable when this occurs.

Amplifier instability in any form is always undesirable and can be corrected by feedback (called *neutralization*) or by changes in the input/output tuning networks. Generally, the changes involve *introducing some slight mis-*

match to improve stability. Although the neutralization and tuning circuits are relatively simple, the equations for determining stability (or instability) and impedance matching are long and complex. As a general rule, such equations are best solved by computer-aided design methods.

In an effort to cut through the maze of information and complex equations, we discuss all steps involved in RF-amplifier design. Armed with this information, you should be able to interpret datasheets, or test information, and use the data to design tuning networks that provide stable RF amplification at the frequencies of interest.

With each step we discuss the various alternative procedures and types of information available. Specific design examples are used to summarize the information. On the assumption that you may not be familiar with two-port networks, we start with a summary of the y-parameter system.

4-7.2 Y-Parameters

Impedance (Z) is a combination of resistance (R, the real part) and reactance (X, the imaginary part). Admittance (y) is the reciprocal of impedance and is composed of conductance (g, the real part) and susceptance (jb, the imaginary part). Thus, g is the reciprocal of R, and jb is the reciprocal of X.

To find g, divide R into 1; to find R, divide g into 1. Z is expressed in ohms (Ω). Y, being a reciprocal, is expressed in mhos or millimhos (mmhos). For example, an impedance Z of 50 Ω equals 20 mmhos (1/50 = 0.02; 0.02 mho = 20 mmhos).

A y-parameter is an expression for admittance in the form

$$y_i = g_i + jb_i$$

where g_i = real (conductive) part of input admittance
jb_i = imaginary (susceptive) part of input admittance
y_i = input admittance (the reciprocal of Z_i)

The term $y_i = g_i + jb$ expresses the y-parameter in *rectangular form*. Some manufacturers describe the y-parameter in *polar form*. For example, they give the *magnitude* of the input admittance as y_i and the angle of input admittance as y_i.   Quite often, manufacturers mix the two systems of *vector algebra* on datasheets.

Conversion of Vector-Algebra Forms.   In case you are not familiar with the basics of vector algebra, the following notes summarize the steps necessary to manipulate vector-algebra terms. With this background, you should be able to perform all calculations involved in the simplified design of RF-amplifier networks using y-parameters.

To convert from rectangular to polar form:

1. Find the magnitude from the square root of the sum of the squares of the components:

$$\text{polar magnitude} = \sqrt{g^2 + jb^2}$$

2. Find the angle from the ratio of the component values.

$$\text{polar angle} = \arctan\left(\frac{jb}{g}\right)$$

The angle is leading if the jb term is positive and is lagging if the jb term is negative.

For example, assume that the y_{fs} is given as $g_{fs} = 30$ and $jb_{fs} = 70$. This is converted to polar form by

$$\left|y_{fs}\right| \text{polar magnitude} = \sqrt{(30)^2 + (70)^2} = 76$$

$$\angle y_{fs} \text{ polar angle} = \arctan\left(\frac{70}{30}\right) = 67°$$

Converting from polar to rectangular form:

1. Find the real (conductive, or g) part when polar magnitude is multiplied by the cosine of the polar angle.

2. Find the imaginary (susceptance, or jb) part when polar magnitude is multiplied by the sine of the polar angle.

If the angle is positive, the jb component is also positive. When the angle is negative, the jb component is also negative.

For example, assume that the y_{fs} is given as $\left|y_{fs}\right| = 20$ and $\angle y_{fs} = -33°$. This is converted to rectangular form by:

$$20 \times \cos 33° = g_{fs} = 16.8$$

$$20 \times \sin 33° = jb_{fs} = 11$$

The Four Basic y-Parameters. Figure 4–25 shows the y-equivalent circuit for a FET. A similar circuit can be drawn for a two-junction transistor when analyzing small-signal characteristics.

Note that y-parameters can be expressed with number subscripts or letter subscripts. The number subscripts are universal and can apply to two-junction transistors, FETs, and IC amplifiers. The letter subscripts are most popular on FET datasheets.

The following notes can be used to standardize y-parameter nomenclature. Note that the letter s in the letter subscript refers to common-source operation of a FET amplifier and is equivalent to a common-emitter two-junction amplifier.

y_{11} is *input admittance* and can be expressed as y_{is}.

y_{12} is *reverse transadmittance* and can be expressed as y_{rs}.

y_{21} is *forward transadmittance* and can be expressed as y_{fs}.

y_{22} is *output admittance* and can be expressed as y_{os}.

Input admittance, with Y_L = infinity (a short circuit of the load), is expressed as

$$y_{11} = g_{11} + jb_{11} = \frac{di_1}{de_1} \quad \text{(with } e_2 = 0)$$

This means the y_{11} is equal to the difference in current i_1, divided by the difference in voltage e_1, with voltage e_2 at 0. The voltages and currents involved are shown in Fig. 4–25.

Some datasheets do not show y_{11} at any frequency, but give *input capacitance* instead. If one assumes that the input admittance is entirely (or mostly) capacitive, then the input impedance can be found when the input capacitance is multiplied by 6.28F (F = frequency in hertz) and the reciprocal is taken.

Because admittance is the reciprocal of impedance, admittance is found when input capacitance is multiplied by 6.28F (where admittance is capacitive). For example, if the frequency is 100 MHz and the input capacitance is 8 pF, the input admittance is $6.28 \times (100 \times 10^6) \times (8 \times 10^{-12})$, or about 5 mmhos.

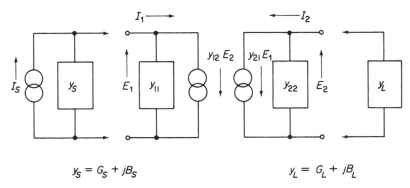

$y_S = G_S + jB_S$ $y_L = G_L + jB_L$

FIGURE 4-25 A y-equivalent circuit (for a FET) with source and load.

This assumption is accurate only if the real part of y_{11} (or g_{11}) is negligible. Such an assumption is reasonable for most FETs, but not necessarily for all two-junction transistors. The real part of two-junction transistor input admittance can be quite large in relation to the imaginary jb_{11} part.

Forward transadmittance, with Y_L = infinity (a short circuit of the load), is expressed as

$$y_{21} = g_{21} + jb_{21} = \frac{di_2}{de_1} \quad \text{(with } e_2 = 0\text{)}$$

This means that y_{21} is equal to the difference in output current i_2 divided by the difference in input voltage e_1, with voltage e_2 at 0. In other words, y_{21} represents the difference in output current for a difference in input voltage.

Two-junction transistor datasheets often do not give any value for y_{21}. Instead, forward transmittance is shown by means of a *hybrid system* of notation using h_{fe} or h_{21} (which means hybrid forward transadmittance with common emitter). No matter what system is used, it is essential that the values of forward transadmittance be considered at the frequency of interest.

Output admittance, with Y_S = infinity (a short circuit of the source or input), is expressed as

$$y_{22} = g_{22} + jb_{22} = \frac{di_2}{de_2} \quad \text{(with } e_1 = 0\text{)}$$

Reverse transadmittance, with Y_S = infinity (a short circuit of the source or input), is expressed as

$$y_{12} = g_{12} + jb_{12} = \frac{di_1}{de_2} \quad \text{(with } e_1 = 0\text{)}$$

y_{12} is usually not considered an important two-junction transistor parameter. However, y_{12} may appear in equations related to RF design.

4-7.3 y-Parameter Measurement

It is obvious that y-parameter information is not always available or in a convenient form. In practical design, it may be necessary to measure the y-parameter using laboratory equipment. The main concern in measuring y-parameters is that the measurements are made under conditions simulating those of the final circuit. For example, if supply voltages, bias voltages, and operating frequency are not identical (or close) to the final circuit, the tests may be misleading.

Although the datasheets for transistors to be used as RF amplifiers usually contain input and output admittance information, it may be helpful to know how this information is obtained. There are two basic methods for measuring the y-parameters of transistors used in amplifier circuits.

One method involves *direct measurement* of the parameter (such as measuring changes in output for corresponding changes in input). The other method uses *tuning substitution* (where the transistor is tuned for maximum transfer of power, and the admittances of the tuning circuits are measured). We summarize both methods in the following paragraphs.

Direct Measurement of y_{fs} (y_{21}).   Figure 4–26 shows a typical test circuit for direct measurement of y_{fs} (which may be listed as y_{21}, g_m, or even g_{fs}. Although a FET is shown, the same circuit can apply to any single-input device, such as a two-junction transistor.

The value of R_L must be such that the drop is negligible of I_{DSS}. Just as important, the value of R_L must be such that the operating voltage point (V_{DS} in the case of FET) is correct for a given supply voltage (V_{DD}) and operating current (I_D). For example, if I_D is 10 mA, V_{DD} is 20 V, and V_{DS} is 15 V, R_L must drop 5 V at 10 mA. Thus, the value of R_L is 5 V/0.01 A = 500 Ω.

During test, the signal source is adjusted to the frequency of interest. The amplitude of the signal source V_{IN} is set to some convenient number such as 1 V or 100 mV. The value of y_{fs}(y_{21}) is calculated from the equation of Fig. 4–26 and is expressed in mhos (or millimhos and micromhos, in practical terms). As an example, assume that the value of R_L is 1000 Ω, V_{IN} is 1 V, and V_{OUT} is 8 V. The value of y_{fs} is 8/(1 × 1000) = 0.008 mho = 8 mmho = 8000 μmho.

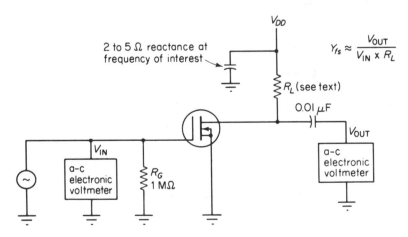

FIGURE 4–26   Test circuit for direct measurement of y_{fs} as a simple number.

Direct Measurement of $y_{os}(y_{22})$. Figure 4–27 shows a typical test circuit for direct measurement of y_{os} (which may be listed as y_{22}, g_{os}, g_{22}, or even r_d, where $r_d = 1/y_{os}$. Some datasheets give y_{os} as a complex number with both the real (g_{os}) and imaginary (b_{os}) values shown by means of curves.

The value of R_S must be such as to cause a negligible drop (so that V_{DS} can be maintained at the desired level, with given V_{DD} and I_D. During test, the signal source is adjusted to the frequency of interest. Both V_{out} and V_{DS} are measured, and the value of y_{os} is calculated from the equation of Fig. 4–27.

Direct Measurement of y_{is} (y_{11}). Although y_{is} is not generally a critical factor for FETs, it is necessary to know the values of y_{is} to calculate impedance-matching networks for FET RF amplifiers (as discussed in Sec. 4-9). If it is necessary to establish the imaginary part (b_{is}), use an admittance meter or R_X meter.

Direct Measurement of $y_{rs}(y_{12})$. Again, although y_{rs} is not generally a critical factor for FETs, it is necessary to know the values of y_{rs} to calculate impedance-matching networks. Although the real part of g_{rs} remains at zero for all conditions and at all frequencies, the imaginary part b_{rs} does vary with voltage, current, and frequency. (The reverse susceptance varies and, under the right conditions, can produce undesired feedback from output to input.) This

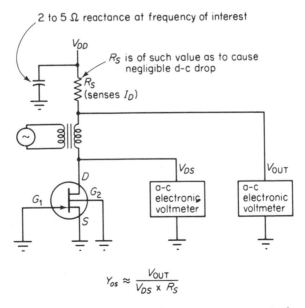

$$Y_{os} \approx \frac{V_{OUT}}{V_{DS} \times R_S}$$

FIGURE 4-27 Test circuit for direct measurement of y_{os} as a simple number.

condition must be accounted for in the design of RF amplifiers to prevent the feedback from causing oscillation. If it is necessary to establish the imaginary part (b_{rs}), use an admittance meter or R_X meter.

Tuning-substitution Measurement of y-Parameters.   Figure 4–28 is a typical test circuit for measurement of y-parameters using the tuning-substitution method. Although a two-junction transistor is shown, the circuit can be adapted for use with FETs.

During test, the transistor is placed in a test circuit designed with variable components to provide wide tuning capabilities. This is necessary to ensure correct matching at various power levels. The circuit is tuned for *maximum power gain* at each power level for which admittance information is desired.

After the test amplifier has been tuned for maximum power gain, the d-c power, signal source, circuit load, and test transistors are disconnected from

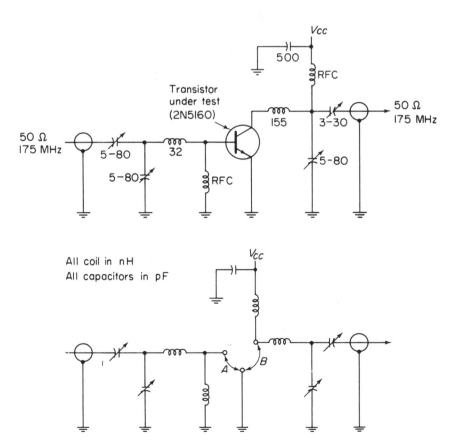

FIGURE 4-28   Test circuit for measurement of y-parameters using the tuning-substitution method.

the circuit. For total circuit impedance to remain the same, the signal-source and output-load circuit connections are terminated at the characteristic resistances (typically 50 Ω).

After the substitutions are completed, complex admittances are measured at the base- and collector-circuit connections of the test transistor (points A and B, respectively, of Fig. 4-28) using a laboratory admittance meter.

Note that the transistor input and output admittances are the *conjugates* of the base-circuit connection and the collector-circuit connection admittances, respectively. For example, if the base-circuit connection (point A) admittance is 8 + j3, the input admittance of the transistor is 8 − j3.

In some systems of two-junction transistor RF-amplifier design, the networks are calculated on the basis of input/output resistance and capacitance, instead of admittance (although admittances are often used to determine stability before going into design of the RF network). Such a system is described in Sec. 4-8. In this system, the admittances measured in the circuit of Fig. 4-28 are converted to resistance and capacitance.

Admittances are expressed in mhos (or millimhos, mmhos). Resistance can be found by dividing the real part of the admittance into 1. Capacitance can be found by dividing the imaginary part of the admittance into 1 (to find reactance); then the reactance is used in the equation $C = 1/(6.28 \, FX_c)$ to find the actual capacitance.

4-7.4 Amplifier Stability Factors

Two factors are used to determine the potential stability (or instability) of transistors in RF amplifiers. One factor is known as the Linvill C factor; the other is the Stern k factor. Both factors are calculated from equations requiring y-parameter information (to be taken from datasheets or by actual measurement at the frequency of interest).

The main difference between the two factors is that the Linvill C factor assumes that the transistor is not connected to a load. The Stern k factor includes the effect of a specific load.

The Linvill C factor is calculated from

$$C = \frac{y_{12}y_{21}}{2g_{11}g_{22} - R_e(y_{12}y_{21})}$$

where $R_e(y_{12}y_{21})$ is the real part of $y_{12}y_{21}$.

If C is less than 1, the transistor is unconditionally stable. That is, using a conventional (unmodified) circuit, no combination of load and source admittance can cause oscillation. If C is greater than 1, the transistor is potentially unstable. That is, certain combinations of load and source admittance can cause oscillation.

The Stern k factor is calculated from

$$k = \frac{2(g_{11} + G_S)(g_{22} + G_L)}{y_{12}y_{21} + R_e(y_{12}y_{21})}$$

where G_S and G_L are source and load conductance, respectively. (G_S = 1/source resistance; G_L = 1/load resistance.)

If k is greater than 1, the amplifier circuit is stable (the opposite of Linvill). If k is less than 1, the amplifier is unstable. In practical design, a Stern k factor of 3 or 4 should be used, rather than 1, to provide a safety margin. This accommodates parameter and component variations (particularly with regard to bandpass response).

Note that both equations are fairly complex and require considerable time for solution. In practical work, computer-aided design techniques are used for stability equations.

Some manufacturers provide alternative solutions to the stability and load-matching problems, usually in the form of datasheet graphs, such as shown in Fig. 4–29. Note that the device is unconditionally stable at frequencies above 250 MHz. At frequencies below about 50 MHz, the device becomes highly unstable.

4-7.5   Solutions to Amplifier Stability Problems

There are two basic design solutions to the problem of unstable RF amplifiers. One solution involves *neutralization*, while the other solution uses a *mismatch* between tuning networks and source or load.

Neutralization.   When an RF amplifier is neutralized, part of the output is fed back (after the output is shifted in phase) to the input so as to cancel oscillation. With neutralization, an amplifier can be matched perfectly to the source and load.

Neutralization is sometimes known as a *conjugate match*. In a perfect conjugate match the transistor input and source, as well as the transistor output and load, are matched resistively, and all reactance is tuned out. Neutralization requires extra components and can create a problem when frequency is changed.

Mismatch Solution.   This solution involves introducing a specific amount of mismatch into either the source or load tuning networks so that any feedback is not sufficient to produce instability or oscillation. The mismatch solution, sometimes known as the *Stern solution*, requires no extra components, but does produce a reduction in gain.

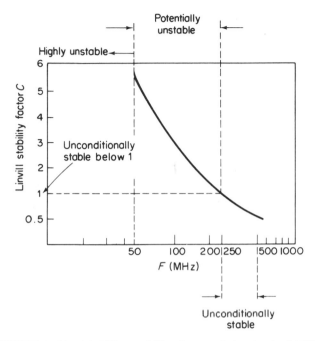

FIGURE 4-29   Linville stability factor C for typical FET.

Comparison of Stability Solutions.   Figure 4–30 shows a comparison of the neutralized and mismatch solutions. The higher gain curve represents neutralized operation (also called unilateralized gain in some literature). The lower gain curve represents the power gain when the Stern k factor is 3.

Assume that the frequency of interest is 100 MHz. If the amplifier is matched directly to the load (perfect conjugate match) without regard to stability (or using neutralization to produce stability), the top curve applies and the power gain is about 38 dB. If the amplifier is matched to a load and source where the Stern k factor is 3 (resulting in a mismatch with the actual load and source), the lower curve applies, and the power gain is about 29 dB.

General Power Gain Equation.   The upper curve of Fig. 4–30 is found by the general power-gain equation

$$G_P = \frac{\text{power delivered to load}}{\text{power delivered to input}}$$

$$= \frac{(y_{21})^2 \, G_L}{(Y_L + y_{22})^2[(R_e \, y_{11} - y_{12}y_{21})/(y_{22} + Y_L)]}$$

G_P applies to circuits with no external feedback and to circuits with external feedback (neutralization), provided the composite y-parameters of both

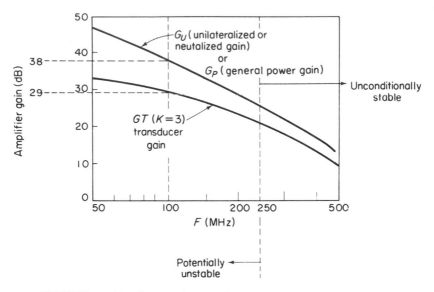

FIGURE 4-30 Comparison of neutralized and mismatch stability solutions.

transistor and feedback networks are substituted for the transistor y-parameters in the equation.

Transistor Gain Equation.   The lower curve of Fig. 4–30 is found by the transducer gain equation

$$GT = \frac{\text{power delivered to load}}{\text{maximum power available from source}}$$

$$= \frac{4G_S G_L (y_{21})^2}{[(y_{11} + Y_S)(y_{22} + Y_L) - y_{12}y_{21}]^2}$$

The transducer gain expression includes input mismatch. The lower curve of Fig. 4–30 assumes that the input mismatch produces a Stern k factor of 3. (That is, the tuning circuit networks are adjusted for admittances that produce a Stern k factor of 3.) The transducer gain expression considers the input and output network as part of the source and load.

Modifying Admittances.   With either gain expression, the input and output admittances of the transistor are modified by the load and source admittances.

The *input admittance* of the transistor is given by

$$Y_{IN} = y_{11} - \frac{y_{12}y_{21}}{y_{22} + y_L}$$

The *output admittance* of the transistor is given by

$$Y_{OUT} = y_{22} - \frac{y_{12}y_{21}}{y_{11} + Y_S}$$

At low frequencies, the second term in the input and output admittance equations is not particularly significant. Above about 30 mHz, the second term makes a significant contribution to the input and output admittances.

Simplifying the Equations.   The imaginary parts of Y_S and Y_L (B_S and B_L, respectively) must be known before values can be calculated for power gain, transducer gain, input admittance, and output admittance. Exact solutions for B_S and B_L almost always consist of time-consuming complex algebraic manipulations.

To find fairly good simplifying approximations for the equations, let $B_S = -b_{11}$ and $B_L = -b_{22}$, so that

$$GP = \frac{(y_{21})^2 G_L}{(G_L + g_{22})^2 \, R_e[y_{11} - y_{12}y_{21}/(g_{22} + G_L)]}$$

$$G_T = \frac{4G_S G_L (y_{21})^2}{(g_{11} + G_S g_{22} + G_L - y_{12}y_{21})^2}$$

$$Y_{IN} = y_{11} - \frac{y_{12}y_{21}}{g_{22} + G_L}$$

$$Y_{OUT} = y_{22} - \frac{y_{12}y_{21}}{g_{11} + G_L}$$

MAG and MUG.   The other gain expressions sometimes found on the datasheets of transistors used in RF amplifiers are maximum available gain, or MAG, and maximum usable gain, or MUG.

MAG is usually applied as the gain in conjugately matched, neutralized circuits and is expressed as

$$MAG = \frac{(y_{21})^2 \, R_{IN} R_{OUT}}{4}$$

where R_{IN} and R_{OUT} are the input and output resistances, respectively, of the transistor.

An alternative MAG expression is

$$MAG = \frac{(y_{21})^2}{4R_e(y_{11})R_e(y_{22})}$$

where $R_e(y_{11})$ is the real part (g_{11}) of the input admittance, and $R_e(y_{22})$ is the real part (g_{22}) of the output admittance.

MUG is usually applied as the *stable gain* that may be realized in a *practical* (neutralized or unneutralized) RF amplifier. In a typical unneutralized circuit, MUG is expressed as

$$\text{MUG} = \frac{0.4y_{21}}{6.28F \times \text{reverse transfer capacitance}}$$

MAG and MUG are often omitted on datasheets for two-junction transistors. Instead, gain is listed as h_{fe} at a given frequency. This is supplemented with graphs that show available power output at given frequencies with a given input.

4–7.6 *Neutralized Solution for RF Amplifiers*

There are several methods for neutralization of RF amplifiers. The most common method is the *capacitance-bridge* technique, as shown in Fig. 4–31a. Capacitance-bridge neutralization becomes more apparent when the circuit is

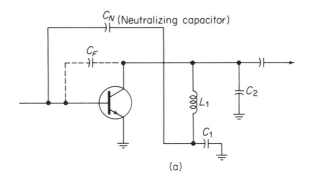

(a)

C_F = Reverse capacitance of transistor

$$C_N \approx C_F \times \left(\frac{C_1}{C_2}\right)$$

$$C_1 \gg C_2$$

(b)

FIGURE 4–31 Capacitance-bridge neutralization circuit.

redrawn as shown in Fig. 4–31b. The condition for neutralization is that $I_F = I_N$ (the neutralization current I_N must be equal to the feedback current I_F in amplitude, but of opposite phase).

The equations normally used to find the value of the feedback neutralization capacitor are long and complex. However, for practical work, if the value of C1 is made quite large in relation to C2 (at least four times), the value of C_N can be found by

$$C_N = C_F \times (C_1/C_2)$$

where C_F is the reverse capacitance of the transistor.

In simple terms, the value of C_N is approximately equal to the value of reverse capacitance times the ratio of C_1/C_2. For example, if reverse capacitance (sometimes listed as collector-to-base capacitance) is 7 pF, C_1 is 30 pF, and C_2 is 3 pF, the C_1/C_2 ratio is 10, and $C_N = 10 \times 7$ pF $= 70$ pF.

4-7.7 The Stern Solution

A stable design with a potentially unstable transistor is possible without external feedback (neutralization) by proper choice of source and load maintenances. This can be seen by inspection of the Stern k factor equation (Sec. 4–7.4). G_S and G_L can be made large enough to yield a stable circuit, regardless of the degree of potential instability. Using this approach, a circuit stability factor (typically k = 3) is selected, and the Stern k factor equation is used to arrive at values of G_S and G_L that produce the desired k.

Of course, the actual G of the source and load cannot be changed. Instead, the input and output tuning circuits are designed as if the actual values are changed. This results in a mismatch and a reduction in power gain, but does produce the desired degree of stability.

To get a particular circuit stability factor, the designer may choose any of the following combinations of matching and mismatching for G_S and G_L to the transistor input and output conductances, respectively:

G_S matched and G_L mismatched

G_L matched and G_S mismatched

Both G_S and G_L mismatched

Other performance requirements or practical considerations often dictate the decision on which combination to use. For example, it may not be practical to mismatch to some extreme value of G_S or G_L.

Once G_S and G_L are chosen, the remainder of the design may be completed using the relationships that apply to the amplifier without feedback. Power gain and input/output admittances may be computed using the appropriate equations (Sec. 4–7.5).

Simplified Stern Approach.   Although the basic procedure may be adequate in many cases, a more systematic method of source and load admittance determination is desirable for designs that demand maximum power gain per degree of circuit stability. Stern has analyzed this problem and developed equations for computing the best G_S, G_L, B_S, and B_L for a particular circuit stability factor (Stern k factor). Unfortunately, these equations are very complex and become quite tedious when used frequently. The complete Stern solution is best applied by computer!

Programs have been written to provide essential information for transistors used as RF amplifiers, including the effects of various specific sources and loads. These programs permit the designer to experiment with theoretical circuits in a matter of seconds.

When a Stern solution must be obtained without the aid of a computer, it is best to use one of the many shortcuts that have been developed over the years. The following shortcut is by far the simplest and most widely accepted, yet provides an accuracy close to that of the computer solutions.

1. Let $B_S = -b_{11}$ and $B_L = -b_{22}$, as in the case of the Sec. 4–7.5 equations. This permits the designer to closely approximate the exact Stern solution for Y_S and Y_L, while avoiding the most complex and time-consuming portion of the computations. Also, the circuit can be designed with tuning adjustments for varying B_S and B_L, thereby creating the possibility of getting the true B_S and B_L (by experiment) for maximum gain (and as accurately as if all the Stern equations had been solved).

2. Mismatch G_S to g_{11} and G_L to g_{22} *by an equal ratio*. That is, find a ratio that produces the desired Stern k factor; then mismatch G_S to g_{11} (and G_L to g_{22}). For example, if the ratio is 4:1, make G_S four times the value of g_{11} (and G_L four times the value of g_{22}).

If the mismatch ratio, R, is defined as $R = G_L/g_{22} = G_S/g_{11}$ the R may be computed for any particular circuit stability k factor using the equation

$$R = \sqrt{k\,\frac{y_{21}y_{12} + R_e(y_{12}y_{21})}{2g_{11}g_{22}} - 1}$$

As an example, assume that it is desired to mismatch input and output circuits to produce a Stern k factor of 4, using a transistor with the following characteristics: $y_{21}y_{12} = 0.5$, $g_{11} = 5.0$, $g_{22} = 0.05$, and $R_e(y_{12}y_{22}) = 0.2$ (all values in mmhos).

$$R = 4 \cdot \sqrt{\frac{0.5 + 0.7}{(2)(5)(0.05)}} - 1 = 1.37$$

Using the value of 1.37 for R, and the equation

$$1.37 = \frac{G_S}{g_{11}} = \frac{G_L}{g_{22}}$$

then

$$G_S = (1.37)(0.05)(10^{-3}) = 6.85 \text{ mmho}$$

and

$$R_S = \frac{1}{G_S} = 147 \ \Omega$$

$$G_L = (1.37)(0.05)(10^{-3}) = 0.0685 \text{ mmho}$$

and

$$R_L = \frac{1}{G_L} = 14{,}600 \ \Omega$$

The shortcut Stern method may be advantageous if the source and load admittances and power gains for several different values of k are desired. Once the R for a particular k has been determined, the R for any other k may be quickly found from the equation

$$R = \frac{(1 + R_1)^2}{(1 + R_2)^2} = \frac{k_1}{k_2}$$

where R_1 and R_2 are values of R corresponding to k_1 and k_3, respectively.

Stern Solution with Datasheet Graphs.   It is obvious that the Stern solution, even with the shortcut method, is somewhat complex. For this reason, some manufacturers have produced datasheet graphs that show the best source and load admittances for a particular transistor over a wide range of frequencies. Figures 4–32 and 4–33 are examples of such graphs.

Figure 4–32 shows the real (G_S) and imaginary (B_S) values that produce maximum gain, but with a stability (Stern k) factor of 3 at frequencies from 50 to 500 MHz. Figure 4–33 shows corresponding information for G_L and B_L.

To use these figures, simply select the desired frequency and note where the corresponding G and B curves cross the frequency line. For example, assuming a frequency of 100 MHz, $Y_L = 0.35 - j2.1$ mmho, and $Y_s = 1.3 - j4.4$ mmho.

If the tuning circuits are designed to match these admittances (rather than the actual admittances of the source and load), the circuit is stable, but with

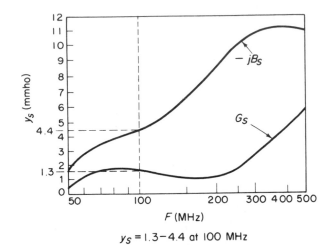

$y_S = 1.3 - 4.4$ at 100 MHz

FIGURE 4-32   Best source admittance, $Y_S = G_S + jB_S$.

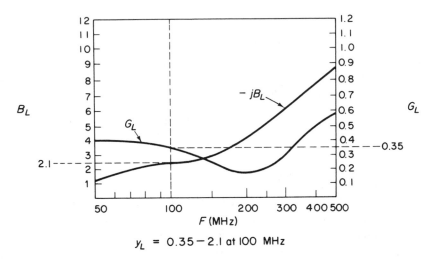

$y_L = 0.35 - 2.1$ at 100 MHz

FIGURE 4-33   Best load admittance, $Y_L = G_L + jB_L$.

reduced gain. Use the transducer gain expression G_T of Sec. 4–7.5 to find the resultant power gain.

4-7.8   Systematic RF-Amplifier Design

A review of the two-port (y-parameter) network design method may be helpful at this point. A design example is given in Sec. 4–9.

Determining Transistor Stability.   The first step in designing RF amplifiers with y-parameters is to find the potential instability of the transistor. This involves extracting the transistor y-parameters from the datasheet or determining y-parameters from actual test, and then plugging the y-parameters into the Linvill C and/or Stern k equations to find potential stability or instability.

Use the Linvill C factor where source and load impedances are not involved (or known). Use the Stern k factor when load and source impedances are known. As a practical matter, it is usually more convenient to go directly to the Stern k factor, since this serves as a starting point if the circuit must be modified to produce stability.

Selecting a Stability Method.   If the transistor is not unconditionally stable, decide on a course of action to ensure circuit stability. Usually, this involves going to neutralization or to some form of mismatching input/output tuning circuits. Mismatching is by far the most popular course of action.

If the transistor is unconditionally stable, without neutralization or mismatch, the design can proceed without fear of oscillation. Under these circumstances, the usual object is to get maximum gain by matching the tuning circuits to the actual source and load.

Determining Source and Load Admittances.   Find the source and load admittances based on gain and stability considerations (together with practical circuit limitations). If the transistor is potentially unstable at the frequency of interest with actual source and load impedances, use another source and load that guarantees a certain degree of amplifier stability.

This involves the Stern solution described in Sec. 4–7.7. If optimum source and load impedances are given on transistor datasheets, use these as a first choice. As a second choice, use computer-aided design techniques to get a Stern solution for the desired stability and gain. If neither of these two is available, use the shortcut Stern technique.

It is a good idea to check circuit stability (Stern k factor) even when an unconditionally stable transistor is found by the Linvill C factor. A transistor may be stable without a load or with certain loads, but not stable with some specific load.

Once the optimum source and load admittances are selected, verify that the required gain is available. In practical terms, it is possible to mismatch almost any RF amplifier sufficiently to produce a stable circuit. However, the resultant power gain may be below that required. In that case, a different transistor must be used (or a lower gain accepted!).

Designing the Networks.   Design appropriate networks (input/output tuning circuits) to provide the desired (or selected) source and load admittances. First, the networks must be resonant at the desired frequency. (That is, inductive and capacitive reactance must be equal at the selected frequency.) Second, the network must match the transistor to the load and source.

It is sometimes difficult to get a desired source and load because of tuning-range limitations, excess network losses, component limitations, and the like. In such cases, the source and load admittances are a compromise or trade-off between desired performance and practical limitations. Generally, this trade-off involves a sacrifice of gain to get stability.

4-8 DESIGNING RF AMPLIFIERS WITH LARGE-SIGNAL PARAMETERS

As discussed at the beginning of Sec. 4–7, it is possible to design the tuning networks for RF amplifiers without using a full set of y-parameters or admittances. Instead, the networks are designed using the input/output capacitances and resistances of the transistor. Often, the capacitance/resistance information is available on datasheets in the form of graphs. This is especially true for transistors designed for use as RF power amplifiers. Before we get into the design calculations, let us discuss the amplifier circuits, particularly the tuning networks.

4-8.1 Typical RF Power Amplifier and Multipliers

Figure 4–34 shows the schematics of typical RF power amplifiers. The same basic circuits can be used as frequency multipliers. However, in a multiplier circuit, the output must be tuned to a multiple of the input. A multiplier may or may not provide amplification. Usually, most of the amplification is supplied by the final amplifier stage, which is not operated as a multiplier. That is, the input and output of the final stage are at the same frequency. A typical RF transmitter has three stages: an oscillator to generate the basic signal frequency, an intermediate stage that provides amplification and/or frequency multiplication, and a final stage for power amplification.

4-8.2 Design Considerations for RF Power Amplifiers and Multipliers

All design considerations in Chapter 1 and 2 apply to RF power amplifiers and multipliers. Of particular importance are interpreting datasheets, determining parameters at different frequencies, and temperature-related design problems. In addition to these basic design considerations, the following problems must be considered.

Tuning Controls.   Note that the circuit of Fig. 4–34a has two tuning controls (variable capacitors in this case) in the output network, while the network circuit of Fig. 4–34b has only one adjustment control. The circuit of Fig. 4–34a is typical for power amplifiers, where the output is tuned to the resonant fre-

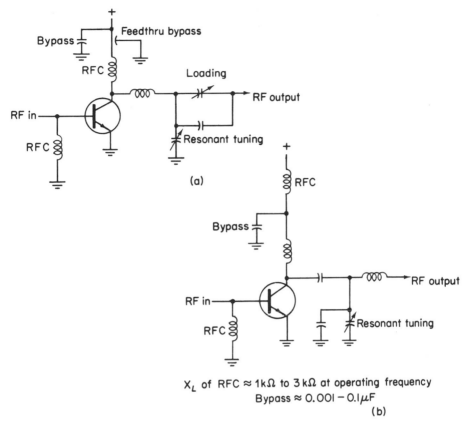

FIGURE 4-34 Typical RF power-amplifier and multiplier circuits.

quency by one control and adjusted for proper impedance match by the other control (often known as the *loading control*). In practice, both controls affect tuning and loading (impedance matching). The circuit of Fig. 4–34b is typical for multipliers or intermediate amplifiers where the main concern is tuning to the resonant frequency.

Parallel Capacitors. Note that the variable capacitors are connected in parallel with fixed capacitors in both networks. This parallel arrangement serves two purposes. First, there is a minimum fixed capacitance in case the variable capacitor is adjusted to minimum value. In some cases, if minimum capacitance is not included in the network, a severe mismatch can occur when the variable capacitor is at minimum (possibly resulting in damage to the transistor). The second purpose for a parallel capacitor is to reduce required capacitance rating (and thus the physical size) of the variable capacitor.

Midrange Capacitors.   When designing networks such as shown in Fig. 4–34, use a capacitor with a midrange capacitance equal to the desired capacitance. For example, if the desired capacitance is 25 pF (to produce resonance at the normal operating frequency), use a variable capacitor with a range of 1 to 50 pF. If such a capacitor is not readily available, use a fixed capacitor of 15 pF in parallel with a 15-pF variable capacitor. This provides a capacitance range of 16 to 30 pF, with a midrange of about 23 pF. Of course, the maximum capacitance range depends on the required tuning range of the circuit. (A wide frequency range requires a wide capacitance range.)

Bias.   The transistors remain cut off until a signal is applied. Therefore, the transistors are never conducting for more than 180° (half a cycle) of the 360° input signal cycle. In practice, the transistors conduct for about 140° of the input cycle, either on the positive half or negative half, depending on the transistor type (NPN or PNP). No bias, as such, is required for this class of operation.

Grounded Emitter.   The emitter is connected directly to ground. In those transistors where the emitter is connected to the case (typical in many RF power transistors), the case can be mounted on a chassis that is connected to the ground side of the supply voltage. A direct connection between emitter and ground is of particular importance in high-frequency operation. If the emitter is connected to ground through a resistance (or a long lead), an inductive or capacitive reactance can develop at high frequencies, resulting in undesired changes in the network.

RFC Connections.   The transistor base is connected to ground through an RF choke (RFC). This provides a d-c return for the base, as well as RF signal isolation between base and emitter or ground. The transistor collector is connected to the supply voltage through an RFC and (in some cases) through the coil portion of the resonant network. The RFC provides d-c return, but RF signal isolation, between collector and power supply.

When the collector is connected to the power supply through the resonant network, the coil must be capable of handling the full collector current. For this reason, final (power) amplifier networks should be chosen so that collector current does not pass through the coil (such as Fig. 4–34a). The circuit of Fig. 4–34b should be used for power applications where the current is low (such as in an intermediate amplifier).

RFC Ratings.   The ratings for RFCs are sometimes confusing. Some manufacturers list a full set of characteristics: inductance, d-c resistance, a-c resistance, Q, current capability, and nominal frequency range. Other manufacturers give only one or two of these characteristics. A-c resistance and Q usually depend on frequency. A nominal frequency-range characteristic is a helpful, but usually not critical, design parameter.

All other factors being equal, the d-c resistance should be at a minimum for any circuit carrying a large amount of current. For example, a large d-c resistance in the collector of a final power amplifier can result in a large voltage drop between power supply and collector.

Usually, the selection of a trial value for an RFC is based on a trade-off between inductance and current capability. The minimum current capacity should be greater (by at least 10%) than the maximum anticipated direct current. The inductance depends on operating frequency. As a trial value, use an inductance that produces a reactance between 1000 and 3000 Ω at the operating frequency.

Bypass Capacitors.   The power-supply circuits of power amplifiers and multipliers must be bypassed, as shown in Fig. 4–34. The feed-through bypass capacitors are used at higher frequencies where the RF circuits are physically shielded from the power-supply and other circuits. The feed-through capacitor permits direct current to be applied through a shield, but prevents RF from passing outside the shield (RF is bypassed to the ground return). As a trial value, use a total bypass capacitance range of 0.001 to 0.1 μF.

Checking Bypass Capacitors and RFCs.   From a practical standpoint, the best test for adequate bypass capacitance and RFC inductance is the presence of RF signals on the power-supply side of the d-c voltage line. If RF signals are present on the power-supply side of the line, the bypass capacitance and/or the RFC inductance are not adequate. (A possible exception is where the RF signals are being picked up because of inadequate shielding.)

If the shielding is good and RF signals are present in the power supply, increase the bypass capacitance value. As a second step, increase the RFC inductance. Of course, circuit performance must be checked with each increase in capacitance or inductance value. For example, too much bypass capacitance can cause undesired feedback and oscillation; too much RFC inductance can reduce amplifier output and efficiency. The procedures for the measurement of RF signals are described in Chapter 9.

Amplifier Efficiency.   A class C RF amplifier has a typical efficiency of about 65% to 70%. That is, the RF power output is 65% to 70% of the d-c input power. To find the required d-c input power, divide the desired RF power output by 0.65 or 0.7. For example, if the desired RF output is 50 W, the d-c input power is 50/0.7, or about 70 W.

Since the collector of an RF amplifier is at a d-c potential approximately equal to the power supply (slightly less because of a drop across the RFC and/or coil), divide the input power by the power-supply voltage to find the collector current. For example, with a d-c input of 70 W and a 28-V power supply, the collector current is about 2.7 A.

Transistor Characteristics.   It is obvious that the transistors must be capable of handling the full power-supply voltage at the collectors and that the current and/or power rating is greater than the maximum calculated values. Likewise, the transistor must be capable of producing the necessary power output at the operating frequency (as discussed in Chapter 1).

It is also obvious that the transistors must provide the necessary power gain at the operating frequency. Likewise, the input power to an amplifier must match the desired output and gain. For example, assume that a 50-W, 50-MHz transmitter is to be designed and that transistors with a power gain of 10 are available. Generally, a transistor oscillator produces less than 1-W output. Therefore, an intermediate amplifier is required to deliver an output of 5 W to the final amplifier. The intermediate amplifier requires about 7-W d-c input (50/7 = 7). Assuming a gain of 10 for the intermediate amplifier, an input of 0.5 W is required from the oscillator.

Intermediate Amplifier Efficiency.   When an intermediate amplifier is also used as a frequency multiplier, the efficiency drops from the 65% to 70% value. As a guideline, the efficiency of a second-harmonic amplifier (output at twice the input frequency) is 42%; third harmonic, 28%; fourth harmonic, 21%; and fifth harmonic, 18%. Therefore, if an intermediate amplifier is to be operated at the second harmonic and produce 5-W RF power output, the required d-c input power is about 12 W (5/0.42 = 12).

Power Gain with Frequency Multiplication.   Another problem to be considered in frequency multiplication is that power gain (as listed on the datasheet) may not remain the same as when amplifier input and output are at the same frequency. Some datasheets specify power gain at the basic frequency and then derate the power gain for second harmonic operation. As a guideline, always use the minimum power-gain factor when calculating power input and output values.

4-8.3 Resonant Network Design

Now we come to the most critical design consideration for RF power amplifiers: the resonant network. This network must be resonant at the desired frequency. (Inductive and capacitive reactance must be equal at the selected frequency.) Second, the network must match the transistor output impedance to the load.

Generally, an antenna load impedance is on the order of 50Ω, while the output impedance of a typical transistor at radio frequencies is a few ohms. In the case where one amplifier feeds into another amplifier, the network must match the output impedance of one transistor to the input impedance of another transistor. Any mismatch can result in a loss of power between stages or to the final load.

Transistor impedance (both input and output) has both resistive and reactive components (Sec. 4–7) and therefore varies with frequency. To design a resonant network for the output of a transistor, it is necessary to know the output reactance (usually capacitive), the output resistance at the operating frequency, and the output power. (These factors are sometimes known as the large-signal parameters.) It is also necessary to know the input resistance and reactance of a transistor at a given frequency and power when designing the resonant network of the stage feeding into the transistor.

Generally, the input resistance, the input capacitance, and the output capacitance of RF power amplifier transistors are shown by means of graphs similar to Fig. 4–35. The reactance can then be found using the corresponding frequency and capacitance. For example, the output capacitance shown on the graph of Fig. 4–35 is about 15 pF at 80 MHz. This produces a capacitive reactance of about 130 Ω at 80 MHz. The reactance and resistance can then be combined to find impedance, as shown in Fig. 4–35.

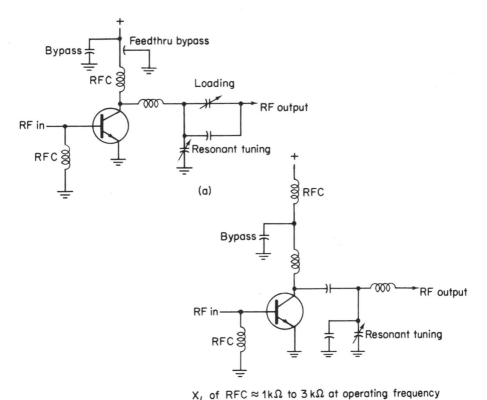

X_L of RFC ≈ 1kΩ to 3 kΩ at operating frequency
Bypass ≈ 0.001 – 0.1 μF

(b)

FIGURE 4-35 Typical RF power-amplifier transistor characteristics.

Input and output transistor impedances are generally listed on datasheets in parallel form. That is, the datasheets assume that the resistance is in parallel with the capacitance. However, some networks require that the impedance be calculated in series form. It is therefore necessary to convert between series and parallel impedance forms. The necessary equations are listed in Fig. 4–35. The output resistance of RF power transistors is usually not shown on datasheets, but may be calculated using the equation of Fig. 4–35.

Typical RF Power Amplifier Tuning Networks.   Figures 4–36 through 4–40 show five typical resonant networks, together with the equations necessary to find the component values. Any of the resonant networks can be used as the tuning networks for RF amplifiers and/or multipliers. Note that the network of Fig. 4–36 is similar to that of Fig. 4–34a, while Fig. 4–38 is similar to Fig. 4–34b (except for the power connection).

The resistor and capacitor shown in the box labeled "transistor to be matched" represent the *complex output impedance* of a transistor. When the network is to be used with a final amplifier, the resistor labeled R_L is the antenna impedance or other load. When the network is used with an intermediate amplifier, R_L represents the input impedance of the following transistor. It is therefore necessary to calculate the input impedance of the transistors being fed by the network, using the data and equations of Fig. 4–34.

The complex impedances are represented in series form in some cases and parallel form in others, depending on which form is the most convenient for

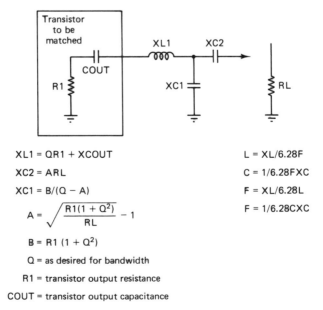

$$XL1 = QR1 + XCOUT$$

$$XC2 = ARL$$

$$XC1 = B/(Q - A)$$

$$A = \sqrt{\frac{R1(1 + Q^2)}{RL}} - 1$$

$$B = R1\,(1 + Q^2)$$

Q = as desired for bandwidth

R1 = transistor output resistance

COUT = transistor output capacitance

$$L = XL/6.28F$$

$$C = 1/6.28FXC$$

$$F = XL/6.28L$$

$$F = 1/6.28CXC$$

FIGURE 4-36   RF network where R1 is less than RL.

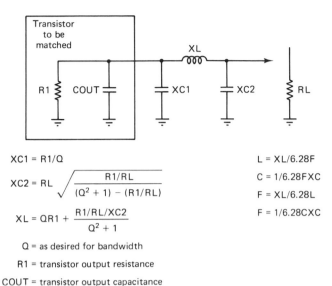

XC1 = R1/Q

$$XC2 = RL \sqrt{\frac{R1/RL}{(Q^2 + 1) - (R1/RL)}}$$

$$XL = QR1 + \frac{R1/RL/XC2}{Q^2 + 1}$$

Q = as desired for bandwidth

R1 = transistor output resistance

COUT = transistor output capacitance

L = XL/6.28F

C = 1/6.28FXC

F = XL/6.28L

F = 1/6.28CXC

FIGURE 4-37 RF network where R1 is about equal to RL.

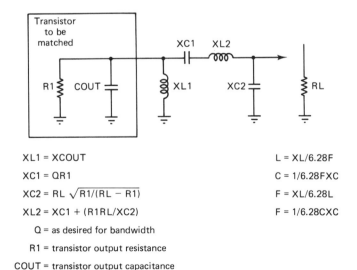

XL1 = XCOUT

XC1 = QR1

XC2 = RL $\sqrt{R1/(RL - R1)}$

XL2 = XC1 + (R1RL/XC2)

Q = as desired for bandwidth

R1 = transistor output resistance

COUT = transistor output capacitance

L = XL/6.28F

C = 1/6.28FXC

F = XL/6.28L

F = 1/6.28CXC

FIGURE 4-38 RF network where R1 is very small in relation to RL.

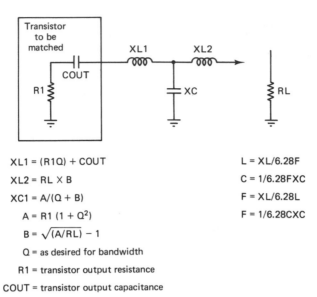

XL1 = (R1Q) + COUT

XL2 = RL × B

XC1 = A/(Q + B)

A = R1 (1 + Q²)

B = $\sqrt{(A/RL)} - 1$

Q = as desired for bandwidth

R1 = transistor output resistance

COUT = transistor output capacitance

L = XL/6.28F

C = 1/6.28FXC

F = XL/6.28L

F = 1/6.28CXC

FIGURE 4-39   Alternate RF network where R1 is very small in relation to RL.

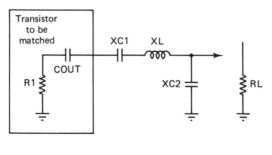

XL = XC1 + (R1RL/XC2) + XCOUT

XC1 = QR1

XC2 = RL $\sqrt{R1/(RL - R1)}$

Q = as desired for bandwidth

R1 = transistor output resistance

COUT = transistor output capacitance

L = XL/6.28F

C = 1/6.28FXC

F = XL/6.28L

F = 1/6.28CXC

FIGURE 4-40   RF network where R1 is very large or very small in relation to RL.

network calculation. The resultant impedance of the network, when terminated with a given load, must be equal to the conjugate of the impedance in the box.

For example, assume that the transistor has a series output impedance of 7.33 − j3.87. That is, the resistance (real part of impedance) is 7.33 Ω, while the capacitive reactance (imaginary part of impedance) is 3.87. For maximum power transfer from the transistor to the load, the load impedance must be the conjugate of the output impedance, or 7.33 + j3.87.

If the transmitter is designed to operate into the typical 50-Ω load (antenna), the network must transfer the (50 + j0)-Ω transmitter load to the 7.33 + j3.87 transistor load. In addition to performing this transformation, the network provides harmonic rejection (unless a harmonic is needed in a multiplier stage), low loss, and provisions for adjustment of both loading and tuning.

Each of the networks has advantages and disadvantages. The following is a summary of the five resonant networks.

R1 Less than RL.   The network of Fig. 4–36 applies to most RF power amplifiers and is especially useful where the series real part of the transistor output impedance (R1) is less than 50 Ω. With a typical 50-Ω load, the required reactance for C1 rises to an impractical value when R1 is close to 50 Ω.

R1 Equal to RL.   The network of Fig. 4–37 (often called a *pi network*) is best suited where the parallel resistor R1 is high (near the value of RL, typically 50 Ω). If the network of Fig. 4–37 is used with a low value of R1, the inductance of L1 must be very small, while C1 and C2 become vary large (beyond practical limits).

R1 Very Small in Relation to RL.   The networks of Figs. 4–38 and 4–39 produce practical values for C and L, especially where R1 is very low. The main limitation for the networks of Figs. 4–38 and 4–39 is that R1 must be substantially lower than RL. These networks, or variations thereof, are often used with intermediate stages where a low output impedance of one transistor is matched to the low input impedance of another transistor.

R1 Very Large or Very Small in Relation to RL.   The network of Fig. 4–40 (often called a *tee network*) is best suited where R1 is much less or much greater than RL.

4-8.4 Design Example

Assume that it is desired to design a network similar to that of Fig. 4–36. The network must match the output of a transistor to a 50-Ω antenna. The transistor shows an output capacitance of 200 pF at the operating frequency (obtained from the datasheet).

The required output is 50 W and 50 MHz with a 28-V power supply. It is assumed that the transistor produces the required output at the operating frequency, with the available signal, as discussed in Sec. 4-8.2.

With a value of 28 V and an output of 50 W, the equations of Fig. 4-35 show the parallel value of R1 as $28^2/(2 \times 50) = 7.84 \ \Omega$.

With an output capacitance of 200 pF, and an operating frequency of 50 MHz, the equations of Fig. 4-35 show the reactance of Cout as

$$\frac{1}{6.28 \times (50 \times 10^6) \times (200 \times 10^{-12})} = 16 \ \Omega$$

The combination of these two values results in a parallel output impedance of $7.84 - j16$.

Usually, the datasheet gives the output capacitance in parallel form with R1. For the network of Fig. 4-36, the values of R1 and Cout must be converted to series form.

Using the equations of Fig. 4-35, the equivalent series output impedance is

$$Rseries = \frac{7.84}{1 + (7.84/16)^2} = 6.32 \ \Omega \quad (R1)$$

$$Xseries = 6.32 \times \frac{7.84}{16} = 3.1 \ \Omega \quad (XCout)$$

The combination of these two values results in a *series output* impedance of $6.32 - j3.1$.

Using the equations of Fig. 4-36 and assuming a Q of 10 for simplicity, the *reactance values* for the network are

$$XL = 10 \times (6.32) + 3.1 = 66.6 \ \Omega$$

$$A = \sqrt{\frac{6.32(1 + 10^2)}{50}} - 1 = 3.3$$

$$XC2 = 3.3 \times 50 = 165 \ \Omega$$

$$B = 6.32(1 \times 10^2) = 638.32$$

$$XC1 = \frac{638.32}{10 - 3.3} = 95 \ \Omega$$

Using the equations of Fig. 4-36, the corresponding *inductance and capacitance values* are

$$L1 = \frac{66.6}{6.28 \times (50 \times 10^6)} = 0.21 \ \mu H$$

$$C1 = \frac{1}{6.28 \times (50 \times 10^6) \times (95)} = 33 \text{ pF}$$

$$C2 = \frac{1}{6.28 \times (50 \times 10^6) \times (165)} = 19 \text{ pF}$$

If C1 and C2 are variable, the values obtained should be the midrange values.

4-9 DESIGNING RF AMPLIFIERS WITH y-PARAMETERS

This section describes the design of tuning networks for RF amplifiers using the y-parameters discussed in Sec. 4–7. Note that the design procedure is similar to that described in Sec. 4–8, but requires that a full set of y-parameters be available (either in the form of datasheet graphs or obtained by actual test of the transistor, as described in Sec. 4–7.3).

Assume that the circuit of Fig. 4–41 is to be operated at 100 MHz. (Although a FET is shown in Fig. 4–41, the design procedures also apply to two-junction transistors.) The source and load impedance are both 50 Ω. The characteristics of the FET are given in the datasheet. The problem is to find the optimum values of C1 through C4, as well as L1 and L2.

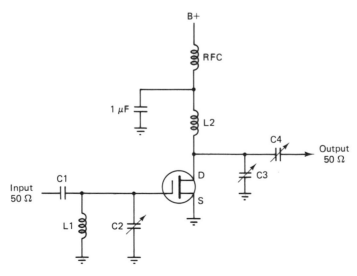

FIGURE 4-41 Common-source FET RF amplifier.

4-9.1   FET Characteristics

At 100 MHz, y-parameters for the FET are

$$y_{is} = y_{11} = 0.15 + j3.0$$

$$y_{fs} = y_{21} = 10 - j5.5$$

$$y_{os} = y_{22} = 0.04 + j1.7$$

$$y_{rs} = y_{12} = 0 - j0.012$$

From Fig. 4–29, the Linvill stability factor C is 2.0. A Linvill C factor greater than 1 indicates that the device (a FET in this case) is potentially unstable. Neutralization or mismatching is necessary to prevent oscillation. Mismatching is used in this example.

Figure 4–30 shows that for a circuit stability (Stern k) factor of 3.0 the transducer gain is about 29 dB. The load and source admittances for the required mismatch and gain (found in Figs. 4–32 and 4–33) are

$$Y_L = 0.35 - j2.1 \text{ mmho}$$

$$Y_S = 1.30 - j4.4 \text{ mmho}$$

4-9.2   Output Matching

The first step is to design networks that match the FET to the load and source of 50 Ω. The calculations are easier to perform if the admittances of Figs. 4–32 and 4–33 are converted to impedances. Let us start by matching the FET output impedance to the load.

The 50-Ω load impedance must be transformed to the optimum load for the FET ($Y_L = 0.35 - j2.1$). This transformation can be performed by the network shown in Fig. 4–42a. In effect, RL is in series with C4. The 50 Ω must be transformed to

$$RL = \frac{1}{GL} = \frac{1}{0.35 \times 10^{-3}} \approx 2860 \ \Omega$$

As discussed, the true value of RL is not changed. However, the network is designed as if the value of RL is changed to the optimum load for the FET.

The series capacitive reactance (for C4) required for this matching is

$$XC4_{series} = RS \left[\sqrt{\frac{RP}{RS}} - 1 \right]$$

where RP is the transformed parallel resistance and RS is the true series resistance. Therefore,

$$XC4_{series} = 50 \left[\sqrt{\frac{2860}{50}} - 1 \right] \approx 372 \ \Omega$$

The capacitance for C4 that provides this reactance at 100 MHz is

$$C4 = \frac{1}{6.28F \times C4} = \frac{1}{6.28(10^8)(327)} \approx 4.3 \ pF$$

The parallel equivalent of this capacitance is needed for determining the bandwidth and resonance later in design.

$$XC4_{parallel} = XC4_{series} \left[1 + \left(\frac{RS}{XC4_{series}} \right)^2 \right]$$

$$= \frac{RP}{XC4_{series}/RS}$$

$$= \frac{2860}{372/50} \approx 382 \ \Omega$$

The equivalent parallel capacitance for C4 ($C4_{parallel}$) is therefore about 4.2 pF at 100 MHz.

4-9.3 Output Network Design

Figure 4–42b shows an equivalent circuit for the output tank after transformation of the load. Since the resistance across the output circuit is fixed by the parallel combination of ROUT and RL (after transformation), the desired bandwidth of the output tank is determined by C3.

As discussed, the output admittance YOUT of the FET is not equal to Y_{os} under all conditions. Only when the input is terminated in a short circuit or the feedback admittance is zero does YOUT equal y_{os}. When y_{os} is not zero and the input is terminated with a practical source admittance, the true output admittance is found from

$$YOUT = y_{os} - \frac{y_{fs}y_{rs}}{y_{is} + Y_S}$$

$$= 0.04 + j1.7 - \frac{(10 - j5.5)(0 - j0.012)}{(0.15 + j3.0) + (1.3 - j4.4)}$$

$$= -0.066 + j2.05 \ mmho$$

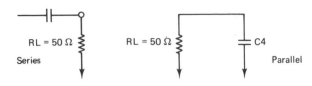

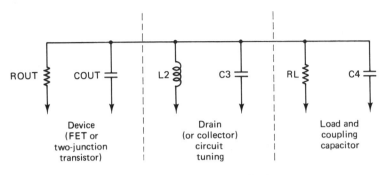

(a) Output impedance transformation

(b) Equivalent output circuit

FIGURE 4-42   Impedance transformation and equivalent output circuit of RF amplifier.

Therefore,

$$ROUT = \frac{1}{GOUT} = \frac{1}{-0.066 \times 10^{-3}} = -15.2 \text{ k}\Omega$$

$$COUT = \frac{BOUT}{6.28F} = \frac{2.05 \times 10^{-3}}{6.28(10^8)} \approx 3.2 \text{ pF}$$

Note that the negative output impedance indicates the instability of the unloaded amplifier.

Now the total impedance across the tank can be calculated:

$$R_{total} = \frac{1}{GOUT + GL}$$

$$= \frac{1}{(-0.066 \times 10^{-3}) + (0.35 \times 10^{-3})} \approx 3.52 \text{ k}\Omega$$

Since the output impedance is several times higher than the input impedance of the FET, amplifier bandwidth depends primarily on output loaded Q. For a bandwidth of 5 MHz (3-dB points),

$$C_{total} = \frac{1}{6.28 R_{total} \ (BW)}$$

$$= \frac{1}{6.28(3.52 \times 10^3)(5 \times 10^6)} \approx 9 \ pF$$

Therefore,

$$C3 = C_{total} - COUT - C4_{parallel} = 9.0 - 3.2 - 4.2 = 1.6 \ pF$$

The output inductance (L2) that resonates with C_{total} at 100 MHz is 280 nH. This completes output network design. (Which do you prefer, this or Sec. 4-8?)

4-9.4   Input Network Design

The input network can be calculated in the same manner and yields the following results:

$$Y_S = 1.30 - j4.4$$

$$XC1_{series} = 190 \ \Omega$$

Therefore,

$$C1 = 8.4 \ pF$$

$$XC1_{parallel} = 203 \ \Omega$$

Therefore,

$$C1_{parallel} = 7.8 \ pF$$

$$YIN = y_{is} = \frac{y_{fs} y_{rs}}{y_{os} + YL} = 0.25 + j4.52 \ mmho$$

Therefore,

$$RIN = 14 \ k\Omega, \qquad CIN = 7.2 \ pF, \qquad R_{total} = 950 \ \Omega$$

The bandwidth of the input tuned circuit is chosen to be 10 MHz. Therefore,

$$C_{total} = 17 \ pF \quad and \quad L1 = 150 \ nH$$

$$C2 = 17 - 7.2 - 7.8 = 2 \ pF$$

4-9.5 Bypass Capacitor

This completes design of the tuned circuits (resonant networks). It is important that the circuit be well bypassed to ground at the signal frequency, since only a small impedance to ground may cause instability or loss in gain. To be on the safe side, select a bypass capacitor value where the reactance is about 1 or 2 Ω at the operating frequency. A 1-μF capacitor provides less than 1-Ω reactance at 100 MHz.

5

DIRECT-COUPLED
AND COMPOUND AMPLIFIERS

The amplifiers discussed in Chapters 1 through 4 have one major limitation in certain applications. The amplifiers discussed thus far cannot amplify direct currents (direct voltages). Direct currents are not passed by coupling capacitors or transformers. Equally important, amplifiers using transformers and/or coupling capacitors are not well suited for amplification at low frequencies.

As discussed in Chapter 3, coupling capacitors form a voltage divider with the input impedance of the following stage. Such dividers increase attenuation of the signal as frequency decreases. In the case of transformers, the impedance offered at low frequencies virtually shorts low-frequency signals (since inductive reactance decreases with frequency).

For these reasons, direct-coupled amplifiers are used if direct currents and/or low-frequency signals must be amplified. Direct-coupled amplifiers, also known as DC amplifiers or direct-current amplifiers, permit a signal to be fed directly to the transistor without any coupling device. Direct-coupled amplifiers are often used in present design between ICs (in those applications where there must be no low-frequency signal attenuation or where a direct signal must be passed from one IC to another).

5-1 BASIC DIRECT-COUPLED AMPLIFIER THEORY

Figure 5–1 shows the basic circuit of a direct-coupled amplifier using a PNP transistor in a common-emitter configuration. Resistors R1 and R2 form a voltage divider to forward bias the emitter–base junction of the transistor. Resistor RE is the emitter stabilizing resistor, and CE is the bypass capacitor.

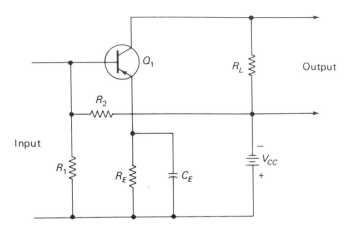

FIGURE 5-1   Basic direct-coupled amplifier.

Resistor RL is the collector-load resistor. The voltage drop across RL, caused by the flow of collector current, is the output voltage.

If the input voltage increases the forward bias of the emitter–base junction, the emitter current is increased. This, in turn, increases the collector current and the output voltage. If the input voltage reduces the forward bias, the collector current is reduced, as is the output voltage.

The range of input voltages that may be applied to a direct-coupled transistor is small. The forward bias must not be increased to the point where the transistor operates in the saturated region, nor can the bias be reduced so that the transistor is cut off.

5-1.1   Direct-Coupled Amplifier with Two Stages of Like Transistors

Figure 5–2 shows a two-stage direct-coupled amplifier using two PNP transistors in a common-emitter configuration. Resistors R1 and R2 form a voltage divider to forward bias the emitter–base junction of Q1. Resistor RE1 is the emitter-stabilizing resistor for Q1, and CE1 is the bypass capacitor.

Resistor RL1 serves the double purpose of collector-load resistor for Q1 and voltage-dropping resistor to forward bias the emitter–base junction of Q2. Resistor RE2 is the emitter-stabilizing resistor for Q2, and CE2 is the bypass capacitor. Resistor RL2 is the collector-load resistor for Q2.

The amplifier of Fig. 5–2, where both transistors are PNP (or NPN), has a tendency to become unstable. For example, if the collector current of Q1 varies because of power-supply or temperature changes, Q2 amplifies these changes and adds changes (because of possible variations in Q2 characteristics). For this reason, it is difficult to cascade more than two stages of direct coupling where both transistors are of the same type (PNP or NPN).

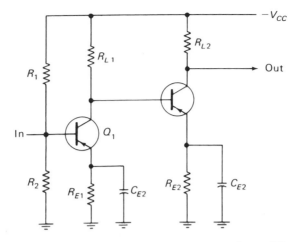

FIGURE 5-2   Two-stage direct-coupled amplifier.

5-1.2   *Direct-Coupled Complementary Amplifier*

Figure 5-3 shows a more stable direct-coupled amplifier. Here, Q1 is an NPN transistor with the collector coupled directly to the base of Q2, a PNP transistor. Such an arrangement is known as a *complementary amplifier.*

The forward bias for the emitter–base junction of Q1 and the reverse bias for the collector–base junction are taken from the + Vcc supply in the normal manner. The emitter of Q2 is connected to + Vcc through RE2 (which is bypassed by CE2). The collector resistor RL1, the impedance of Q1, and resistor RE1 form a voltage divider across the Vcc supply.

With such an arrangement, the base of Q2 is less positive (or more negative) than the emitter. Thus, the emitter–base junction of Q2 (PNP) is forward

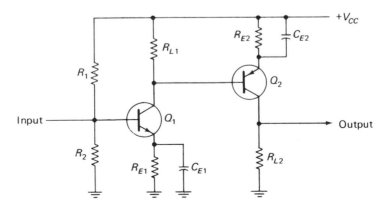

FIGURE 5-3   Stable direct-coupled amplifier.

biased. The collector of Q2 is grounded through the output-load resistor RL2. Since the negative terminal of the power supply (– Vcc) is also grounded, reverse bias for the collector–base junction of Q2 is established.

5-1.3 Three-Stage Complementary Amplifier

Figure 5–4 shows a three-stage complementary amplifier. Note that the increased stabilization for any complementary amplifier occurs because any change in collector current (due to temperature, power supply variation, and so on) is opposed by an equal change in collector current in the *following stage*. Of course, NPN and PNP amplifiers (transistors) must be used alternately, as shown in Fig. 5–4.

5-1.4 Direct Coupling between Unlike Stages

Figure 5–5 shows how to use direct coupling between unlike stages (between a common-emitter amplifier and an emitter follower in this case). Here, the collector of Q1 is direct coupled to the base of Q2, which is connected in a common-collector (emitter-follower) configuration.

Forward bias for the emitter–base junction of Q2 is taken from the power supply through RL, which also serves as the collector load for Q1. Capacitor C places the collector of Q2 at ground, as far as signal is concerned. Output is taken across RE2, which is the emitter resistor for Q3. The circuit of Fig. 5–5 is particularly useful when the output must be at a low impedance. The value of RE2 sets the approximate output impedance of the circuit.

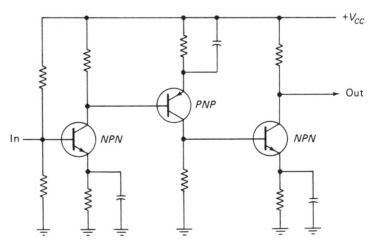

FIGURE 5-4   Three-stage complementary amplifier.

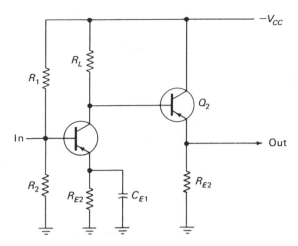

FIGURE 5-5 Direct coupling between unlike stages.

5-2 PRACTICAL DIRECT-COUPLED TWO-JUNCTION TRANSISTOR AMPLIFIERS

Figure 5-6 is the working schematic of a two-stage direct-coupled complementary amplifier. Figure 5-6 also shows the equations for finding approximate component values and gives some typical component values and currents.

Note that two- and three-stage amplifiers, similar to that shown in Fig. 5-4, are available commercially in IC or packaged form. It is generally easier to design with these packages or ICs rather than with individual components, since impedance relationships, Q point, and so forth, are calculated by the IC or package manufacturer. Also, the datasheets supplied with the ICs and packages provide information regarding source voltage, gain, impedances, and the like.

The datasheet information can be followed to adapt the IC or package for a specific application. However, in some cases it is necessary to select values of components external to the IC (or between ICs). For this reason, and since it may be necessary to design a multistage direct-coupled amplifier (with individual components) for some special application, the following design considerations and examples are provided.

5-2.1 Design Considerations

Note that the circuit elements for transistor Q1 in Fig. 5-6 are essentially the same as for the circuits of Fig. 3-11. Also note that the same circuit arrangement is used for transistor Q2, except that RA and RB are omitted (as is the coupling capacitor between stages).

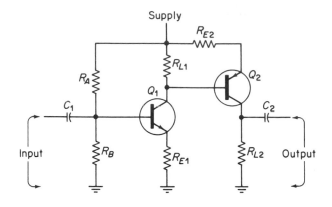

Input impedance $\approx R_B$

Output impedance $\approx R_{L2}$

Q_1 voltage gain $\approx \dfrac{R_{L1}}{R_{E1}}$

Q_2 voltage gain $\approx \dfrac{R_{L2}}{R_{E2}}$

Overall voltage gain $\approx 70\,\%$ of combined Q_1 and Q_2 voltage gains

$R_L > 5\,R_E$

$R_L \approx 10\,R_E$

$R_B \approx 10\,R_E$

$R_B < 20\,R_E$

Q_2 collector voltage $\approx 0.5 \times$ supply voltage (as adjusted by R_A)

FIGURE 5-6 Two-stage direct-coupled complementary amplifier.

The design considerations for the circuit of Fig. 5–6 are the same as those for Fig. 3–11, except for the following:

1. The input impedance of the complete circuit is approximately equal to RB. The output impedance is set by RL2.

2. The value of C1 depends on the low-frequency limit and the value of RB. The value of C2 depends on the low-frequency limit and the value of input resistance for the following stage (or for the load).

5-2.2 Design Example

Assume that the circuit of Fig. 5–6 is to be used as a voltage amplifier. The desired output is 7 V (peak to peak) with a 500-Ω impedance. The input is 100 mV, also from a 500-Ω impedance. (This requires a voltage gain of 70.) The low-frequency limit is 30 Hz, with a high-frequency limit of 100 kHz. Minimum distortion is desired. (The circuit should not be overdriven.) A source of 20 V is specified. The circuit is not battery operated. A 2N3568 silicon NPN is to be used for the input; a 2N3638 is to be used for the output. Both transistors are time-tested devices.

Supply Voltage and Operating Point.   The supply voltage, 20 V, is specified. Since this is about three times the desired output of 7 V, there should be no distortion. First establish the Q point for the output stage. The collector voltage should be about one-half the source, or $20/2 = 10$ V, as the Q point.

Load Resistance and Collector Current.   The value of RL2 is specified as 500 Ω. Use the nearest standard value of 510 Ω. With a 10-V drop across RL2, the Q2 collector current is 10/510, or 19.6 mA (rounded off to 20 mA).

Emitter Resistance, Current, and Voltage.   When two stages are direct coupled in the stabilized circuit of Fig. 5–6, the overall gain is about 70% of the combined gain for each stage. Since the required voltage gain is 70, the combined gains should be 100 (or a gain of 10 for each stage).

To provide a voltage gain of 10 in the output stage, the value of RE2 should be one tenth of RL2, or $510/10 = 51$ Ω. The current through RE2 is the collector current of 20 mA, plus the base current. Assuming a circuit gain of 10, the base current is 20/10, or 2 mA. The combined currents through RE2 are $20 + 2$, or 22 mA. This produces a drop of 1.12 V across RE2 (rounded off to 1 V).

Output Base Voltage and Input Collector Current.   The base of Q2 should be 0.5 V from the emitter voltage. Since Q2 is PNP, the base should be more negative (or less positive) than the emitter. The emitter of Q2 is at $+19$ V (20 V $-$ 1 V). Therefore, the base of Q2 should be $+18.5$ V. This sets the collector voltage for Q1 at the Q point.

Input Stage Resistances and Current.   The value of RB (input impedance) is specified as 500 Ω. Use the nearest standard value of 510 Ω. As shown in Fig. 5–6, RB should not be greater than 20 times or less than 5 times RE1. An RB/RE1 ratio of 10 is nominal for reasonable stability. Therefore, RE1 should be 510/10, or 51 Ω. With the RE1 value established and a stage voltage gain of 10 desired, the value of RL1 should be 51×10, or 510 Ω.

With a 1.5-V drop across RL1, the collector current is 1.5/510, or 2.94 mA (rounded off to 3 mA). The current through RE1 is the collector current of 3 mA, plus the base current. Assuming a gain of 10, the base current is 3/10, or 0.3 mA. The combined currents through RE1 are $3 + 0.3$ mA, plus the 2-mA base current of Q2, or $3 + 0.3 + 2 = 5.3$ mA. This produces a drop of about 270 mV across RE1.

The base voltage of Q1 is about 0.5 V higher than the emitter voltage, or $0.5 + 0.270 = 0.770$ V. With a 0.770-V drop across RB, the current through RB is 0.770/510, or about 1.5 mA.

The value of RA should be sufficient to drop the 20-V source to 19.230 V so that the base is 0.770 V above ground. The current through RA is the current through RB of 1.5 mA, plus the base current of 0.3 mA, or

1.5 + 0.2 = 1.8 mA. The resistance required to produce a 19.230-V drop with 1.8 mA is 19.230/1.8, or about 10.7 kΩ. Use an 11-kΩ standard resistance as the trial value.

Adjust RA for the desired collector Q-point voltage at Q2 rather than Q1. The final adjustment of RA is made for a distortion-free 7-V output signal at the collector of Q2 (with a 100-mV input signal applied to Q1).

Input and Output Capacitors.   Even though there is direct coupling between the stages, coupling capacitors are necessary at the input and output, unless the external device is provided with capacitors for isolation of the direct current.

The value of C1 forms a high pass filter with RB. The high limit of 100 kHz can be ignored. The low-frequency limit of 30 Hz requires a capacitance value of $1/(3.2 \times 30 \times 510)$, or about 20 μF, for an approximate 1-dB drop. The value of C2 is found in the same way, except the resistance value R must be the load resistance. Since the load and input impedances are the same, the capacitance value of C2 is the same as C1. The voltage values of C1 and C2 should be 1.5 times the maximum voltage involved, or $20 \times 1.5 = 30$ V.

Transistor Selection.   In this case, the transistors are specified in the design example. However, it is always wise to check basic transistor characteristics against the proposed circuit parameters *before* making any connections (even though the two transistors involved are classic devices).

The maximum collector voltages for the 2N3568 and 2N3638 are 80 and 25 V, respectively, both well above the 20-V source.

The powers dissipated for the 2N3568 and 2N3638 are 55 and 200 mW, respectively, both well below the 300-mW datasheet limit.

Minimum betas for the 2N3568 and 2N3638 are 40 and 20, respectively, both well above the required 10.

Increasing Gain with Emitter Bypass.   If gain must be increased, the emitter resistor can be bypassed. Amplifier gain then depends on transistor beta, rather than circuit values. However, as discussed in Chapter 3, bypassing the emitter creates problems for low-frequency signals. Since one of the prime reasons for using direct coupling is to amplify low-frequency signals, emitter bypassing may defeat the advantage of direct coupling. Likewise, if the circuit is to amplify direct-current signals, the emitter bypass is of little value.

Alternate Bypass Configuration.   Figure 5-7 shows an alternative method for bypassing the emitters of a direct-coupled amplifier with one capacitor. In effect, collector-load resistor RL2 is broken up into two resistances (RL2A and RL2B) and connected in series with the emitter of Q1 as a form of feedback. The bypass capacitor C3 is connected between the junction of RL2A and RL2B.

With this configuration, the signal is bypassed, as is part of the feedback. This alters input impedance as well as voltage gain. In effect, voltage gain is

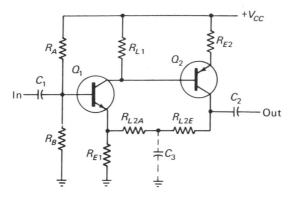

FIGURE 5-7 Alternative method for bypassing the emitters of a direct-coupled amplifier.

sacrificed for increased input impedance, resulting in increased stability. However, the low-frequency signal-attenuation problem caused by C3 still exists.

5-3 PRACTICAL DIRECT-COUPLED FET AMPLIFIERS

Both IGFETs and JFETs can be used as direct-coupled amplifiers. However, IGFETs are especially well suited to direct-coupled applications. Since the gate of an IGFET acts essentially as a capacitor, rather than a diode junction, no coupling capacitor is needed between stages. For a-c signals, this means that there are no low-frequency cutoff problems (in theory). In practical design, the input capacitance can form an RC filter with the source resistance and produce some low-frequency attenuation.

Figure 5–8 is the working schematic of an all-IGFET, three-stage amplifier. Note that all three IGFETs are of the same type, and all three drain resistors (R1, R2, R3) are the same value. This arrangement simplifies design. At first glance it may appear that all three stages are operating at zero bias. However, when ID flows, there is some drop across the corresponding drain resistor, producing a voltage at the stage drain and an identical voltage at the gate of the next stage.

The gate of the first stage is at essentially the same voltage as the drain of the last stage because of feedback resistor RF. There is no current drain through RF, with the possible exception of reverse gate current (which can be ignored for practical purposes).

Operating Point.   To find a suitable operating point for the amplifier, it is necessary to trade off between desired output voltage swing, IGFET characteristics, and supply voltage. For example, assume that an output swing of 7 V peak to peak is desired, and the supply voltage is 24 V. Further assume that the ID is about 0.55 mA when VGS is 7 V.

A suitable operating point is 7 V to accommodate the 7-V output swing without distortion. (The swing is from 3.5 to 10.5 V around the 7-V point.) This requires an 18-V drop from the 24-V supply. With 0.55-mA ID and an 18-V drop, the values of R1, R2, and R3 are about 33 kΩ.

Gain.   The overall gain depends on the relationship of the gain without feedback and the feedback resistance RF. Gain without feedback is set by Yfs and the value of R1/R2/R3. For example, assume a Yfs of 1000 μmho (0.001 mho); the gain of each stage is 33 (33,000 × 0.001 = 33). With each stage at a gain of 33, the overall gain (without feedback) is about 36,000.

To find the value of RF, divide the gain without feedback by the desired gain. Multiply the product by 100. Then multiply the resultant product by the value of R1. For example, assume a desired gain of 3000 (the gain without feedback is 36,000); 36,000/3000 = 12; 12 × 100 = 1200; 1200 × 33,000 = 39.6 MΩ (use the nearest standard of 40 mΩ).

Input Impedance.   The input impedance depends on the relationship of gain to feedback resistance. The approximate impedance is RF/gain. Since the gain depends on Yfs, input impedance is subject to variation from FET to FET and with temperature.

5–3.1 *Direct-Current FET Amplifier*

The circuit of Fig. 5–8 requires one coupling capacitor at the input. This is necessary to isolate the input gate from any direct-current signal or voltage that may appear at the input. This makes the circuit of Fig. 5–8 unsuitable as a direct-current amplifier. The coupling capacitor forms an RC filter with the input resistance. However, since resistance is high, a 0.01-μF coupling capacitor produces less than 1-dB drop, even at frequencies of a few hertz.

The circuit can be converted to a direct-current amplifier when the coupling capacitor is replaced by a series resistor Rin, as shown in Fig. 5–9.

The considerations concerning operating point are the same for both circuits. However, the series resistance must be terminated at a d-c level equivalent to the operating point. For example, if the operating point is − 7 V, point A must be at − 7 V. If point A is at some other d-c level, the operating point is shifted.

The relationships of input impedance, RF, and gain still hold. However, input impedance is about equal to Rin. Therefore, gain is about equal to the ratio of RF/Rin. This makes it possible to control gain by setting the RF/Rin ratio. Of course, the gain cannot exceed the gain-without-feedback (open loop) factor, no matter what the ratio of RF/Rin (closed loop). As a general rule,

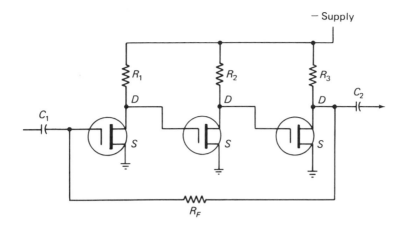

Gain without $R_F \approx (R_1 \times y_{fs})\,(R_2 \times y_{fs})\,(R_3 \times y_{fs})$

$$Z_{in} \approx \frac{R_F}{gain}$$

FIGURE 5-8 All-IGFET, three-stage, direct-coupled amplifier.

the greater the ratio of open-loop gain to closed-loop gain the greater is the circuit stability.

As an example, assume that the open-loop gain is 36,000, and the desired gain is 6000. This requires a ratio of 6 to 1.

As another example, assume that the desired gain is 5000, RF is 40 mΩ, and the open-loop gain is 36,000. 5000 is considerably less than 36,000, so the circuit is well capable of producing the desired gain with feedback. To find the value of Rin,

$$Rin \approx \frac{40 \times 10^6}{5 \times 10^3} \approx 8 \times 10^3 \qquad (8000 \ \Omega)$$

5-3.2 Amplification from Grounded Sources

The circuit of Fig. 5–9 requires that the signal source be at a d-c level equal to the operating point. In many cases, it is necessary to amplify d-c signals at the zero or ground level. This can be done using a depletion-mode JFET at the input, as shown in Fig. 5–10.

Feedback is introduced by connecting the sources of both Q1 and Q3 to a common-source resistor R2. The source of Q2 is not provided with a resistor, but there is some bias on Q2 produced by the ID drop across R3. The input impedance of the Fig. 5–10 circuit is set by the value of R1 and the gate–drain capacitance of Q1.

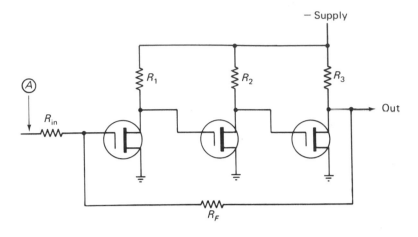

$$\text{Gain} \approx \frac{R_F}{R_{in}}$$

Input impedance $\approx R_{in}$ Output impedance $\approx R_3$

FIGURE 5-9   Converting to a direct-current amplifier.

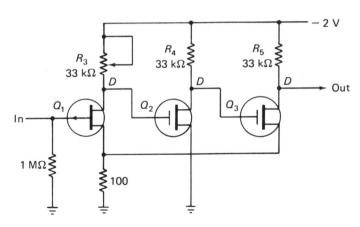

FIGURE 5-10   Amplifying signals from zero or ground level.

Gain can be sacrificed for stability by increasing the value of R2. With the values shown and typical FETs, the gain should be on the order of 3000 to 5000. The bias and operating point for Q2 and Q3 are set by R3, shown as 33 kΩ. In practice, the value of R3 is approximated by calculation and then adjusted for a desired operating point at the output (drain) of Q3.

5-4 FET/TWO-JUNCTION DIRECT-COUPLED AMPLIFIERS

In certain applications, FET stages and two-junction transistors can be combined to form amplifiers (sometimes known as hybrid amplifiers). The classic example is where a single FET stage is used at the input, followed by two two-junction amplifier stages. Such an arrangement takes advantage of both the FET and two-junction transistor characteristics.

A FET is essentially a voltage-operated device, permitting large voltage swings with low currents. This makes it possible to use high resistance values (resulting in high impedances) at the input and between stages. In turn, these high resistance values permit use of low-value coupling capacitors and eliminate the need for bulky, expensive electrolytic capacitors.

If operated at the OTC point, the FET is highly temperature stable, tending to make the overall amplifier equally stable. However, FETs typically operate at low currents (except for certain power FETs) and are usually considered as low-power devices.

Two-junction transistors are essentially current-operated devices, permitting large currents at about the same voltage levels as the FET. Thus, with equal supply voltages and signal-voltage swings, the two-junction transistor can supply much more current gain (and power gain than the FET).

Since currents are high, the impedances (input, interstage, and output) must be low in two-junction transistor amplifiers. This requires large-value coupling capacitors if low frequencies are involved. The low impedances also place a considerable load on devices feeding the amplifier, particularly if the devices are high impedances. (There is also considerable mismatch at the input.) On the other hand, a low output impedance is often a desirable characteristic for an amplifier.

When a FET is used as the input stage, the amplifier input impedance is high. This places a small load on the signal source and allows use of a low-value input coupling capacitor (if required). If the FET is operated at the 0TC point, the amplifier input is temperature stable. (Generally, the input stage is the most critical in regard to temperature stability.) When two-junction transistors are used as the output stages, the output impedance is low, and current gain (as well as power gain) is high.

FET/two-junction amplifiers can be direct-coupled or capacitor-coupled, depending on requirements. The direct-coupled configuration offers the best low-frequency response, permits d-c amplification, and is generally simpler (uses fewer components). The capacitor-coupled version permits a more stable design and eliminates the voltage-regulation problem common to all direct-coupled amplifiers. (That is, a direct-coupled amplifier cannot distinguish between changes in signal level and changes in power-supply level. This problem is discussed further in Sec. 5–5.)

The FET can be combined with any of the classic two-stage two-junction transistor amplifier combinations. The most common combinations are the Darlington pair (for no voltage gain, but high current gain and low output impedance) and the NPN/PNP complementary amplifier pair (for both voltage gain and current gain). (The Darlington pair is discussed in Sec. 5-6.)

5-4.1   FET Input, Two-Junction Output Amplifier

Figure 5-11 shows a direct-coupled amplifier using a FET input stage and a two-junction transistor pair as the output. Note that *local feedback* is used in the FET stage (provided by RS), as well as *overall feedback* (provided by R4).

The design considerations for the FET portion of the circuit are essentially the same as described in Chapter 3, with certain exceptions. Input impedance is set by the value or R2, as usual. Output impedance is set by the combination of R4 and RS. However, since RS is quite small in comparison to R4, the output impedance is essentially equal to R4.

The FET stage gain is set by the ratio of RL to RS, plus the 1/Yfs factor. However, since RS is quite small, the FET gain is set primarily by the ratio of RL to 1/Yfs. The gain of the two-junction transistor pair is set by the beta of the two transistors and the feedback. Thus, the gain can only be estimated.

Note that the drop across R3 is the normal base–emitter drop of a transistor (about 0.5 to 0.7 V for silicon and 0.2 to 0.3 V for germanium). The drop across RL is twice this value. Thus, for a typical silicon transistor, the base

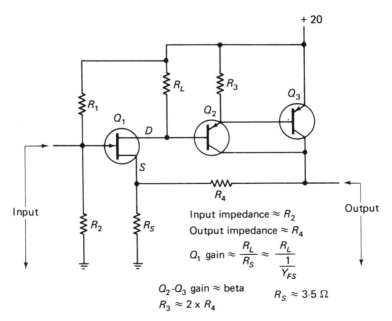

FIGURE 5-11   FET/two-junction direct-coupled amplifier.

of Q2 and the drain of Q1 operate at about 1 V removed from the supply. In a practical experimental circuit, RL must be adjusted to give the correct bias for Q2 (and operating point for Q1). The same is true for R3. However, as a first trial value, R3 should be about twice the value of R4.

Design starts with a selection of ID for the FET. If maximum temperature stability is desired, use the 0TC level of ID. This usually requires a fixed bias, as described in Chapters 1 and 3. If temperature stability is not critical, the FET can be operated at zero bias by omitting R1. There is some voltage developed across RS. However, since RS is small, the VGS is essentially zero, and the ID is set by the 0VGS characteristics of the FET.

With the value of ID set, select a value of RL that produces approximately a 1- to 1.5-V drop to bias Q2.

The input impedance is set by R2, with the output impedance set by R4. The value of R3 is about twice that of R4. The value of RS is less than 10 Ω, typically on the order of 3 to 5 Ω.

As a brief design example, assume that the circuit of Fig. 5–11 is to provide an input impedance of 1 MΩ, an output impedance of 500 Ω, and maximum gain. Temperature stability is not critical. Under these conditions, the values of R2 and R4 are set at 1 MΩ and 500 Ω, but the voltage drop across RS can be ignored. FET Q1 operates at VGS = 0, for practical purposes. Assume that ID is 0.2 mA under these conditions. With a required drop of 1.5 V and a 0.2-mA ID, the value of RL is about 7.5 k Ω. Since R4 is 500 Ω, R3 should be 1000 Ω.

The key component in setting up this circuit is RL. With the circuit operating in experimental form, adjust RL for the desired Q-point voltage at the output (collectors of Q2 and Q3).

5–4.2 Nonblocking Direct-Coupled Amplifier

Generally, a direct-coupled amplifier does not require any coupling capacitors. One exception is a coupling capacitor at the input to isolate the amplifier from direct current (when the signal is composed of both direct current and alternating current). When a coupling capacitor is used at the gate of a JFET (or at the base of some two-junction transistors), a condition known as *blocking* can possibly occur.

Blocking is produced by the fact that the gate junction of a JFET is similar to that of a diode. That is, the diode acts to rectify the incoming signal. If a capacitor is connected in series with the diode (gate junction), large signals can charge the capacitor.

On one half-cycle, the diode is forward biased and charges rapidly. On the opposite half-cycle, the diode is reverse biased and discharges slowly. If the signal and charge are large enough, the amplifier can be biased at or beyond cutoff until the capacitor discharges. Thus, the amplifier can be blocked to incoming signals for a period of time.

One method of eliminating the blocking problem is to use an IGFET at the input. Such a circuit is shown in Fig. 5–12, where an IGFET drives a two-junction transistor pair. Since the gate of an IGFET acts essentially as a capacitor (rather than a diode junction), there is no rectification of the signal and no blocking.

The input impedance is set by Rin, with the output impedance set by R7. With the resistance ratios as shown by the equation of Fig. 5–12, the voltage gain is about 10 when capacitor C2 is out of the circuit and about 1000 when C2 is in the circuit. Keep in mind that feedback is reduced or removed (and gain increased) when C2 is in the circuit, since C2 functions to bypass feedback signals to ground. With C2 removed, the full feedback is applied, and gain is minimum (stability is maximum).

If it is desired to operate the amplifier at some gain level between 10 and 1000, use a resistance in series with C2 (shown in phantom as R8).

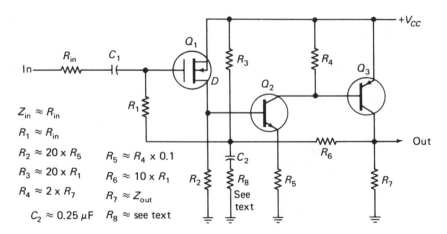

FIGURE 5-12   Nonblocking direct-coupled amplifier.

5-5 STABILIZATION PROBLEMS IN DIRECT-COUPLED AMPLIFIERS

One main problem with any direct-coupled amplifier is that the circuit responds the same way to a change in power-supply voltage (due, for example, to temperature drift) as to a change in the d-c signal level. Although feedback can stabilize the gain of a capacitor-coupled amplifier to a point where the gain is set almost entirely by the resistors in the feedback network, the level of a direct-coupled amplifier is not so easily stabilized.

Temperature-drift problems can be stabilized (or the effects of temperature drift minimized) in several ways. If FETs are involved, one or more of the FET stages can be operated at the 0TC point. If two-junction transistors are used, thermistors and/or temperature-compensating diodes can be used, as discussed in Chapter 1.

Complementary circuits tend to be more stable in the presence of temperature and power-supply variations. However, none of these methods can compensate for a constantly changing or drifting power supply (from whatever cause) when the signal is also direct current. For that reason, several techniques are used to stabilize direct-coupled amplifiers.

5-5.1 Chopper Stabilization

A widely used method for circumventing the voltage-change problems of direct-coupled amplifiers is to convert the d-c signal to an equivalent a-c signal (through modulation). The a-c signal is amplified in a gain-stabilized a-c amplifier and then reconverted to direct current (through demodulation). During amplification the signal is not affected by drift.

One method used to convert the d-c signal to an a-c signal is to switch the amplifier input alternately to both sides of a transformer, as shown in Fig. 5–13. This periodically inverts the polarity of the signal applied to the amplifier. (Note that although mechanical switches are shown the switching system is solid-state in present-day amplifiers, similar to the SHR and SHL signals described in Sec. 3–14.6.)

Another pair of contacts at the output establishes the ground level for a storage capacitor in series with the output. The output storage capacitor becomes charged to a level corresponding to the amplitude of the output square wave. Synchronous detection preserves the polarity of the input voltage and recovers both positive and negative voltages with the correct polarity.

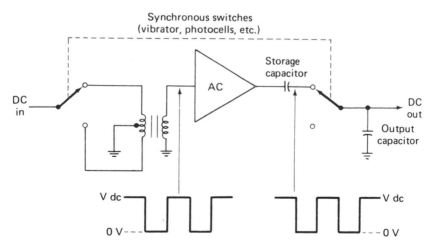

FIGURE 5-13 Basic chopper stabilization technique.

The synchronous modulation and demodulation is known as *chopping*, or *chopper stabilization*. A direct-coupled amplifier with a chopper circuit offers drift-free amplification of low-level signals in the microvolt region.

One problem with any type of modulated amplifier (chopper stabilized) is frequency response. If the input is pure direct current, there is no problem using modulation. However, if the input is very low frequency alternating current, the modulating frequency must be higher than the signal frequency. If not, the input signal waveform can be distorted or completely lost.

For example, if both the input signal and modulating frequency are 100 Hz, and if the amplifier input is shorted at the same instant as the positive swing of the input signal, the amplifier sees only the negative portion of the input. As a guideline, the modulating frequency should be *four times that of the highest a-c input signal to be amplified*.

Figure 5–14 shows another method of chopper stabilization. Here, a chopper-modulated amplifier is used to correct the direct-coupled amplifier for voltage change or drift. Note that the input signal direct current is amplified through a conventional d-c amplifier. A portion of the amplified output is tapped off from a divider network and compared with the original input signal.

The divider network reduces the output by the same amount that is amplified by the d-c amplifier. Therefore, the divider output should be equal to the input level at the summing point. Any difference at the summing point (caused by voltage change, drift, and the like) is amplified through the modulated amplifier and then applied to the main-channel d-c amplifier as *negative feedback* to cancel the drift.

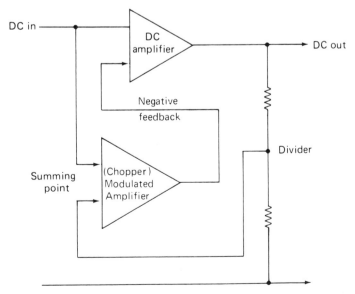

FIGURE 5-14 Chopper-modulated technique for direct-coupled amplifiers.

5-5.2 Dual Amplification

Techniques other than modulation can be used when it is necessary to amplify both direct and alternating current (up to about 100 kHz). One such method is shown in Fig. 5–15, where two parallel amplifiers are used. One amplifier is direct coupled for direct-current and low-frequency alternating current. The other amplifier is a conventional a-c amplifier. Appropriate networks separate the two frequency bands.

For example, direct current is rejected from the a-c amplifier branch by coupling capacitors, but appears at the d-c amplifier input and output as a charge across capacitors C1 and C2. High-frequency alternating current is attenuated at the d-c amplifier input by resistors R1 and R2. Note that feedback is used in both branches to assure uniform gain. Such an amplifier provides stable gain (1% or better) from direct current up to about 100 kHz.

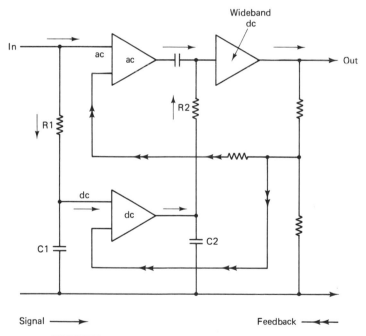

FIGURE 5-15 Dual-amplification technique.

5-6 DARLINGTON COMPOUNDS

Figure 5–16 shows the basic Darlington circuit (known as the Darlington compound) together with two practical versions of the circuit. As shown, the Darlington compound is an emitter follower (or common collector) driving a sec-

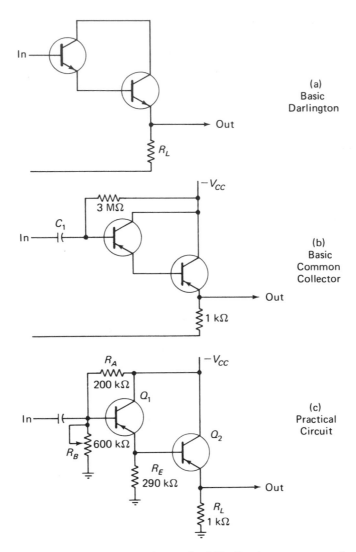

(a)
Basic
Darlington

(b)
Basic
Common
Collector

(c)
Practical
Circuit

FIGURE 5-16 Basic and practical Darlington compounds.

ond emitter follower. Going back to the basic amplifier theory of Chapters 1 and 3, an emitter follower provides no voltage gain, but can provide considerable power gain.

The main reason for using a Darlington compound (especially in audio work) is to produce high current (and power) gain. For example, Darlington compounds (possibly in IC form) are often used as audio drivers to raise the power of a signal from a voltage amplifier to a level suitable to drive a final power amplifier. Darlingtons are also used as a substitute for a driver section (or to eliminate the need for a separate driver).

5-6.1 *Darlingtons as Basic Common Collectors (Emitter Followers)*

When the Darlington is used as a common collector, as shown in Fig. 5–16, the output impedance is about equal to the load resistance RL. The input impedance is approximately equal to beta^2 × RL. The current gain is about equal to the average beta of the two transistors, squared. However, in most common-collector circuits, power gain is of primary concern. That is, the designer is interested in how much the signal power can be increased across a given output load.

For example, assume that the value of RL (in Fig. 5–16c) is 1000 Ω and that the average beta is about 15. This produces an input impedance of about 225 kΩ (15^2 × 1000) and an output impedance of 1000 Ω. Now assume that a 2.5-V signal is applied at the input, and an output of 2 V appears across RL. This input power is 2.5^2/225 kΩ = 0.028 mW. The output power is 2^2/1 kΩ = 4 mW. The power gain is 4/0.028 = 140.

5-6.2 *Darlingtons as Basic Common-Emitter Amplifiers*

Darlington compounds can be used as common-emitter amplifiers to provide voltage gain. This can be done by adding a collector resistor to any of the circuits in Fig. 5–16 and taking the output from the collector rather than the emitter.

With such an arrangement, Q1 becomes a common-collector driving Q2, which appears as a common-emitter amplifier. The entire circuit then appears as a common-emitter amplifier and can be used to replace a single transistor. Such an arrangement is often used where high voltage gain is used.

A more practical method of using a Darlington as a common-emitter amplifier is to eliminate RB and RE (of Fig. 5–16c), ground the emitter of Q2, and transfer RL to the collector of Q2. Such an arrangement is shown in Fig. 5–17, where the circuit is stabilized by the collector feedback through RA (which holds both collectors at a potential somewhat less than 0.5 V from the base of Q1). Note that both collectors are at the same voltage and that this voltage is about equal to two base–emitter voltage drops (or about 1.5 V for two silicon transistors).

With the current of Fig. 5–17, the current gain is about equal to the ratio of RA/RL. Both the input and output voltage swings are somewhat limited in the Fig. 5–17 circuit. The input is biased at approximately 1 to 1.5 V. However, voltage gains of 100 (or more) are possible, since input impedance (or resistance) is approximately equal to the ratio of RA/current gain, or equal to RL. (With input and output impedances approximately equal to RL, the voltage gain follows the current gain.)

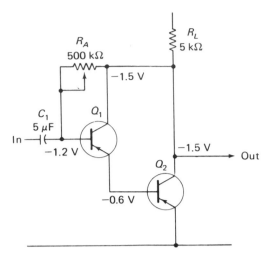

FIGURE 5-17   Darlington compound with collector feedback and common-emitter output.

5-6.3   *Multistage Darlingtons*

Darlington compounds need not be limited to two transistors. Three (and even four) transistors can be used in the Darlington circuit. A classic example of this is the General Electric circuit of Fig. 5–18 (available from many manufacturers in IC form). This circuit is essentially a common-collector and common-emitter Darlington, followed by a common-emitter amplifier.

With R1 out of the circuit, both the input and output impedances are set by RL. (In practice, the input impedance is slightly higher than RL, typically

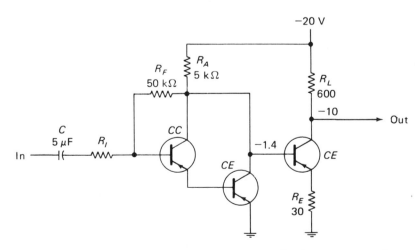

FIGURE 5-18   CC–CE–CE multistage Darlington compound.

about 700 to 800 Ω.) With R1 removed, the voltage gain is about 1000. When R1 is used, the input impedance is approximately equal to R1, and the voltage gain is reduced accordingly. For example, if R1 is 10 kΩ, the 1000 voltage gain drops to about 50.

5-7 SPECIAL DIRECT-COUPLED AMPLIFIER CIRCUITS

In addition to the basic direct-coupled amplifier circuits described thus far in this chapter, three special direct-coupled amplifiers are in common use: the emitter-coupled circuit, the cascode amplifier, and the power-complementary amplifier.

5-7.1 *Emitter-Coupled Amplifier*

Figure 5-19 is the schematic of a basic emitter-coupled amplifier. This circuit, or one of the many variations, is similar to that of the *phase inverter* or *phase splitter* described in Sec. 3-13 in that two 180° out-of-phase signals or outputs can be taken from one point. Unlike the single-stage inverter of Fig. 3-13, the emitter-coupled amplifier can be used when a low-voltage output must be amplified to drive a push–pull output stage.

The emitter resistor RE is common to both Q1 and Q2, which are biased at the Q point (typically at about one-half the supply voltage) in the normal

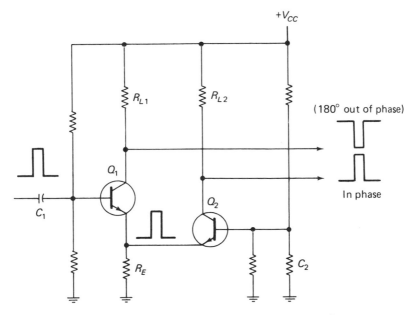

FIGURE 5-19 Basic emitter-coupled amplifier.

manner. The signal is applied to Q1 and appears in amplified, but phase-inverted form at the collector of Q1. An in-phase signal is also developed across RE and appears at the emitters of both Q1 and Q2 simultaneously.

Since the base of Q2 remains fixed by the bias, signal variations on the emitter of Q2 produce an in-phase signal at the collector of Q2. If the values of RL1 and RL2 are equal, the two output collector signals are identical in amplitude, but opposite in phase. Capacitor C2 provides a signal path for the base of Q2.

The values for components of the Fig. 5–19 circuit are found using the guidelines established in Chapters 1 and 3. However, remember that RE *passes twice the current* usually found in a single-stage amplifier, all other factors being the same. This is because the emitter–base and emitter–collector currents of both transistors must pass through RE. Thus, RE is normally one-half the value of a single-stage design for a given base–emitter voltage relationship.

5–7.2 Cascode Amplifier

The basic cascode amplifier is shown in Fig. 5–20. The cascode amplifier is a form of direct coupling used primarily in RF circuits (Chapter 4). The cascode circuit is quite popular in vacuum-tube amplifiers, but is less widely used in solid-state RF equipment.

The input signal is fed to the base of Q1, which is a conventional common-emitter amplifier. Variations in signal at the base of Q1 appear as amplified variations in voltage at the collector of Q1.

The collector of Q1 is coupled directly to the emitter of Q2, which operates as a common-base amplifier. The base is grounded as far as the signal is concerned through C2. A grounded or common-base amplifier does not require neutralization, as discussed in Chapter 4.

One problem in solid-state cascode amplifiers is the bias network. The collector of Q1 and the emitter of Q2 are at the same potential. Generally, this potential is less than the supply voltage, but can still be well above ground. To function properly, the base of Q2 should be within about 1 V of the emitter. This bias is provided by the network of R1 and R2.

5–7.3 Power-Complementary Amplifier

The basic complementary amplifier described in Secs. 5–1 and 5–2 can be used in audio power applications when a third power transistor is added. Such an arrangement is shown in Fig. 5–21, where a PNP/NPN pair is followed by a power PNP transistor.

Using the values shown and assuming that Q3 is capable of handling about 5 W, the circuit of Fig. 5–21 can deliver about 5 W of stable power at audio frequencies. Although the voltage gain is low (about 10), the power gain is high (about 100,000), since there is a large difference between input and output impedances.

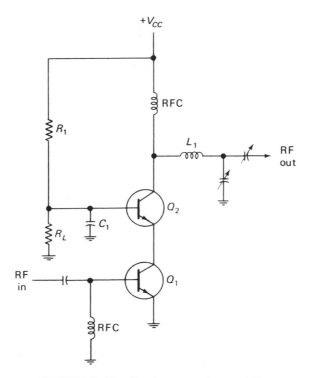

FIGURE 5-20 Basic cascode amplifier.

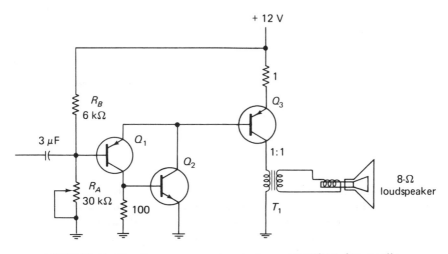

FIGURE 5-21 Power complementary amplifier for audio frequencies.

The input impedance is approximately equal to the value of RB. The output impedance is equal to the impedance of the transformer. If the loudspeaker has an 8-Ω impedance, and T1 has a 1 : 1 ratio, the output impedance is 8 Ω.

The circuit is adjusted for class A operation by the value of RA. In practice, the value shown (30 kΩ) is used as a starting point and is adjusted for the desired class A operation and/or a specific Q point (collector voltage of Q3) under no-signal conditions. Note that Q3 must be operated with a heat sink (Chapter 2) since the power dissipation is in excess of 5 W.

5-8 SERIES-OUTPUT AND COMPLEMENTARY AMPLIFIERS

Three classic direct-coupled circuits are often found in discrete-component amplifiers: the series output, the quasi-complementary, and the full complementary. In present design, these circuits are usually found in IC form (or have been replaced by similar circuits in IC form), so no detailed design examples are given here. However, the circuits are included for reference.

5-8.1 Series-Output Amplifier

Figure 5–22 shows two typical series-output amplifiers used in audio systems. One configuration requires two power supplies, but omits the coupling capacitor to the load. This configuration provides better low-frequency response (since there is no capacitor), but can be inconvenient because of the two power supplies. The configuration with a single power supply has reduced low-frequency response since the coupling capacitor forms a high-pass filter with the load resistance.

Either configuration of series output has two drawbacks. A phase inverter (Sec. 3–13) is required to drive the series-output stage, even if gain is not required. Also, an additional driver stage (or possibly an emitter-coupled amplifier, Sec. 5–7.1) may be necessary to bring the power up from the output of voltage amplifier to a level required by the series-output stage.

These problems are overcome by means of a *complementary* circuit, either quasi-complementary or full complementary.

5-8.2 Quasi-Complementary Amplifier

Figure 5–23 shows a quasi-complementary output. This circuit consists of a Darlington compound using NPN transistors and a direct-coupled complementary pair using an input PNP and an output NPN. Both base signals can be in phase (although the signals are often at different voltage levels) so that a phase inverter or emitter-coupled amplifier is not needed.

For example, if the input is positive going, Q1 is forward biased, as is Q2. A positive-going input at Q3 reverse biases Q3, since Q3 is PNP. This pro-

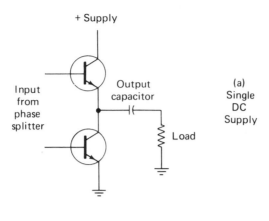

Input from phase splitter

+ Supply

Output capacitor

Load

(a)
Single
DC
Supply

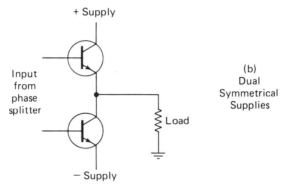

Input from phase splitter

+ Supply

Load

− Supply

(b)
Dual
Symmetrical
Supplies

FIGURE 5-22 Typical series-output amplifiers.

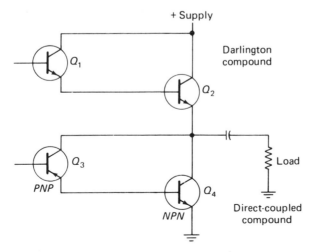

+ Supply

Q_1

Q_2

Darlington compound

Q_3

PNP

Q_4

NPN

Load

Direct-coupled compound

FIGURE 5-23 Quasi-complementary amplifier.

duces a negative-going output from Q3 to Q4 (an NPN) and reverse biases Q4. When the signal at the bases of Q1 and Q3 is negative going, the condition is reversed (Q3 is reverse biased, Q4 is forward biased).

5-8.3 Full-Complementary Amplifier

Figures 5–24 and 5–25 show two versions of the full-complementary output. Either version has an advantage over the quasi-complementary circuit in that both halves of the circuit are identical. This makes it easier to match both halves (for positive and negative signals) to minimize distortion that could be caused by uneven amplification of the signal.

The circuit of Fig. 5–25 uses two Darlington compounds and is often called a *dual-Darlington output*. The amplifier of Fig. 5–24 is used when *power gain* is needed.

Phase inversion is not needed for either circuit. A positive-going input forward biases Q1 and Q2 and reverse biases Q3 and Q4. A negative-going input produces the opposite results.

As in the case of series-output circuits (Sec. 5–8.1), the load-coupling capacitor can be omitted if two power supplies are used (one positive and one negative with respect to ground or common). The trade-off between the convenience of two power supplies, versus improved low-frequency response, must be decided by circuit requirements.

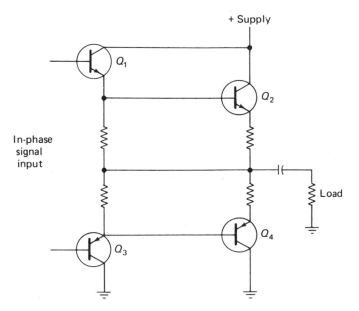

FIGURE 5–24 Full-complementary amplifier with dual-Darlington output.

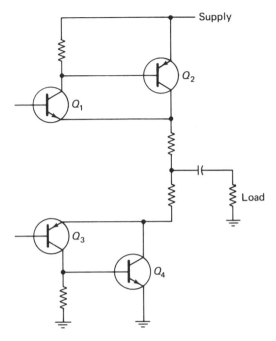

FIGURE 5-25 Full-complementary amplifier with direct-coupled output.

As a guideline, a 2000-μF capacitor working into a 4-Ω load (such as a 4-Ω loudspeaker) produces an approximate 3-dB drop at 20 Hz (a 10-V output drops to 7 V).

5-9 DIRECT-COUPLED IC AMPLIFIERS

Most ICs use some form of direct coupling. It is difficult to fabricate a capacitor on an IC chip, and it is next to impossible to form a coil or transformer within any IC. These problems are eliminated in ICs by direct coupling between stages.

Direct-coupling IC amplifiers operate at all frequency ranges, from direct current to RF. Of course, coils, capacitors, and transformers are external components connected to the IC. The transistors and resistors associated with the amplifier circuit are part of the IC.

Many IC amplifiers use the *differential-amplifier* principle. For that reason, we describe some typical direct-coupled IC amplifiers in Chapter 6.

6

DIFFERENTIAL AMPLIFIERS

The differential amplifier is similar to the emitter-coupled amplifier (Fig. 5–19), except that the two output signals are the result of a *signal difference* between the two inputs. In a theoretical differential amplifier, no output is produced when the signals at the inputs are identical. That is, an output is produced *only when there is a difference in signals at the input.*

One main use for differential amplifiers is as the input stage for an *operational amplifier,* or op-amp, as is described in Chapter 7. Another use for a differential amplifier in laboratory work is that of an amplifier for meters, oscilloscopes, recorders, and the like. Such instruments are operated in areas where many signals may be radiated (power-line radiation, stray signals from generators, and the like). Test leads connected to the input terminals pick up these radiated signals, even when the leads are shielded.

If a single-ended input is used, the undesired signals are picked up and amplified along with the desired signal input. If the amplifier has a differential input, both leads pick up the *same radiated signal* at the same time. Since there is no difference between the radiated signals at the two inputs, there is no amplification of the undesired inputs.

Signals common to both inputs (such as radiated signals) are known as *common-mode signals.* The ability of a differential amplifier to prevent conversion of a common-mode signal into a difference signal (which produces an output) is expressed as the *common-mode rejection ratio* (CMR or CMRR).

6-1 BASIC DIFFERENTIAL-AMPLIFIER THEORY

Figure 6-1 shows a basic differential amplifier. The circuit responds differently to common-mode signals than to single-ended signals.

A common-mode signal (such as power-line pickup) drives both bases in phase with equal-amplitude a-c voltages, and the circuit acts as though the transistors are in parallel to cancel the output. In effect, one transistor cancels the other.

Non-common-mode signals are applied to either of the bases (Q1 or Q2). The *inverting input* is applied to the base of transistor Q2, and the *noninverting input* is applied to the base of Q1. With a signal applied only to the inverting input and the noninverting input grounded, the output is an amplified and inverted version of the input. For example, if the input is a positive pulse, the

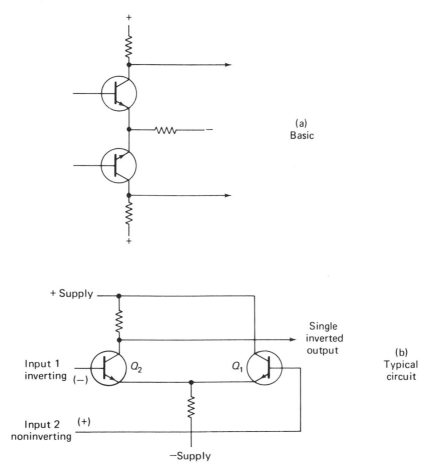

(a)
Basic

(b)
Typical
circuit

FIGURE 6-1 Basic differential amplifier.

output is a negative pulse. If the noninverting input is used, with the inverting input grounded, the output is an amplified version of the input (but without inversion).

The emitter resistor introduces emitter feedback to both transistors simultaneously. This reduces the common-mode signal gain without reducing the differential signal gain in the same proportion.

Figure 6–2 shows a more practical differential amplifier (packaged in IC form). The circuit is basically a single-stage differential amplifier (Q2 and Q4) with input emitter followers (Q1 and Q5) and *constant-current source* Q3 in the emitter-coupled leg. Note that the single emitter resistor of the Fig. 6–1 circuit is replaced by Q3 in the circuit of Fig. 6–2 and that direct coupling (Chapter 5) is used throughout the IC circuits.

The use of a transistor such as Q3 is typical for many differential amplifiers, particularly those found in op-amps (as the first stage). The circuit

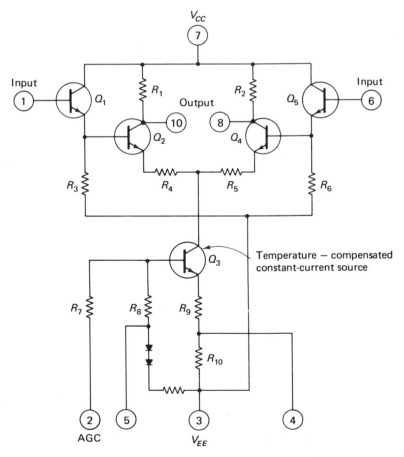

FIGURE 6-2   Differential amplifier in IC form.

of Q3 is known as a *temperature-compensation constant-current source*. All current for the differential amplifier is fed through Q3 (an NPN) connected between the emitters of the differential amplifier and the negative power supply (VEE).

If there is an increase in current, a larger voltage is developed across the current-source Q3 emitter resistor. This larger voltage acts to reverse bias the base/emitter junction, thus reducing current through Q3. Since all current for the differential amplifier is passed through Q3, current to the amplifier is also reduced. If there is a decrease in a current, the opposite occurs, and the amplifier current increases. Thus, the differential amplifier is maintained at a constant current level.

Transistor Q3 is also *temperature compensated* by the diodes connected in the base–emitter bias network. These diodes have the same (approximate) temperature characteristics as the base–emitter junction and offset any change in Q3 base–emitter current flow that results from temperature change.

The circuit of Fig. 6–2 has an input impedance of about 100 kΩ (for each input and a gain of about 30 dB at frequencies up to about 1 MHz. The useful frequency response can be increased by means of external resistors and coils.

The degenerative-feedback resistors R4 and R5 in the emitter-coupled point of transistors increase the linearity of the circuit. The low-frequency output impedance between each output (terminals 8 and 10) and ground is essentially the value of the collector resistor R1 and R2 in the differential stage.

6-2 COMMON-MODE DEFINITIONS

The terms *common mode* and *common-mode rejection* are used frequently in differential-amplifier applications. All manufacturers do not agree on the exact definition of common-mode rejection.

One manufacturer defines common-mode rejection (CMR, or sometimes listed as CMrej) or the common-mode rejection ratio (CMRR) as the ratio of differential gain (usually large) to common-mode gain (usually a fraction). That is, the amplifier may have a large gain of differential signals (different signals at each input terminal, or one input terminal grounded and the opposite input terminal with a signal), but little gain (or possibly a loss) of common-mode signals (same signal at both terminals).

Another manufactuere defines CMR as the relationship of *change* in the output voltage to the *change* in the input common-mode voltage producing the change, divided by the open-loop gain (amplifier gain without feedback).

For example, using the latter definition, assume that the common-mode input (applied to both terminals simultaneously) is 1 V, the resultant measured output is 1 mV, and the open-loop gain is 100. The CMR is then

$$\frac{\text{output/input}}{\text{open-loop gain}} = \text{CMR}$$

$$\frac{0.001/1}{100} = 100,000 = 100 \text{ dB}$$

Another method to calculate CMR is to divide the output signal by the open-loop gain to find an *equivalent differential input signal*. Then the common-mode input signal is divided by this equivalent differential input signal. Using the same figures as in the previous CMR calculation,

$$\frac{0.001}{100} = 0.00001, \quad \text{equivalent differential input signal}$$

$$\frac{1}{0.00001} = 100,000 \quad \text{or} \quad 100 \text{ dB}$$

No matter what basis is used for calculation, CMR is an indication of the *degree of circuit balance* of the differential stages, since a common-mode input signal should be amplified identically in both halves of the circuit. A large output for a given common-mode input is an indication of large imbalance or poor CMR. If there is an imbalance, a common-mode signal becomes a differential signal after passing the first stage (or at the output of a single-stage differential amplifier).

As with amplifier gain, CMR usually decreases as frequency increases. However, as a guideline, the CMR of any differential amplifier should be *at least 20 dB greater than the open-loop gain at any frequency* (within the limits of the amplifier).

6-3 FLOATING INPUTS AND GROUND CURRENTS

Since a differential amplifier is sensitive only to the difference between two input signals (in theory), the signal source need not be grounded and can be *floating*. Therefore, differential amplifiers are often used in test-equipment applications where the signal source is from a bridge (such as a bridge-type transducer) and the power supply is grounded.

The differential amplifier also allows injection of a fixed d-c voltage into either channel to permit establishment of a new voltage reference level at the output (some point other than 0 V). This is commonly referred to as *zero suppression* and is discussed further in Chapter 7.

A floating-input circuit can create problems. When the input is floating, cable shielding between the amplifier and signal source may be connected to chassis ground rather than to signal ground. However, both a-c and d-c voltages can exist between two widely separated earth grounds, causing current to flow. (Such currents are known as *ground currents,* and the circuits producing the current flow are known as *ground loops.*

6-3.1 Basic Ground-Current Condition

Figure 6–3 shows the basic ground-current condition (where a bridge-type transducer is used with a differential transistor, in this example). Note that the signal source is connected to the transducer earth ground (*local ground* or *physical ground,* as the ground may sometimes be called). This ground point is connected to the amplifier ground through the cable shielding. The amplifier ground is connected to one of the differential inputs through the internal capacitance (represented as Cd) of the amplifier, even though there may be no d-c connection between ground and the input terminal of a floating-input amplifier.

The same differential input terminal is connected to the signal source through the signal leads and the transducer elements (bridge resistors, in this example). Thus, the a-c ground currents are mixed with the signal currents. This can result in an imbalance of the differential amplifier. Also, radiated signals picked up by the shield appear as undesired differential signals, rather than common-mode signals, and produce an undesired output.

6-3.2 Guard-Shield Technique

Many methods are used to minimize the ground-current condition. One method is shown in Fig. 6–4. Here, a guard shield is placed around the input circuits of the differential amplifier. This not only shields the amplifier from radiated signals, but also provides an electrostatic shield to break the internal capacitance Cd into two series capacitances Cda and Cdb. A much higher impedance is then

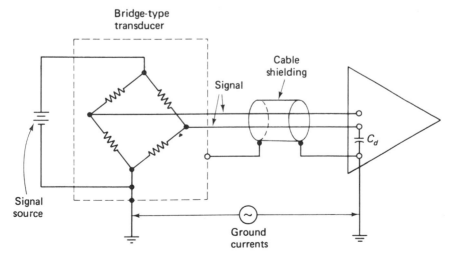

FIGURE 6-3 Basic ground-current condition.

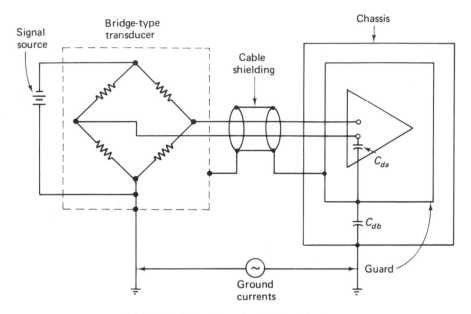

FIGURE 6-4 Guard-shield technique.

presented to the flow of a-c ground signals. This type of amplifier is termed a *floated-input* and *guarded* amplifier.

6-4 DIFFERENTIAL INPUTS AND OUTPUTS

Differential amplifiers can have differential inputs with differential outputs, differential inputs with single-ended output, or single-ended input with differential output, depending on the particular application. The differential input is usually required where there are high common-mode signals or where the source is floating (no reference to ground). The differential output is required when both amplitude and polarity must be considered.

A *laboratory differential voltmeter* is a classic example of this latter requirement. Such meters must amplify very small differential signals in the presence of large common-mode signals. The output of the meter amplifier is fed to the meter movement or to a digital readout, where both the amplitude and polarity of the differential signal are indicated (as a plus or minus voltage readout), but common-mode signals are rejected.

A differential output amplifier can be made using two amplifiers. However, to be really effective both amplifiers must be *identical* in characteristics (gain, temperature, drift, and so on). This is best done using a dual-channel amplifier, in differential form, with both channels fabricated on the same semiconductor chip in the same IC. We discuss both discrete-component and IC differential amplifiers in the following paragraphs.

6-4.1 Typical Discrete-Component Differential Amplifier

Figure 6–5 shows a differential amplifier used in a typical discrete-component laboratory instrument. Note that FETs are used at the input (for high input impedance), whereas the output uses two-junction emitter followers (for low output impedance). Direct coupling (Chapter 5) is used throughout to ensure good low-frequency response.

In either design or troubleshooting of any differential amplifier, a consideration of major importance is that both halves of the amplifier be *electrically symmetrical.* That is, the input impedance at point A (Fig. 6–5) should be exactly equal to the input impedance at point B.

Resistors R1/R3 form an input attenuator, with an impedance equal to R1 + R3. The voltage attenuation is equal to the ratio R3/(R1 + R3). Similarly, R2 and R4 form a division ratio of R4/(R2 + R4). The sum of R1 + R3 is typically about 1 MΩ.

Resistors R5 and R6 (typically in the range of 100 kΩ) limit the input current to protect Q1 and Q2. Capacitors C1 and C2 couple a-c signals directly to the gates of Q1 and Q2 and thus provide high-frequency peaking. (At high frequencies, the impedance offered by R5 and R6 could reduce the input signals.)

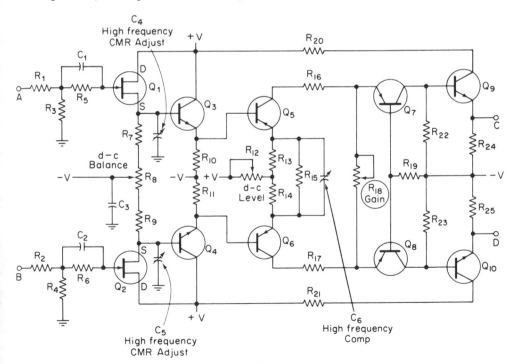

FIGURE 6-5 Differential amplifier in a discrete-component laboratory instrument.

6-4.2 *Typical IC Differential Amplifier*

Figure 6–6 shows a differential amplifier in IC form. Such an IC can replace the majority of circuits shown in Fig. 6–5. Of course, the adjustment controls in Fig. 6–5 must be external to the IC of Fig. 6–6. For example, the high-frequency CMR adjust capacitors C4 and C5 of the Fig. 6–5 circuit can be connected to pins 1 and 2 of the Fig. 6–6 circuit.

In the Fig. 6–6 circuit, transistors Q2 and Q4 amplify (differentially) the input signal applied between pins 9 and 10. The emitter-follower input buffering, provided by Q1 at the noninverting input and Q3 at the inverting input, is used to get the high input impedance to minimize the input capacitance and to reduce the dependence of offset and impedance variations on device current gains.

Temperature-Compensated Current Source. Transistor Q5 serves as the temperature-compensated current source, used primarily to provide a high common-mode rejection ratio. A common-mode feedback loop is incorporated around the input stage to provide Q-point operating stability for the input devices and thus ensure a more predictable input common-mode voltage swing. The base of Q7 and collector of Q8 are wire bonded to pins 1 and 7 (as are the base of Q9 and collector of Q10 to pins 2 and 6) to provide for open-loop frequency compensation capacitors (or for external adjustment controls).

Level Shifting. A resistive level shifter follows the input stage to maintain the desired voltage levels at the overall IC output. (Level shifters are discussed further in Sec. 6–7.) The second differential-gain stage also contains emitter-follower buffering to minimize losses in the resistive level shifter and loading of the first-stage collectors.

Output Impedance and Swing. Low output impedance and large output swing are provided at the output by differential emitter followers Q15/Q17. To improve peak negative swing of the amplifier, a current source Q18 (rather than the usual emitter resistor) is used to bias the emitter follower.

Stabilization of Output. Stabilization of the output Q point is obtained by a local common-mode feedback loop incorporating a separate differential amplifier stage Q13/Q14. This circuit holds average voltage output at ground potential (for split-supply operation) and at VCC/2 (for single power-supply operation).

Drift. Changes in output Q point (also known as *no-signal drift* or *zero drift)* are minimized by temperature-compensation diodes D2/D3 in the base circuits of the feedback differential amplifier.

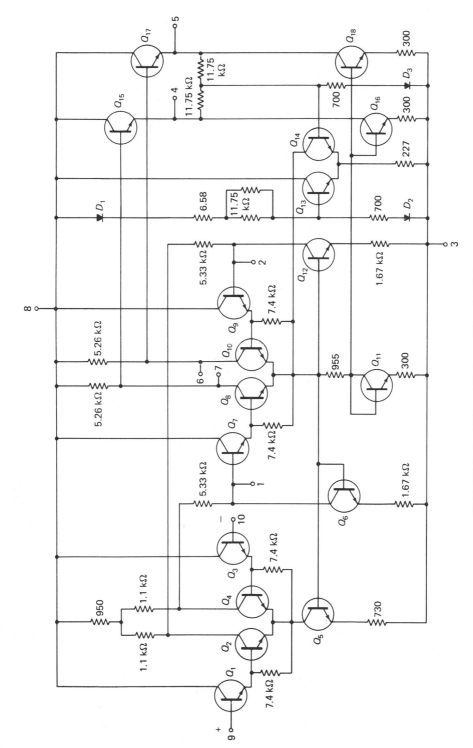

FIGURE 6-6 Differential amplifier in IC form.

257

Additional Applications. The IC differential amplifier of Fig. 6–6 is not limited to laboratory instruments. The IC can be used wherever a high-gain, wide-frequency-range differential amplifier is required.

Figure 6–7 shows the connections for both balanced-input and unbalanced-input operation. The recommended values for the external resistors and capacitors, together with corresponding gain and frequency ranges, are given in the IC datasheet. The IC can also be used as a single-ended op-amp (Chapter 7) if desired.

Balanced input - differential output

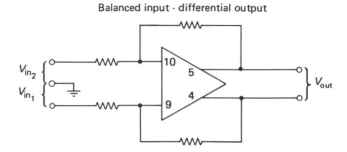

Single-ended input - differential output

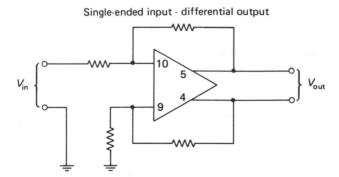

FIGURE 6-7 Connections for balanced input and unbalanced input.

6-5 ADJUSTMENT CONTROLS IN DIFFERENTIAL AMPLIFIERS

Note that there are many adjustment controls in the circuit of Fig. 6–5. To properly analyze the circuit for design or troubleshooting, the effects of these controls must be known.

For example, in troubleshooting it may be possible to eliminate an apparent major fault simply by adjustment of the controls. The following descriptions are applicable to the specific controls of Fig. 6–5. However, similar controls (possibly identified by different names) appear on many differential amplifiers in laboratory equipment.

6-5.1 Frequency Compensation

Capacitors C4 and C5 (high-frequency CMR adjust) shunt the high-frequency signals to ground. Capacitor C6 (high-frequency compensation) shunts the high-frequency signals between the emitters of Q5 and Q6. These three controls form the high-frequency compensation adjustments for the amplifier. When properly adjusted, the high-frequency signals are attenuated so that the response is flat over the desired range.

6-5.2 Balance

The d-c balance potentiometer R8 compensates for any inherent difference between the two halves of the input circuit. It is impossible to match the two halves perfectly. Also, component values change with age and thus produce an imbalance.

Generally, R8 is a very temperature stable potentiometer and is mechanically secure so as not to lose the setting with any vibration. (Sometimes the potentiometer shaft is provided with a locknut to prevent a change in setting due to vibration.)

6-5.3 DC Level and Gain

The function of R12 (d-c level) and R18 (gain) are often confused. Potentiometer R12 *sets the level of the d-c voltage at the output,* whereas R18 *sets the overall gain of the amplifier.* It is quite possible for the amplifier to operate at the correct output level, but not to provide the necessary gain, and vice versa.

6-5.4 Temperature Drift

To prevent temperature changes from causing the amplifier d-c levels to drift, transistors Q1 through Q4 are mounted on the same heat sink. (This problem is eliminated in an IC.) Sometimes Q1 through Q4 are enclosed by a metal can so that all remain at the same temperature. In some cases, Q5 and Q6 may also share the same heat sink, but usually are separated from Q1 through Q4. However, transistors Q1 through Q4 are invariably *matched pairs* and should be replaced (for troubleshooting) or selected (for design) as such (in discrete-component differential amplifiers).

6-5.5 Cascode Amplifiers

Transistors Q5 and Q7 (as well as corresponding transistors Q6 and Q8) form a cascode amplifier (Sec. 5–7.2), with the result that the input to Q5 is voltage, whereas the output of Q5 into Q7 is current. The output of Q7 is voltage again, since the input impedance of Q7 is quite low, whereas the Q7 output impedance is quite high. Transistors Q9 and Q10 are emitter followers used to reduce the high impedance from Q7 to a low output impedance.

6-6 FAILURE PATTERNS FOR DIFFERENTIAL AMPLIFIERS

The four most common failures of a differential amplifier are loss of gain (voltage or current), poor common-mode rejection, d-c imbalance, and output-signal drift (dc). All these conditions are discussed in the following paragraphs. Of course, in IC equipment, remember that the IC must be replaced even if only one fault is traced to the IC and all other circuit performance is good.

6-6.1 Loss of Gain (Voltage or Current)

If there is a lack of gain in a particular design, the cause is usually one of incorrect component values (low RL/RE ratio for example), or possibly a problem of low beta (or poor Yfs in FETs). Such conditions can only be corrected by trial-and-error substitution or in experimental circuits.

In an existing differential amplifier, if a loss of gain is gradual, component aging is usually at fault. Always try to correct a loss-of-gain problem, frequency-response problem, or an improper output level by adjustment. Gradual losses are not common in well-designed equipment. Also, usually only half the circuit loses gain, thus unbalancing the amplifier. Any differential amplifier that requires continual adjustment of the balance control is suspect. If there is a total breakdown or extreme imbalance, check the active elements in the usual manner. Refer to Chapter 10.

6-6.2 Poor Common-Mode Rejection

The most common cause of poor common-mode rejection is unsymmetrical gain. This is one of the reasons for including the CMR adjustment capacitors (C4/C5 of Fig. 6-5).

Always try to correct a CMR problem by adjustment first. Do not forget to include the balance adjustments, since any imbalance is reflected as poor CMR ratio. Another point to remember is that *attenuator probes, attenuator networks,* and the like, must be perfectly matched with regard to both alternating and direct current to get good CMR ratio.

Some laboratory instruments are provided with attenuator networks and probes, usually at the input circuit. Although usually not part of the amplifier, attenuator operation has a direct effect on amplifier performance. Try to avoid using external attenuator in calibrating a differential amplifier for CMR. (Unmatched external calibrators simply add one more unknown.)

6-6.3 Direct-Current Imbalance

Problems of d-c imbalance are almost always because of component aging or mismatched temperature coefficients of symmetrical components. That is, the components can age faster in one-half of one stage, thus unbalancing the entire amplifier circuit. The same applies to an imbalance introduced by a mismatched temperature coefficient.

If many stages are used in the amplifier, start by balancing the *last stage first*. Make the inputs to the last stage equal (tie both halves together, ground both inputs, or apply the same signal to both inputs, depending on the specific circuit) and adjust the balance control *for that stage.*

In many cases, the last stage does not have a balance control, but can be checked for balance. For example, in Fig. 6–5 the voltage drop across R20 should be the same as across R21, the drop across R24 should be the same as R25, and so on.

Continue this checking process, moving toward the first stage and always monitoring the *main output,* until the trouble is located. Usually, this sequence can isolate a badly unbalanced section.

If the circuit is part of an oscilloscope (differential amplifiers are often found in laboratory scopes), an imbalance is easily detected by grounding the two inputs and rotating the scope gain control (often called the *vernier* control). An imbalance shows up in the form of *trace shift* (where the scope trace shifts vertically as the gain control is changed, even though both inputs are at zero).

Trace shift is the result of the d-c signal being developed across the gain control (because of imbalance). The amplifier can then be balanced by adjusting the amplifier d-c balance potentiometer until no trace shift is observed when the gain control is rotated.

6-6.4 Output Signal Drift

A slow drift of any kind at the output is usually linked to a d-c imbalance. Temperature compensation (or a lack thereof) plays a major role in such cases. A well-designed differential circuit is seldom plagued by temperature problems (especially in IC form). However, occasionally, circuits may be overheated, and solid-state components can be partially damaged (changing the temperature coefficients, but not destroying the component).

When replacing parts (for troubleshooting) or selecting parts (for design), always make sure that both halves of the amplifier are well matched in *all characteristics* (gain, temperature coefficient, and so on), especially in the first stage of the amplifier (even if the first stage is an IC). Minor imbalances in the first stage of a differential amplifier are amplified by each stage thereafter, possibly resulting in a major imbalance at the output.

6-7 DIFFERENTIAL AMPLIFIERS AS IC OP-AMPS

As discussed in Chapter 7, op-amps are high-gain direct-coupled circuits where the *gain and frequency response are controlled by external feedback networks.* With these networks, the op-amp can be used to produce a broad range of intricate transfer functions and thus may be adapted for use in many widely different applications.

Although the op-amp was originally designed to perform various mathematical functions (differentiation, integration, analog comparison, and summation) the op-amp (by modification of the feedback network) may be used to provide the broad, flat, frequency-gain response of *video amplifiers* or the peaked response of various types of *shaping amplifiers.* This capability makes the op-amp a most versatile form of amplifier.

6-7.1 Balanced Differential Op-Amp

The most common type of IC op-amp uses a balanced differential-amplifier circuit such as shown in Fig. 6–8. The basic purpose of such a circuit is to produce an output signal that is linearly proportional to the difference between two signals applied to the input. The circuit provides an overall open-loop gain of about 2500.

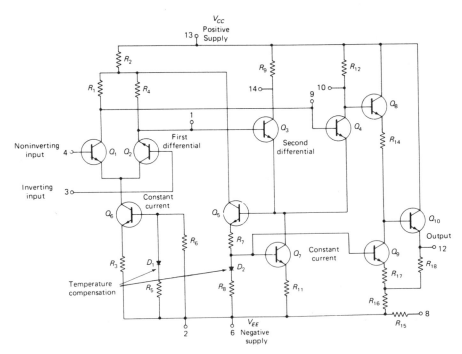

FIGURE 6-8 Balanced differential amplifier in IC op-amp form.

As discussed, in a theoretical differential amplifier, no output is produced when the signals at the inputs are identical. Thus, differential amplifiers are particularly useful for op-amps, because common-mode signals are eliminated or greatly reduced.

6-7.2 Op-Amp Circuit Description

As shown in Fig. 6–8. the op-amp consists basically of two differential amplifiers and a single-ended output circuit in cascade. The pair of cascaded differential amplifiers provides most of the gain.

Inputs. The inputs to the op-amp are applied to the bases of the pair of emitter-coupled transistors Q1 and Q2 in the differential amplifier. With the noninverting input grounded and signals applied only to the inverting input, the output is an *amplified* and *inverted* version of the input.

For example, if the input is a positive pulse, the output is a negative pulse. If the noninverting input is used, with the inverting input grounded, the output is an amplified version of the input (without inversion). Transistors Q1 and Q2 develop driving signals for the second differential amplifier.

Constant Current. Transistor Q6, a constant-current device, is included in the first differential stage to provide bias stabilization for Q1 and Q2. If there is an increase in supply voltage which normally increases the current through Q1/Q2), the reverse bias on Q6 increases. This reduces the Q1/Q2 current and offsets the effect of the initial supply-voltage increases.

Thermal Compensation. Diode D1 provides thermal compensation for the first differential stage. If there is an increase in operating temperature of the IC (which normally increases the current through Q1/Q2), there is a corresponding current flow through D1, since the diode is fabricated on the same silicon chip as the transistors. The increase in D1 current also increases the reverse bias of Q6, thus offsetting the initial change in temperature.

Second Differential Amplifier. The emitter-coupled transistors Q3 and Q4 in the second differential amplifier are driven push–pull by the outputs from the first differential amplifier. Bias stabilization for the second differential amplifier is provided by current transistor Q7. Compensating diode D2 provides the thermal stabilization for the second differential amplifier and also for the current transistor Q9 in the output stage.

Negative Feedback. Transistor Q5 develops the negative feedback to reduce common-mode error signals that are developed in the emitters of transistors Q3/Q4. Because the second differential-amplifier stage is driven push–pull, the signal at this point (the emitters) is zero when the first differential-amplifier stage and the base–emitter circuits of the second stage are matched, and there is no common-mode input.

A portion of any common-mode, or error, signal that appears at the emitters of transistors Q3/Q4 is developed by transistor Q5 across resistor R2 (the common-collector resistor for transistors Q1, Q2, and Q5) in the proper phase to reduce the error. The emitter circuit of transistor Q5 also reflects a portion of the same error signal into current transistor Q7 in the second differential stage so that the initial error signal is further reduced.

Supply-Voltage Variations.   Transistor Q5 also develops feedback signals to compensate for common-mode effects produced by variations in the supply voltage. For example, a decrease in the voltage at the emitters of Q3/Q4. This negative-going change in voltage is reflected by the emitter circuit of Q5 to the bases of current transistors Q7/Q9. Less current then flows through these transistors.

The decrease in collector current of Q7 results in a reduction of the current through Q3/Q4, and the collector voltages of these transistors tend to increase. This tendency partially cancels the decrease that occurs with the reduction of the positive supply voltage. The partially canceled decrease in collector voltage of Q4 is coupled directly to the base of Q8 and is transmitted by the emitter of Q8 to the base of Q10. At this point the decrease in voltage is further canceled by the increase in the collector voltage of Q9. Because of the feedback stabilization provided by Q5, the IC op-amp has high common-mode rejection, excellent open-loop stability, and low sensitivity to power-supply variation.

Output.   The output from the op-amp is taken from the emitter of Q10 so that the d-c level of the output signal is substantially lower than that of the differential-amplifier output at the collector of Q4. In this way, the circuit shifts the output d-c level to the same level as the input (when no signal is applied).

6-7.3 DC Level-Shifting Problems

In any cascade direct-coupled amplifier, either discrete component of IC, the d-c level rises through successive stages toward the supply voltage. In linear ICs, the d-c voltage builds up through the NPN stages in the positive direction and must be shifted negatively if large output signal swings are required. For example, if the supply voltage is 10 V and the output is at 9 V under no-signal conditions, the maximum output voltage swing is limited to less than 1 V.

With multistage high-gain ICs, such as op-amps and many other multifunction amplifiers (which use external feedback), it is especially important to provide for compensation of the d-c level shift. Such amplifiers must have equal (and preferably zero) input and output d-c levels so that the d-c coupling of the feedback connection does not shift any bias point. For example, if the input is at 0 V, but the output is at 3 V, this 3 V is reflected back through an external feedback resistor and changes the input to a 3-V level.

The use of an output stage, such as shown in Fig. 6–8, is a commonly used technique to prevent a shift in d-c level between the output and input of an IC. Transistor Q8 operates as an input buffer, and transistor Q9 is essentially a current sink for Q8. The shift in d-c level is accomplished by the voltage drop across resistor R14 (produced by the collector current of transistor Q9). The emitter of the output transistor Q10 is connected to the emitter of Q9. Feedback through R18 results in a decrease of the voltage drop across R14 for negative-going output swings and an increase in this voltage drop for positive-going output swings.

The circuit of Fig. 6–8 provides many connections to internal circuit components. For example, feedback may be coupled from the output to the input to stabilize gain (or to compensate for common-mode effects that result from variations in supply voltages). Also, the additional connections (such as at the collectors of Q1, Q2, Q3, Q4, and Q9) provide a variety of input/output points for the *phase-lead* and *phase lag* compensation techniques described in Chapter 7.

7

OPERATIONAL AMPLIFIERS

The designation operational amplifier or op-amp was originally adopted for a series of high-performance direct-coupled amplifiers used in analog computers. These amplifiers were used to perform mathematical operations in analog computation (summation, scaling, subtraction, integration, and so on). Today, the availability of inexpensive IC amplifiers has made the packaged op-amp useful as a replacement for any low-frequency amplifier.

Because such packaged circuits are readily available in great variety at low cost, no detailed circuit-design procedures are given here. Instead, we concentrate on how to select external components to perform a given function, how to connect external power sources, and how to interpret IC op-amp datasheets.

Most of the basic design information for a particular op-amp can be obtained from the datasheet. Likewise, a typical datasheet describes a few specific applications for the op-amp. However, op-amp datasheets generally have two weak points. First, the datasheets do not show how the listed parameters relate to design problems. Second, the datasheets do not describe the great variety of applications for which a basic op-amp can be used.

In any event, it is always necessary to interpret op-amp datasheets. Each manufacturer has a separate system of datasheets. It is impractical to discuss all datasheets here. Instead, we discuss typical information found on op-amp datasheets and see how this information affects simplified design.

7-1   BASIC IC OP-AMP

IC op-amps generally use several differential stages in cascade to provide both common-mode rejection (Chapter 6) and high gain. Differential amplifiers require both positive and negative power supplies and have two inputs to provide for in-phase or out-of-phase amplification.

Many op-amp applications require that the output be fed back to the input through a resistance or impedance. The output is fed back to the negative or inverting input to produce degenerative feedback (and thus produce the desired gain and frequency response).

As in any amplifier, the signal shifts in phase through an op-amp when passing from input to output. The amount of phase shift depends on frequency. When the phase shift approaches 180°, the signal adds to (or cancels out) the 180° feedback phase shift. Thus, the feedback is in phase with the input (or nearly so) and causes the amplifier to oscillate.

This condition of phase shift with increased frequency limits the bandwidth of an op-amp. The condition can be compensated for by adding a *phase-shift network* (usually an RC circuit, but sometimes a single capacitor).

7-1.1   *Typical IC Op-Amp Circuit*

Figure 7-1 shows the circuit diagram and equivalent circuit (or symbol) as they appear on the datasheet of a typical IC op-amp. This circuit is a three-stage amplifier, with the first stage a differential-in, differential-out amplifier designed for high gain and high-common-mode rejection and input overvoltage protection. The input diodes prevent damage to the circuit should the input terminals be accidentally connected to the power-supply leads (or other undesired high-voltage source).

The second stage is a differential-in, single-ended-output amplifier with low gain and high-common-mode rejection. Common-mode feedback is used from the second stage back to the first stage, thus providing more control of a common-mode input signal.

Using these two differential-amplifier stages and a common-mode feedback, a typical common-mode rejection of 110 dB is obtained. The third stage is a single-ended amplifier, providing high gain, voltage translation to a ground reference, output current-drive capabilities, and output short-circuit protection.

7-1.2   *Connecting a Power Source to an IC*

Typically, op-amps require connection to both a positive and a negative power supply. This is because most op-amps use one or more differential amplifiers. When two power supplies are required, the supplies are usually equal or sym-

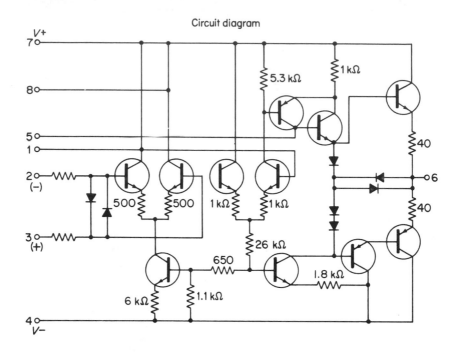

Circuit diagram

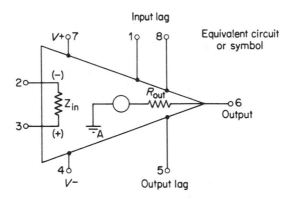

FIGURE 7-1 Typical IC op-amp circuit and symbol.

metrical (such as $+6$ V or -6 V, $+12$ V and -12 V). This is the case with the op-amp of Fig. 7-2, which normally operates with $+12$ and -12 V.

It is possible to operate an op-amp that normally requires two supplies from a single supply by means of special circuits (external to the op-amp). Such circuits are discussed at the end of this section.

IC Op-Amp Power Connection. Unlike most discrete transistor circuits in which it is usual to label one power-supply lead positive and the other negative without specifying which (if either) is common to ground, it is necessary

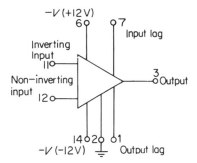

FIGURE 7-2 Typical op-amp with symmetrical 12-V power supplies.

that all IC op-amp power-supply voltages be referenced to a common or ground (which may or may not be physical or equipment ground).

As in the case of discrete transistors, manufacturers do not agree on power-supply labeling for IC op-amps. For example, the circuit shown in Fig. 7–2 uses $+V$ to indicate the positive voltage and $-V$ to indicate the negative voltage. Another manufacturer might use the symbols VEE and VCC to represent negative and positive, respectively. As a result, the op-amp datasheet must be studied carefully before applying any power source.

Typical Op-Amp Power-Supply Connections. Figure 7–3 shows typical power-supply connections for an op-amp. The protective diodes shown are recommended for any power-supply circuit in which the leads could be accidentally reversed. The diodes permit current to flow only in the appropriate direction. The op-amp of Fig. 7–3 requires two power sources (of 12 V each) with the positive lead of one and the negative lead of the other tied to ground or common.

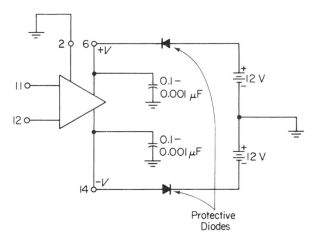

FIGURE 7-3 Typical power-supply connections for op-amp.

The two capacitors shown in Fig. 7–3 provide for decoupling (signal bypass) of the power supply. Usually, disc ceramic capacitors are used. The capacitors should always be connected as close to the op-amp terminals as is practical, not at the power-supply terminals. This is to minimize the effects of lead inductance.

It is particularly important to decouple each op-amp where two or more op-amps share a common voltage supply. A guideline for op-amp power-supply decoupling capacitors is to use values between 0.1 and 0.001 μF.

In addition to the capacitor shown in Fig. 7–3, some op-amp layouts may require additional capacitors on the power lines. The main problem with op-amp power-supply connections is undesired oscillation because of feedback. Most modern op-amps are capable of producing high gain at high frequencies. If there is a feedback path through a common power-supply connection, oscillation occurs.

In the case of IC op-amps, which are physically small, the input, output, and power-supply terminals are close, creating an ideal condition for undesired feedback. To make the problem worse, most op-amps are capable of passing frequencies other than those specified on the datasheet.

For a example, an op-amp to be used in the audio range (say up to 20 kHz with a power gain of 20 dB) could possibly pass a 10-MHz signal. This higher-frequency signal could be a harmonic of signals in the normal operating range and, with sufficient gain, could feed back and produce undesired oscillation.

When laying out any op-amps, particularly IC op-amps in experimental form, always consider the circuit to be RF, even though the op-amp is not supposed to be capable of RF operation and the circuit is not normally used with RF.

Grounding Metal IC Op-Amp Cases.   The metal case of an IC op-amp shown in Fig. 7–3 is connected to terminal 2 and to no other point in the internal circuit. Thus, terminal 2 can and should be connected to equipment ground, as well as to the common or ground of the two power supplies.

The metal cases of some IC op-amps may be connected to a point in the internal circuit. If so, the case is at the same voltage as the point of contact. For example, the case might be connected to pin 14 of the op-amp shown in Figs. 7–2 and 7–3. If so, the case is below ground (or "hot") by 12 V. If the case is mounted directly on a metal chassis that is at ground, the op-amp and power supply will be damaged. Of course, not all IC op-amps have metal cases. Likewise, not all metal cases are connected to the internal circuits. However, this point must be considered before using a particular IC op-amp.

Calculating Current Required for Op-Amps.   The datasheets for op-amps usually specify a nominal operating voltage (and possibly a maximum operating voltage), as well as a "total device dissipation." These figures can be used to calculate the current required for a particular op-amp. Use the simple d-c Ohm's law, and divide the power by the voltage to find the current. However, certain points must be considered.

First, use the actual voltage applied to tne op-amp. The actual voltage should be equal to the nominal operating voltage, but in no event higher than the maximum voltage.

Second, use the total device dissipation. The datasheet may also list other power dissipations, such as "device dissipation," which is defined as the power dissipated by the op-amp (with output at zero and no load). The other dissipation figures are always smaller than the total power dissipation.

Power-Supply Tolerances.   Typically, op-amps operate satisfactorily with ± 20% power supplies. These tolerances apply to actual operating voltage, not to maximum voltage limits. The current (or power consumed) varies proportionately.

Power-supply ripple and regulation are both important. Generally, solid-state power supplies with filtering and full feedback regulation are recommended, particularly for high-gain op-amps. Ideally, ripple (and all other noise) should be 1% or less.

The fact that an op-amp generally requires two power supplies (because of the differential amplifiers) creates a particular problem with regard to *off-set*. For example, if the + V supply is 20% high and the − V supply is 20% low, there is an unbalance and offset, even though the op-amp circuits are perfectly balanced.

The effects of operating an op-amp beyond the voltage tolerance are essentially the same as those experienced when the op-amp is operated at temperature extremes. That is, a high power-supply voltage causes the op-amp to overperform, whereas low voltages result in underperformance. A low voltage usually does not result in damage to the op-amp, as is the case when operating the op-amp beyond the maximum rated voltage.

Single-Power-Supply Operation.   An op-amp is generally designed to operate from symmetrical positive and negative power-supply voltages. This results in a high common-mode rejection capability as well as good low-frequency operation (typically a few hertz down to direct current). If the loss of very low frequency operation can be tolerated, it is possible to operate op-amps from a single power supply, even though designed for dual supplies. Except for the low-frequency loss, the other op-amp characteristics should be unaffected.

The following notes describe a technique that can be used with most op-amps to permit operation from a single power supply, with a minimum of design compromise. The same maximum ratings that appear on the datasheet are applicable to the op-amp when operating from a single-polarity power supply and must be observed for normal operation.

The technique described here is generally referred to as the split-zener method. The main concern in setting up for single-supply operation is to *maintain the relative voltage levels*. With an op-amp designed for dual-supply operation, there are three reference levels: $+V$, 0, and $-V$. For example, if the datasheet calls for $\pm 12 - V$ supplies, the three reference levels are $+12$ V, 0 V, and -12 V.

For single-supply operation, these same reference levels can be maintained using $++V$, $+V$, and ground (that is, $+24$ V, $+12$ V, and 0 V), where $++V$ represents a voltage level double that of $+V$. This is shown in Fig. 7–4, where the op-amp is connected in the split-zener mode. Note that there is no change in the relative voltage levels even though the various op-amp terminals are at different voltage levels (with reference to ground). Terminal 14 (normally connected to the -12-V supply) is at ground. Terminal 2 (normally ground or common) is set at one-half the total zener voltage ($+12$ V). Terminal 6 (normally connected to the $+12$-V supply) is set at the full zener voltage ($+24$ V).

With single supply, the differential input terminals (11 and 12), which are normally at ground in a dual-supply system, must also be raised up one-half

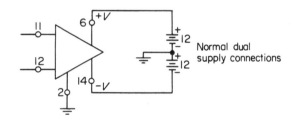

Single-supply connections

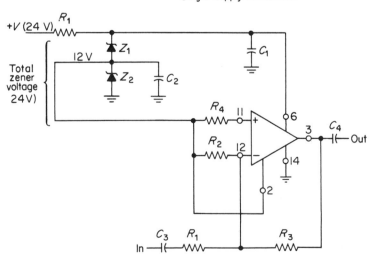

FIGURE 7-4 Connections for single power-supply operation (with ground reference).

the zener voltage ($+12$ V). Under these conditions, the output terminal (3) is at one-half the zener voltage, plus or minus any offset (Sec. 7–3).

To minimize offset errors because of unequal voltage drops caused by the input bias current across unequal resistances, it is recommended that the value of the input offset resistance R4 be equal to the parallel combination of R2 and R3. (The problems of offset correction are discussed fully in Secs. 7–3 and 7–4.)

Any deviation in zener level also contributes to errors in the op-amp output level. Typically, zeners can vary as much as 50 to 100 μV per voltage. However, except in rare cases, this deviation should present no problem in op-amp design.

Note that the op-amp of Fig. 7–4 has a ground reference (terminal 2). Not all op-amps have such terminals. Some op-amps have only $+V$ and $-V$ terminals or leads, even though the two levels are referenced to a common ground. That is, there is no physical ground terminal or lead on the op-amp.

Single Supply without Ground Reference.   Figure 7–5 shows the split-zener connections for single-supply operation for op-amps without ground reference. Here, the input terminals (A and B) are set at one-half the total zener supply voltage: the $-V$ terminal (D) is set at ground, and the $+V$ terminal (F) is set at the full zener voltage ($+24$ V).

Negative Supplies.   Figures 7–4 and 7–5 both show a connection to positive power supplies. Negative supplies can also be used. With a negative supply, the $+V$ terminal is connected to ground and the $-V$ terminal is connected to the total zener supply ($+24$ V), with the input terminals and op-amp ground terminal (if any) connected to one-half the zener supply. Of course, the polarity of the zener diodes must be reversed.

Zener Resistance.   Figures 7–4 and 7–5 both show a series resistance RS for the zener diodes. This is standard practice for zener operation. The approximate or trial value for RS is found by

$$\frac{(\text{maximum supply voltage} - \text{total zener voltage})^2}{\text{safe power dissipation of zeners}}$$

For example, assume that the total zener voltage is 24 V (12 V for each zener), that the supply voltage may go as high as 27 V, and that 2-W zeners are used. Under these conditions

$$\frac{(27 - 24)^2}{2} = 4.5 \ \Omega \text{ for RS}$$

Single-Supply Characteristics.   From a design standpoint, operation of an op-amp with a single supply is essentially the same as with the conventional

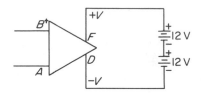

Normal dual-supply connections

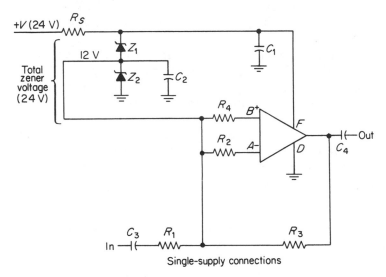

Single-supply connections

FIGURE 7-5 Connections for single power-supply operation (without ground reference).

dual power supply. The following notes describe the basic differences in operational characteristics with both types of supplies.

Single-Supply Phase/Frequency.   The normal op-amp phase/frequency compensation methods are the same for both types of supply. (Phase/frequency compensation is described fully in Sec. 7-3.) The high-frequency limits are essentially the same. The low-frequency limits of an op-amp with single supply are set by the values of C3/C4 (not found on a dual supply).

Capacitors C3/C4 are required for single-supply operation since both the input and output of the op-amp are at a voltage level equal to one-half the total zener voltage (or 12 V in our example). Thus, the op-amp cannot be used as a direct-current amplifier with a single-supply system. In a dual-supply system, the inputs and outputs are at 0 V.

Single-Supply Closed-Loop Gain.   The closed-loop gain (Sec. 7-2) is the same for both types of supply and is determined by the ratio R3/R1.

Single-Supply Decoupling Capacitors.   The values of decoupling capacitors C1 and C2 are essentially the same for both types of supply. However, it may be necessary to use slightly larger values with the single-supply system, since the impedance of the zeners is probably different from that of the power supply (without zeners).

Single-Supply R2/R4 Values.   The value of R2 should be between 50 and 100 kΩ for a typical op-amp. Values of R2 much higher or lower than these limits can result in decreased gain or in an abnormal frequency response. From a practical design standpoint, choose trial values using the guidelines and then run gain and frequency-response tests (Chapter 9).

The value of R4, the input offset resistance (Sec. 7–3), is chosen to minimize offset error from impedance unbalance. As an approximate trial value, the resistance of R4 should be equal to the parallel combination of R2 and R3. That is,

$$R4 \approx \frac{R2R3}{R2 + R3}$$

7-2 DESIGN CONSIDERATIONS FOR FREQUENCY RESPONSE AND GAIN

Most of the design problems for IC op-amps are the result of trade-offs between gain and frequency response (or bandwidth). The open-loop (without feedback) gain and frequency response are characteristics of the basic IC package, but can be modified with external *phase compensation* networks. The closed-loop (with feedback) gain and frequency response depend (primarily) on *external feedback* components.

7-2.1 Inverting and Noninverting Feedback

Figures 7–6 and 7–7 show the two basic op-amp configurations, inverting feedback and noninverting feedback, respectively. The equations on Figs. 7–6 and 7–7 are classic guidelines and do not take into account the fact that open-loop gain is not infinitely high, and the output or load impedance is not infinitely low. As a result, the equations contain built-in inaccuracies and must be used as guides only.

7-2.2 Loop Gain

As shown in Fig. 7–8, loop gain is defined as the ratio of open-loop gain to closed-loop gain. When loop gain is large, the inaccuracies of the equations in Figs. 7–7 and 7–8 decrease.

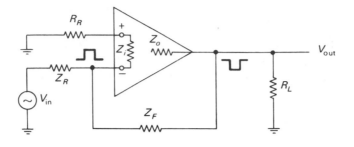

$$V_{out} \approx V_{in} \times \frac{Z_F}{Z_R}$$

closed-loop gain $\approx \dfrac{Z_F}{Z_R}$

loop gain $\approx \dfrac{\text{open-loop gain}}{\text{closed-loop gain}}$

$$Z_{in} \approx Z_R, \qquad Z_{out} \approx \frac{Z_o}{\text{loop gain}}$$

FIGURE 7-6 Theoretical inverting feedback op-amp.

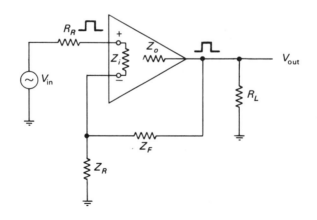

$$V_{out} \approx V_{in} \times \frac{Z_F}{Z_R} + 1$$

closed-loop gain $\approx \dfrac{Z_F}{Z_R} + 1$

loop gain $\approx \dfrac{\text{open-loop gain}}{\text{closed-loop gain}}$

$$Z_{in} \approx Z_i + \frac{\text{open-loop gain} \times Z_i}{\text{closed-loop gain}}$$

$$Z_{out} \approx \frac{Z_o}{\text{loop gain}}$$

FIGURE 7-7 Theoretical noninverting feedback op-amp.

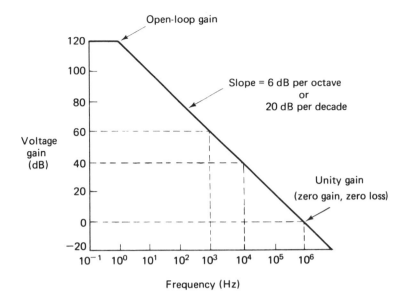

Loop gain $\approx \dfrac{\text{open-loop gain}}{\text{closed-loop gain}}$

FIGURE 7-8 Frequency-response curve of theoretical op-amp.

The relationships in Fig. 7–8 are based on a theoretical op-amp. That is, the open-loop gain rolls off at 6 dB/octave, or 20 dB/decade. (The term 6 dB/octave means that the gain drops by 6 dB each time frequency is doubled. This is the same as a 20-dB drop each time the frequency is increased by a factor of 10.)

If the open-loop gain of an amplifier is as shown in Fig. 7–8, any stable, closed-loop gain can be produced by the proper selection of feedback components, provided the closed-loop gain is less than the open-loop gain. The only concern is a trade-off between gain and frequency response.

For example, if a voltage gain of 40 dB (10^2) is desired, a feedback resistance 10^2 times larger than the input resistance is used. The gain is then flat to 10^4 Hz, and rolloff is 6 dB/octave to unity gain at 10^6 Hz. If 60-dB (10^3) gain is required, the feedback resistance is raised to 10^3 times the input resistance. This reduces the frequency response. Gain is flat to 10^3 Hz (instead of 10^4 Hz), followed by a rolloff of 6 dB/octave down to unity gain.

7-2.3 Open-Loop Frequency and Phase Response

The open-loop frequency-response and phase-shift curves more closely resemble those shown in Fig. 7–9. Here, gain is flat at 60 dB to about 200 kHz and then rolls off at 6 dB/octave to 2 MHz. As frequency increases, rolloff con-

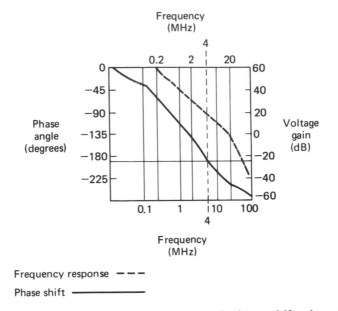

FIGURE 7-9 Frequency-response and phase-shift characteristics of practical IC op-amp.

tinues at 12 dB/octave (40 dB/decade) to 20 MHz (where gain is about unity
or 0) and then rolls off at 18 dB/octave (60 dB/decade).

Some IC op-amp datasheets provide a curve similar to that shown in Fig.
7-9. If the curves are not available, it is possible to test the op-amp and draw
actual response curves (frequency response and phase shift). The necessary procedures are described in Chapter 9.

The sharp rolloff at high frequencies, in itself, is not a problem in op-
amp use (unless the op-amp must be operated at a frequency very near the high
end). However, note that the *phase response* (phase shift between input and
output) changes with frequency. The phase response of Fig. 7-9 shows that a
negative feedback (at low frequencies) can become positive (and cause the
amplifier to be unstable) at high frequencies (possibly resulting in oscillation).
In Fig. 7-9, a 180° phase shift occurs at about 4 MHz. This is the frequency
at which open-loop gain is about +20 dB.

As a guide, when a selected closed-loop gain is equal to or less than the
open-loop gain at the 180° phase-shift point, the amplifier is *unstable*. For example, if a closed-loop gain of 20 dB or less is selected, a circuit with the curves
of Fig. 7-9 is unstable. (Note the point where the −180° phase angle intersects
the phase-shift line. Then draw a vertical line up to cross the open-loop-gain line.)

The closed-loop gain must be *more* than the open-loop gain at the frequency where the 180° phase shift occurs, but *less* than the maximum open-

loop gain. Using Fig. 7–9 as an example, the closed-loop gain must be greater than 20 dB, but less than 60 dB.

7-2.4 Phase-Compensation Methods

Op-amp design problems created by excessive phase shift can be solved by compensating techniques that *alter response* so that excessive phase shifts no longer occur. There are three basic methods of phase compensation.

Closed-Loop Feedback. The closed-loop gain can be altered by means of capacitors and/or inductances in the feedback circuit. These elements change the pure resistance to an impedance that varies with frequency, thus providing a different amount of feedback at different frequencies and a shift in phase of the feedback signal. This offsets the undesired through-amplifier phase shift.

Compensation of phase shift by closed-loop feedback methods is generally not recommended, since such compensation creates impedance problems at both the high- and low-frequency limits. Feedback compensation is not considered here, except where the op-amp is to be used in a *bandpass* function, as described in Secs. 7–12 and 7–13.

Open-Loop Input Impedance. The open-loop input impedance can be altered by means of a resistor and capacitor, as shown in Fig. 7–10. The input impedance of the series C and R decreases as frequency increases, thus altering through-amplifier (open-loop) gain. As shown in Fig. 7–10, this arrangement causes the rolloff to start at a lower frequency, but produces a stable rolloff similar to that of the ideal curve of Fig. 7–8. With the circuit properly compensated, the desired closed-loop gain is produced by selection of external resistors, as described in Sec. 7–2.6.

Open-Loop Phase-Compensation/Rolloff. The open-loop gain can be altered by one of several phase-compensation schemes, as shown in Figs. 7–11, 7–12, and 7–13.

In Fig. 7–11, *known as phase-lead compensation*, the open-loop gain is changed by an external capacitor (usually connected between collectors in one of the high-gain stages). An example of this is given in Sec. 7–2.7.

In Fig. 7–12, *generally referred to as RC rolloff, straight rolloff, or phase-lag compensation*, the open-loop gain is altered by means of an appropriate external RC network connected across a circuit component. With this method, the rolloff starts at the corner frequency produced by the RC network.

In Fig. 7–13, *known as Miller-effect rolloff, or Miller-effect phase-lag compensation*, the open-loop gain is altered by an RC network connected between the input and output of an inverting gain stage in the op-amp.

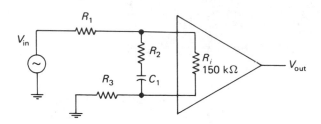

R_i = input impedance of IC

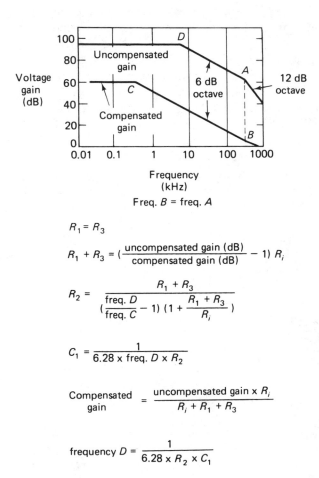

Freq. B = freq. A

$$R_1 = R_3$$

$$R_1 + R_3 = \left(\frac{\text{uncompensated gain (dB)}}{\text{compensated gain (dB)}} - 1 \right) R_i$$

$$R_2 = \frac{R_1 + R_3}{\left(\dfrac{\text{freq. } D}{\text{freq. } C} - 1 \right)\left(1 + \dfrac{R_1 + R_3}{R_i} \right)}$$

$$C_1 = \frac{1}{6.28 \times \text{freq. } D \times R_2}$$

$$\frac{\text{Compensated}}{\text{gain}} = \frac{\text{uncompensated gain} \times R_i}{R_i + R_1 + R_3}$$

$$\text{frequency } D = \frac{1}{6.28 \times R_2 \times C_1}$$

FIGURE 7-10 Frequency-response compensation by modification of open-loop input impedance of op-amp.

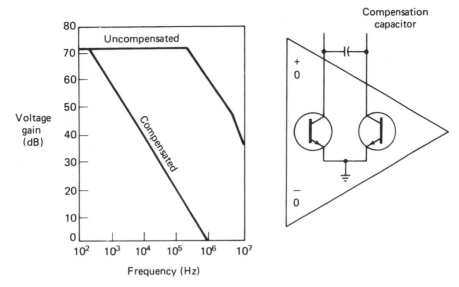

FIGURE 7-11 Frequency compensation with external capacitor (phase-lead compensation).

7-2.5 *Selecting a Phase-Compensation Scheme*

A comprehensive IC op-amp datasheet usually recommends one or more methods for phase compensation and shows the relative merits of each method. Typically, this is done by means of response curves for various values of the compensating network. (An example is given in Sec. 7-2.7.)

The recommended phase-compensation methods and values should be used in all cases. Proper phase compensation of an op-amp is a difficult, trial-and-error job at best. By using the datasheet values, it is possible to take advantage of the manufacturer's test results on production quantities of a given IC op-amp.

If the datasheet is not available or if the datasheet does not show the desired information, it is still possible to design a phase-compensation network using the guideline equations.

The first step in phase compensation (when not following the datasheet) is to connect the op-amp to an appropriate power source as discussed in Sec. 7-2. Then test the op-amp for open-loop frequency response and phase shift as described in Chapter 9.

Draw a response curve similar to Fig. 7-9. On the basis of actual open-loop response and the equations of Fig. 7-10 through 7-13, select trial values for the phase-compensating network. Then repeat the frequency-response and phase-shift tests. If the response is not as desired, change the values as necessary.

A careful inspection of the equations on Figs. 7-10 through 7-13 shows that it is necessary to know certain internal characteristics of the op-amp before an accurate prediction of the compensated frequency response can be found.

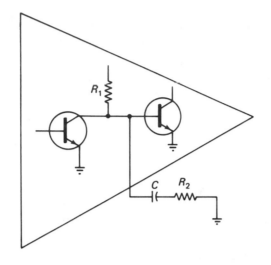

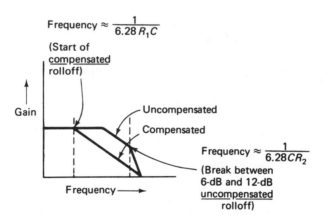

FIGURE 7-12   Frequency-response compensation with external capacitor and resistor (RC rolloff or phase-lag compensation).

For example, in Figs. 7–12 and 7–13, the values of R and C in the compensation network are based on the uncompensated open-loop frequency at which gain changes from a 6-dB/octave drop to a 12-dB/octave drop. This can be found by test of the uncompensated op-amp.

To predict the frequency at which the compensated response starts to roll off (the *corner* frequency), or that gain after compensation, requires a knowledge of internal stage transconductance (or gain) and stage load. This information is usually *not available* and cannot be found by simple test.

An exception to this is modification of the open-loop input impedance, as shown in Fig. 7–10. The only op-amp characteristic required here is the in-

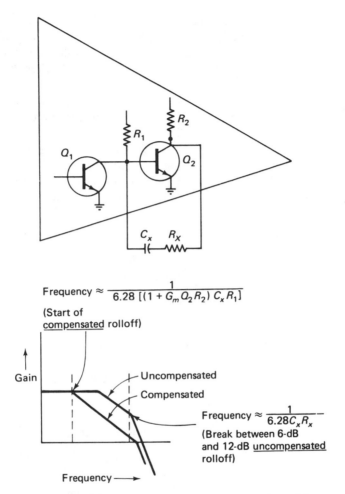

FIGURE 7-13 Frequency-response compensation with external capacitor and resistor (Miller-effect rolloff).

put impedance (or resistance). This is almost always available on the datasheet. If not, input impedance can be found by a simple test, as described in Chapter 9.

7-2.6 Example of Open-Loop Input-Impedance Modification

Figure 7-10 shows the connections, frequency plots, and equations for modification of open-loop input impedance.

The first step is to note the frequency at which the uncompensated rolloff changes from 6 to 12 dB (point A). The compensated rolloff should be zero (unity gain, point B) at the same frequency.

Draw a line up to the left from point B that increases at 6 dB/octave. For example, if point B is at 350 kHz, the line should intersect 35 kHz as the line crosses the 20-dB point. (In a practical circuit, the rolloff starts at a slightly lower frequency, about 28 to 30 kHz at 20-dB gain, since the rolloff point is rounded, rather than a sharp corner.)

Any combination of compensated gain and rolloff starting frequency (point C) can be selected along the line. For example, if the rolloff starts at 10 kHz, the gain is about 30 dB, and vice versa.

Assume that the circuit of Fig. 7-10 is used to produce a compensated gain of 60 dB, with rolloff starting at 280 Hz and dropping to zero (unity gain at 350 kHz). The op-amp to be used has an uncompensated gain of about 94 dB, with a rolloff pattern similar to that of Fig. 7-10. The typical input impedance is 150 Ω.

Using the compensated gain equation of Fig. 7-10, the relationship is

$$60 \text{ dB} = \frac{(94 \text{ dB})(150,000)}{150,000 + R1 + R3}$$

Therefore,

$$R1 + R3 = \left(\frac{94}{60} - 1\right) \times 150,000$$

$$= 0.57 \times 150,000$$

$$= 85,500$$

If R1 = R3, then R1 = R3 = 42,750. The nearest standard value is 43 kΩ. Using the equation of Fig. 7-10, the value of R2 is

$$R2 = \frac{85,500}{\left(\dfrac{5000}{280} - 1\right)\left(1 + \dfrac{85,500}{150,000}\right)}$$

$$= \frac{85,500}{16.85 \times 1.57}$$

$$= 3000 \ \Omega, \quad \text{nearest standard value}$$

The value of C1 is

$$C1 = \frac{1}{(3000)(6.28)(5000)} = 0.01 \ \mu\text{F}, \quad \text{nearest standard value}$$

If the circuit of Fig. 7-10 shows any instability in the open- or closed-loop condition, try increasing the values of R1 and R3 (to reduce gain); then select new values for R2 and C1.

7-2.7 Example of Phase-Lead Compensation

As discussed, phase-lead compensation requires the addition of a capacitor (or capacitors) to the basic op-amp circuit. Generally, the capacitors are external. However, some IC op-amps include an internal capacitor to provide fixed, phase-lead compensation.

Phase-lead compensation requires a knowledge of internal op-amp circuit characteristics. As a result, the manufacturer's datasheet or similar information must be used. The alternative method is to use typical values for the compensating capacitor and test the results. Of course, this is time consuming and may not prove satisfactory, even after tedious testing.

Figure 7-14 shows typical phase-lead compensation characteristics of an IC op-amp. Two of the external compensating capacitors tabulated in Fig. 7-14 are connected from the collectors of the first differential amplifier in the op-amp to ground by means of terminals on the IC.

The following is an example of how to use the curves of Fig. 7-14. Assume that a gain of 60 dB is required, at the highest practical frequency.

First, the intersection of the various gain/frequency curves is followed out along the 60-dB line (horizontally) to the curve for a capacitor value of 0.001 μF (the next-to-highest frequency capacitor). The intersection occurs at about 230 kHz. This means that if a 0.001-μF phase-lead capacitor is used, the op-amp response should be flat at 60 dB (with a 3-dB range) up to a frequency

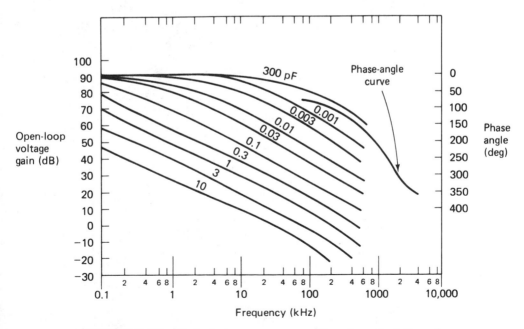

FIGURE 7-14 Typical phase-compensation characteristics of an IC op-amp.

of 230 kHz. At higher frequencies, the op-amp output drops off. Thus, any gain up to 60 dB can be selected by proper choice of feedback and input resistances.

Next, follow the 230-kHz line (vertically) until the line intersects the phase-angle curve. The intersection occurs at about 118°. This means that if a 0.001-μF capacitor is used, and the op-amp is operated at 230 kHz, the phase shift is 118°. Thus, there is a 62° *phase margin* (180°–118°) between input and output (assuming that the input signal is applied to the inverting input in the usual manner). A 62° phase margin should provide very stable operation.

Now assume that it is desired to operate at a higher frequency, but still to provide the 60-dB gain. Follow the 60-dB line (horizontally) to intersect with the 300-pF curve (300 pF is the smallest recommended capacitor). The intersection occurs at about 600 kHz. However, the 600-kHz line intersects the phase-angle curve at about 175°, resulting in a phase margin of about 5° (180° – 175°) and (probably) produces unstable operation. Thus, if a 60-dB gain must be obtained, the capacitor value must be larger than 300 pF.

The curves of Fig. 7–14 show that, for a given gain, a larger value of phase-lead capacitance reduces frequency stability, and vice versa. However, a reduction in frequency increases stability. For a given frequency of operation, capacitor size has little effect on stability, only on gain.

Built-in Phase-Lead Capacitance.   Figure 7–15 shows the characteristics of an IC op-amp with built-in phase-lead capacitance. Note that this rolloff approaches that of the ideal op-amp shown in Fig. 7–8.

Any closed-loop gain (less than open-loop gain) can be selected by feedback resistances in the normal manner. Likewise, any combination of gain/fre-

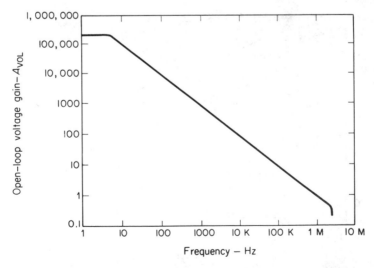

FIGURE 7-15 Open-loop gain versus frequency characteristics of op-amp with internal phase compensation.

quency response can be selected. For example, if it is desired to start the rolloff at 10^4 Hz, choose a feedback resistance ZF = 10,000 (and ZR = 100). The closed-loop gain is then flat to 10^4 Hz and rolls off at 6 dB/octave to unity gain at 10^6 Hz.

Remember that the gain/frequency characteristics of the op-amp in Fig. 7–15 *cannot be changed*. Thus, if a different gain/frequency response is required (say 10^2 gain at 10^5), the op-amp is unsuitable.

7–3 INTERPRETING IC OP-AMP DATASHEETS

The designer must have a thorough knowledge of op-amp characteristics to get the best results from the op-amp in any system. Op-amp characteristics provide a good basis for system design. However, commercial op-amps, either IC or discrete component, are generally designed for specific applications. Although some commercial op-amps are described as general purpose, it is impossible to design an op-amp with truly universal characteristics.

For example, certain op-amps are designed to provide high-frequency gain at the expense of other performance characteristics. Other op-amps provide very high gain or high input impedance in low-frequency applications. IC op-amps, which are fabricated by the diffusion process, can be made suitable for *comparator applications* (where both halves of the differential amplifier circuits must be idential). Likewise, IC op-amps can be processed to provide high gain at low-power-dissipation levels. For these reasons, any description of op-amp characteristics must be of a general nature, unless a specific application is being considered.

Most of the op-amp characteristics required for proper use of the op-amp in any application can be obtained from the manufacturer's datasheet or similar catalog information. There are some exceptions to this rule. For certain applications, it may be necessary to test the op-amp under simulated operating conditions. In using datasheet information or test results, or both, it is always necessary to interpret the information. Each manufacturer has a separate system of datasheets. It is impractical to discuss all datasheet formats here. Instead, we discuss typical information found on op-amp datasheets, as well as test results, and see how this information affects the op-amp user. The procedures for testing these characteristics are given in Chapter 9.

7–3.1 Open-Loop Voltage Gain

The open-loop voltage gain AVOL or AOL is defined as the ratio of a change in output voltage to a change in input voltage at the input of the op-amp. Open-loop gain is always measured without feedback and usually without phase-shift compensation.

Open-loop gain is *frequency dependent* (gain decreases with increased frequency). Typically, gain is flat (within about ± 3 dB) up to a certain frequency (say 0.1 MHz). Then gain rolls off to unity at another higher frequency (say 10 MHz). The open-loop gain is also *temperature dependent* and *dependent on supply voltage*. Generally, gain increases with supply voltage. The effects of temperature on gain are different at different frequencies, but usually gain decreases with increased temperature.

Ideally, open-loop gain should be infinitely high since the primary function of an op-amp is to amplify. In general, the higher the gain, the higher the accuracy of the op-amp transfer function (the relationship of output to input). However, there are practical limits to gain (and there are gain levels at which an increase buys little in the way of increased performance). The true significance of open-loop gain is many times misapplied in op-amp operation, where (in reality) open-loop gain determines *closed-loop accuracy limits* rather than ultimate accuracy.

The numerical values of the open-loop gain (and the bandwidth) of an op-amp are of relatively little importance in themselves. The important requirement is that the open-loop gain must be greater than the closed-loop gain over the frequency of interest.

For example, if a 40-dB op-amp and a 60-dB op-amp are used in a 20-dB closed-loop-gain configuration, and the open-loop gain is decreased 50% in each case (say because of component aging), the closed-loop gain of the 40-dB op-amp varies 9% and that of the 60-dB op-amp varies only 1%.

The *frequency-rolloff characteristics* determine the frequency response of an op-amp (primarily). An 18-dB/octave rolloff is generally considered the maximum slope that can occur in the active region before proper phase compensation becomes extremely difficult or impossible to get. Also, because op-amps have useful application down to and including unity gain, the active region of the op-amp may be considered as the entire-portion frequency characteristic above the 0-dB bandwidth. Thus, a well-designed op-amp should roll off at no greater than 18-dB/octave until below unity gain.

As discussed, open-loop gain can be modified by several compensation methods. A typical op-amp datasheet shows the results of such compensation, usually by means of graphs, such as the one shown in Fig. 7–16.

After compensation is applied, the op-amp can be connected in the closed-loop configuration. The voltage gain is usually not listed as such on op-amp datasheets. However, the datasheet may show some typical gain curves with various ratios of feedback (Fig. 7–16 is an example). If available, such curves can be used directly to select values of feedback components (as well as phase-compensation components).

7–3.2   Phase Shift and Phase Margin

One figure of merit commonly used in evaluating the stability of an op-amp is phase margin. As discussed in Sec. 7–2, oscillation can be sustained if the total phase shift around the loop (from input to output and back to input) can

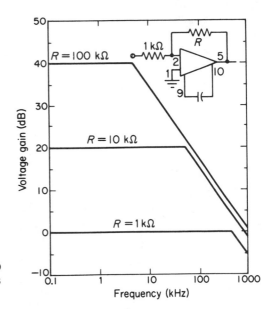

FIGURE   7-16 Closed-loop voltage gain of IC versus frequency.

reach 360° before the total gain around the loop drops below unit (as the frequency is increased). Because an op-amp is often used in the inverting mode, 180° of phase shift is available at the beginning. Additional phase shift is developed by the op-amp because of internal circuit conditions.

Phase margin represents the difference between 180° and the phase shift introduced by the op-amp at the frequency where the loop-gain in unity. A value of 45° phase margin is considered quite conservative to provide a guard against production variations, temperature effects, and other stray factors. This means that the op-amp should not be operated at a frequency where the phase shift exceeds 135° (180° − 45°). However, it is possible to operate op-amps at frequencies where the phase shift is kept within the 160° to 170° region. Of course, the ultimate stability of an op-amp must be established by tests.

7-3.3 Bandwidth, Slew Rate, and Output Characteristics

The bandwidth, slew rate, output voltage swing, output current, and output power of an op-amp are all interrelated. These characteristics depend on frequency and phase compensation. The characteristics also depend on temperature and power supply, but to a lesser extent. Before discussing the interrelationship, let us define each of the characteristics.

Bandwidth.   Op-amp bandwidth is usually expressed in terms of open-loop operation. For example, as shown in Fig. 7-17, an open-loop bandwidth of 800 kHz means that the open-loop gain of the op-amp drops to a value of 3 dB below the flat or low-frequency level at a frequency of 800 kHz.

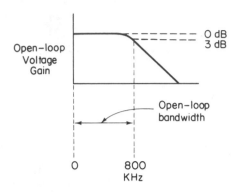

FIGURE 7-17 Bandwidth and open-loop gain relationships.

Frequency Range.   The term frequency range is sometimes used instead of open-loop bandwidth. The frequency range of an op-amp is often listed as "useful frequency range" (such as dc up to 18 MHz). Useful frequency range for an op-amp is similar to the FT or total frequency term used with transistors (Chapter 1). Generally, the high-frequency limit specified for an op-amp is the frequency at which gain drops to unity.

Power Bandwidth.   Op-amp power bandwidth is a more useful characteristic since the term represents the bandwidth of the op-amp in closed-loop operation (connected to a normal load). Power bandwidth is usually given as the peak-to-peak output-voltage capability of the op-amp (working into a given load) across a band of frequencies and usually implies that the output is free of distortion, or that distortion is within limits (such as total harmonic distortion less than 5%).

Power Output.   Op-amp power output is generally listed in terms of power across a given load (such as 250 mW across 500 Ω). However, a power output is usually listed at only one frequency. The same is true of *output current* or *maximum output current* found on some datasheets. (Thus, power bandwidth is the more useful characteristic.)

Output Voltage Swing.   Op-amp output voltage swing is defined as the peak or peak-to-peak output-voltage swing (referred to zero) that can be obtained *without clipping*. A symmetrical voltage swing depends on frequency, load current, output impedance, and slew rate. Generally, an increase in frequency decreases the possible output-voltage swing. Often, the datasheet shows output voltage swing versus frequency and phase compensation.

Slew Rate.   Op-amp slew rate is the maximum rate of change in output voltage, with respect to time, that the op-amp is capable of producing when maintaining linear characteristics (symmetrical output without clipping).

Slew rate is expressed in terms of

$$\frac{\text{difference in output voltage}}{\text{difference in time}} \quad \text{or} \quad \frac{d\text{VO}}{dt}$$

Usually, slew rate is listed in terms of *volts per microsecond*. For example, if the output voltage from an op-amp is capable of changing 7 V in 1 μs, the slew rate is 7. If, after compensation or other change, the op-amp changes a maximum of 3 V in 1 μs, the new slew rate is 3.

Slew rate of an op-amp is the direct function of the phase-shift-compensation capacitor. At higher frequencies, the current required to charge and discharge a compensating capacitor can limit available current to succeeding stages or loads, and thus result in slower slew rates. This is one reason why op-amp datasheets usually recommend the compensation of early stages in the op-amp where signal levels are still small and little current is required.

Slew rate decreases as compensation capacitance increases. This is shown in Fig. 7–18. Where high frequencies are involved, the lowest values of compensation capacitor should be used. Figure 7–19 shows the minimum capacitance values that can be used with different closed-loop gain levels of a particular op-amp. The curves of Figs. 7–18 and 7–19 are typical of those found on op-amp datasheets in which slew rate is of particular importance.

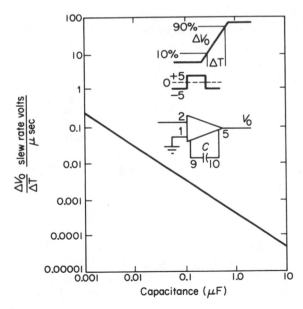

FIGURE 7-18 Slew rate of IC versus rolloff-compensation capacitance.

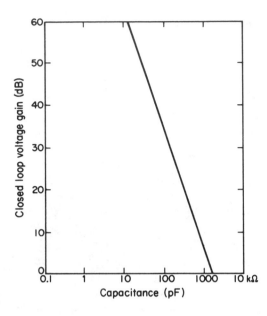

FIGURE 7-19 Closed-loop voltage gain of IC versus minimum rolloff capacitance.

The major effect of slew rate in op-amp applications is on power output. All other factors being equal, a lower slew rate results in lower power output. Slew rate and the term *full power response* of an op-amp are directly related. Full power response is the maximum frequency measured in a closed-loop unity-gain configuration for which rated output voltage can be obtained for a sine-wave signal, with a specified load and without distortion because of slew-rate limiting.

The slew rate versus full-power response relationship can be shown as

$$\text{slew rate in V/s} = 6.28 \times \text{FM} \times \text{EO}$$

where FM is the full-power response frequency in hertz and EO is the peak output voltage (one-half the peak-to-peak voltage).

For example, assume that EO is about 13 V (one-half the peak-to-peak voltage of 26 V) at a frequency of 30 kHz. The slew rate is

$$6.28 \times 30{,}000 \times 13 = 2{,}449{,}200 \text{ V/s} = 2.45 \text{ V/}\mu\text{s}, \quad \text{or simply 2.45}$$

The equation can be turned around to find the full-power response frequency. For example, assume that an op-amp is rated as having a slew rate of 2.5 V/μs (slew rate 2.5) and a peak-to-peak output of 20 V (EO = 20). Find the full-power response frequency FM as follows:

$$\frac{2.5 \text{ V/}\mu\text{s}}{6.28 \times 10} = \frac{2{,}500{,}000}{62.8} = 40{,}000 \text{ Hz} = 40 \text{ kHz}$$

With a constant output load, the power output of an op-amp depends on output voltage. In turn, all other factors being equal, output voltage depends on slew rate. Since slew rate depends on phase-compensation capacitance, op-amp power also depends on compensation.

Some datasheets omit slew rate but provide a graph similar to that of Fig. 7–20. This graph shows the direct relationship between full-power output-frequency and phase-compensation capacitance. For example, with a phase-compensation capacitance of 0.01 μF and a 500-Ω load, the op-amp shown in Fig. 7–2 delivers full-rated output power up to a frequency of about 80 kHz.

7–3.4 Output Impedance

Output impedance is defined as impedance seen by a load at the output of the op-amp, as shown in Fig. 7–21. Excessive output impedance can reduce the gain since, in conjunction with load and feedback resistors, output impedance forms an attenuator network. In general, output impedance of IC op-amps is less than 200 Ω. Generally, input resistances are at least 1 kΩ, with feedback resistances several times higher than 1 kΩ. Therefore, the output impedance of a typical IC op-amp has little effect on gain.

If the IC is serving primarily as a voltage amplifier (which is usually the case), the effect of output impedance is at a minimum. Output impedance has more significant effect in design of *power devices* that must supply large amounts of load current. As shown by the equation of Fig. 7–21, output impedance increases as frequency increases, since open-loop gain decreases.

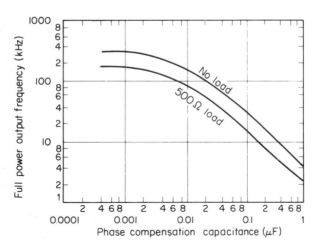

FIGURE 7-20 Frequency for full-power output as a function of phase-compensating capacitance.

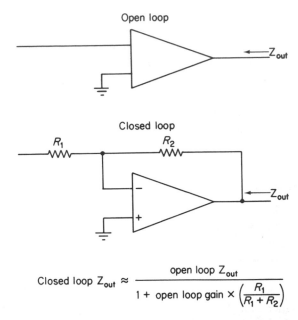

$$\text{Closed loop } Z_{out} \approx \frac{\text{open loop } Z_{out}}{1 + \text{open loop gain} \times \left(\dfrac{R_1}{R_1 + R_2}\right)}$$

FIGURE 7-21 Open- and closed-loop output-impedance relationships.

7-3.5 Input Impedance

Input impedance is defined as the impedance seen by a source looking into one input of the amplifier, with the other grounded, as shown in Fig. 7–22.

The primary effect of input impedance on design is to reduce amplifier loop gain. Input impedance changes with temperature and frequency. Generally, input impedance is listed on the datasheet at 25 °C and 1 kHz.

If input impedance is quite different from the impedance of the device driving the amplifier, there is a loss of input signal because of the mismatch. However, in practical terms, it is not possible to alter the IC input impedance. So, if impedance match is critical, either the IC or the driving source must be changed to effect a match.

7-3.6 Input Common-Mode Voltage Swing

Input common-mode voltage switch VICM is defined as the maxium peak input voltage that can be applied to either input terminal of the amplifier without causing abnormal operation or damage, as shown in Fig. 7–23. Some IC datasheets list a similar term, *common-mode input voltage range*, VCMR. Usually VICM is listed in terms of peak voltage, with positive and negative peaks being equal. VCMR is often listed as a positive and negative voltage of different value (such as +1 V and −3 V).

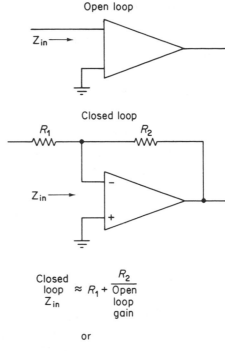

FIGURE 7-22 Open- and closed-loop input-impedance relationships.

$$\begin{array}{l} \text{Closed} \\ \text{loop} \\ Z_{in} \end{array} \approx R_1 + \dfrac{R_2}{\begin{array}{c}\text{Open}\\\text{loop}\\\text{gain}\end{array}}$$

or

$$\begin{array}{l} \text{Closed} \\ \text{loop} \\ Z_{in} \end{array} \approx \underline{\underline{\text{Open loop gain} = \infty}} \;\; R_1$$

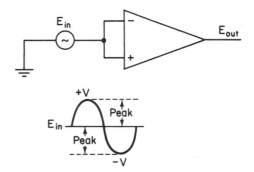

FIGURE 7-23 Input common-mode voltage-swing relationships.

In practical design considerations, either of these parameters limits the differential signal amplitude that can be applied to the amplifier input. As long as the input signal does not exceed the VICM or VCMR values (in either the positive or negative direction), no design problem should be encountered.

Note that some op-amp datasheets also list *single-ended* input voltage signal limits, where the differential input is not to be used.

7-3.7 Common-Mode Rejection Ratio

The common-mode rejection ratio characteristics for an op-amp are essentially the same as for a differential amplifier, as discussed in Sec. 6-2.

7-3.8 Input-Bias Current

Input-bias current is defined as the average values of the two input-bias currents of the op-amp differential input stage, as shown in Fig. 7-24. Input-bias current is a function of the large-signal current-gain of the input stage and decreases as temperature increases.

In use, the only real significance of input-bias current is that the resultant voltage drops across the input resistors can restrict the input common-mode voltage range at higher impedance levels. The input-bias current produces a voltage drop across the input resistors. This voltage drop must be overcome by the input signal.

7-3.9 Input-Offset Voltage and Current

Input-Offset Voltage. Op-amp input-offset voltage is defined as the voltage that must be applied at the input terminals to get zero output voltage, as shown in Fig. 7-25. Input-offset voltage indicates the matching tolerance in the differential-amplifier stages. A perfectly matched amplifier requires zero input voltage to produce zero output voltage.

Typically, input-offset voltage is about 1 or 2 mV for an IC op-amp. The offset voltage from an op-amp can also be defined as the deviation of the output d-c level from the arbitrary input/output level, usually taken as a ground reference when both inputs are shorted together.

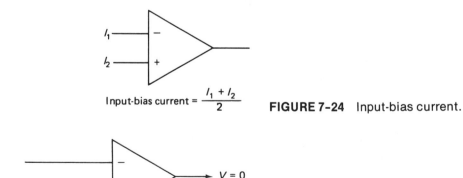

Input-bias current $= \dfrac{I_1 + I_2}{2}$ **FIGURE 7-24** Input-bias current.

FIGURE 7-25 Input-offset voltage.

Input-Offset Current. Op-amp input-offset current is defined as the difference in input-bias current into the input terminals of an op-amp, as shown in Fig. 7–26. Input-offset current is an indication of the degree of matching of the input differential stage. Typically, input-offset current is about 1 or 2 µA for an IC op-amp. The offset current for an op-amp can also be defined as the deviation when the inputs are driven by two identical d-c input-bias-current sources.

Input and Feedback. Offset voltage and currents are usually referred back to the input because the output values depend on feedback. (That is, datasheets rarely list output-offset characteristics.) In normal use, the offset in an op-amp results from a combination of offset voltage and current.

For example, if an op-amp has a 1-mA input-offset voltage and a 1-µA input-offset current, with the inputs returned to ground through 1-kΩ resistors, the total input offset is either zero or 2 mV, depending on the phase relationship between the two offset characteristics.

Need for Offset Minimization. The offset of an op-amp is a d-c error that should be minimized for many reasons, including the following:

1. The use of an op-amp as a true d-c amplifier is limited to signal levels *much greater than the offset.*

2. Comparator applications require that the output voltage be zero (within limits) when the two input signals are equal and in phase.

3. In a d-c cascade, the offset of the first stage determines the offset characteristics of the entire system. Any offset at the input is multiplied by the gain at the output. If the op-amp serves to drive additional amplifiers, the increased offset at the output is multiplied even further. The gain of the entire system must then be limited to a value that is insufficient to cause limiting in the final output stage.

Effect of Input-Offset Voltage. The effect of input-offset voltage on op-amp use is that the input signal must overcome the offset voltage before an output is produced. For example, if an op-amp has a 1-mV input-offset voltage and a 1-mV signal is applied, there is no output. If the signal is increased to 2 mV, the op-amp produces only the peaks.

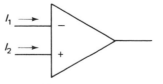

FIGURE 7-26 Input-offset current.

Input-offset current = $I_1 - I_2$ or $I_2 - I_1$

Since input-offset voltage is increased by gain, the effect of input-offset voltage is increased by the ratio of feedback resistance to input resistance, plus unity or 1, in the closed-loop condition. For example, if the ratio is 100 : 1 (for a gain of 100), the effect of input-offset voltage is increased by 101.

Effect of Input-Offset Current.   Input-offset current can be of greater importance than the input-offset voltage when high impedances are used in design. If the input-bias current is different for each input, the voltage drops across the input resistors (or input impedance is not equal). If the resistance is large, there is a large unbalance in input voltage. This condition can be minimized by means of a resistance connected between the noninverting input and ground, as shown in Fig. 7–27.

The value of this resistor R3 should equal the parallel equivalent of the input and feedback resistors R1 and R2, as shown by the equations. In practical design, the trial value for R3 is based on the equations of Fig. 7–27. The value of R3 is then adjusted for minimum voltage difference at both terminals (under normal operating conditions, but with no signal).

Neutralization of Offset.   Some op-amps (particularly IC op-amps) include provisions to neutralize any offset. Typically, an external voltage is applied through a potentiometer to terminals on the op-amp. The voltage is adjusted until the offset, at the input and output, is zero. Often the terminals are connected to the emitters of the first differential-amplifier stage. Figure 7–28 shows a typical external offset null or neutralization circuit.

External Offset Minimizing Circuit.   For op-amps without offset compensation, the effects of input offset can be minimized by an external circuit. Figure 7–29 shows two such circuits, one for inverting and the other for

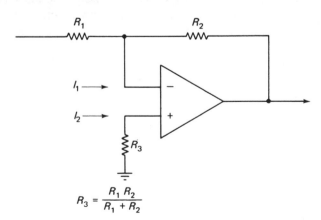

$$R_3 = \frac{R_1 R_2}{R_1 + R_2}$$

FIGURE 7–27   Minimizing input-offset current (and input-offset voltage).

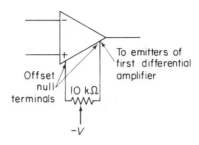

FIGURE 7-28 Typical offset null or neutralization circuit.

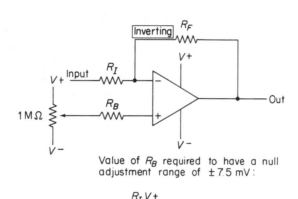

Value of R_B required to have a null adjustment range of ± 7.5 mV:

$$R_B \approx \frac{R_I V+}{7.5 \times 10^{-3}} \text{ assuming } R_B \gg R_I$$

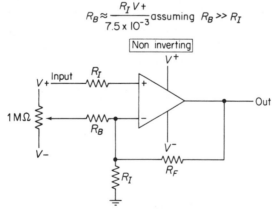

Value of R_B required to have a null adjustment range of ± 7.5 mV:

$$R_B \approx \frac{R_I R_F V+}{(R_I + R_F) \, 7.5 \text{ mV} \times 10^{-3}}$$

$$\text{Assuming } R_B \gg \frac{R_I R_F}{R_I + R_F}$$

FIGURE 7-29 Input-offset minimizing circuits.

noninverting op-amps. The equations shown in Fig. 7–29 assume that resistor RB must be of a value to produce a null range of ±7.5 mV. This is generally sufficient for any op-amp. However, if a different input-offset voltage range is required, simply substitute the desired range for ±7.5 V.

One reason for offset minimization or null is that the input and output d-c levels should be equal, or nearly equal. This condition is desirable to assure that the resistive feedback network can be connected between the input and output without upsetting the differential or the common-mode d-c bias.

Average Temperature Coefficient. The average temperature coefficient of input-offset voltage, listed on some datasheets as TCV, depends on the temperature coefficients of various components within the op-amp. Temperature changes affect stage gain, match of differential amplifiers, and so on, thus changing input-offset voltage. From a design standpoint, TCV need be considered only if the parameter is large and the op-amp must be operated under extreme temperatures. For example, if the input-offset voltage doubles with an increase to a temperature that is likely to be found during normal operation, the higher input-offset voltage should be considered the "normal" value for design.

7-3.10 Power-Supply Sensitivity

Power-supply sensitivity (S + and S −) is defined as the ratio of change in input-offset voltage to the change in supply voltage producing the change, with the remaining supply held constant, as shown in Fig. 7–30. Some IC datasheets list

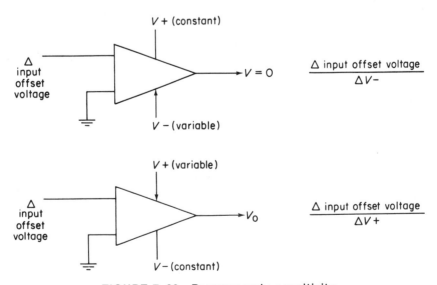

FIGURE 7-30 Power-supply sensitivity.

a similar parameter: *input-offset voltage sensitivity*. In either case, the parameter is expressed in terms of milli- or microvolts per volt (mV/V or μV/V), representing the change (in milli- or microvolts) of input-offset voltage to a change (in volts) of one power supply.

Usually, there is a separate sensitivity parameter for each power supply, with the opposite power supply assumed to be held constant. For example, a typical sensitivity listing is 0.1 mV/V for positive. This implies that with the negative supply held constant the input-offset voltage changes by 0.1 mV for each 1-V change in positive supply voltage.

The effect of power-supply sensitivity (or input-offset voltage sensitivity) is obvious. If an IC has considerable sensitivity to power-supply variation, the power-supply regulation must be increased to provide correct operation with input signal levels.

7-3.11   Input-Noise Voltage

There are many systems for measuring noise voltage in an IC amplifier and equally as many methods used to list the value on the datasheet. Some datasheets omit the values entirely. In general, noise is measured with the IC in the open-loop condition, with or without compensation, and with the input shorted or with a fixed resistance load at the input terminals. The input and/or output voltage is measured with a sensitive voltmeter or scope. Input noise is typically a few millivolts, with output noise usually less than 100 mV (because of the amplifier gain).

Except where the noise voltage is very high or the input signal is very low, amplifier noise can be ignored. Obviously, an amplifier with a 10-μV noise at the input would mask a 10-μV input signal. If the input signal is raised to 1 mV with the same amplifier, the noise is unnoticed. Noise depends on temperature as well as the method of compensation used.

7-3.12   Power Dissipation

An IC amplifier datasheet usually lists two power-dissipation ratings. One value is the *total device dissipation*, which includes any load current. The other is *device dissipation* PD or PT, which is defined as the d-c power dissipated by the IC itself (with output at zero and no load).

The device dissipation must be subtracted from the total dissipation to calculate the load dissipation. For example, if an IC can dissipate a total of 300 mW (at a given temperature, supply voltage, and with or without a heat sink) and the IC itself dissipates 100 mW, the load cannot exceed 200 mW (300 − 100 = 200 mW).

7-4 IC OP-AMP DESIGN EXAMPLE

Figure 7–31 shows a closed-loop op-amp system, complete with external circuit components. The design considerations discussed in Secs. 7–1 through 7–3 apply to the circuit of Fig. 7–31. The following paragraphs provide a specific design example for the circuit.

7-4.1 Op-Amp Characteristics

Assume that the op-amp shown in Fig. 7–31 has the following characteristics:

Supply voltage: $+15$ V and -15 V nominal, ± 19 V maximum

Total device dissipation: 750 mW, derating 8 mW/°C

Temperature range: 0° to $+70$°C

Input-offset voltage: 3 mV typical

Input-offset current: 10 nA nominal, 30 nA maximum

Input-bias current: 100 nA nominal, 200 nA maximum

Input-offset voltage sensitivity: 0.2 mV/V

Device dissipation: 300 mW maximum

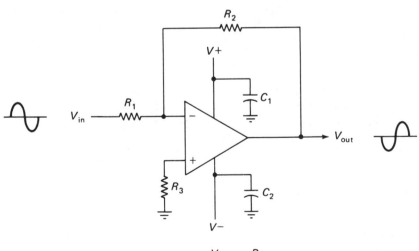

$$\text{Voltage gain} = \frac{V_{out}}{V_{in}} = \frac{R_2}{R_1} \qquad C_1 = C_2 = 0.1 - 0.001 \ \mu F$$

$$R_3 = \frac{R_1 R_2}{R_1 + R_2}$$

$$\text{Peak output voltage capability} = \frac{\text{slew rate (in 1 s)}}{6.28 \times \text{maximum frequency (Hz)}}$$

FIGURE 7-31 Basic IC op-amp system circuit.

Open-loop gain: as shown in Fig. 7–14

Slew rate: 4V/μs at a gain of 1; 6 V/μs at a gain of 10; 33 V/μs at a gain of 100

Open-loop bandwidth: as shown in Fig. 7–14

Common-mode rejection: 94 dB

Output-voltage swing: 23 V (peak-to-peak) typical

Input impedance: 1 MΩ

Output impedance 300 Ω

Input-voltage range: -13 V, $+10$ V

Output power: 250 mW typical

7-4.2 Design Example

Assume that the circuit is to provide a voltage gain of 100 (40 dB), the input signal is 80 mV rms, the input source impedance is not specified, the output load impedance is 500 Ω, the ambient temperature is 25 °C, the frequency range is dc up to 300 kHz, and the power supply is subject to 20% variation.

Frequency–Gain Relationship. Before attempting to calculate any circuit values, make certain that the op-amp can produce the desired voltage gain at the maximum frequency. This can be done by reference to a graph similar to Fig. 7–14. Note that the maximum frequency of 300 kHz intersects the phase curve at about 135°. This allows a phase margin of 45°, which is considered quite conservative. Any phase margin greater than about 25° is usually quite safe.

Note that the maximum frequency of 300 mHz intersects curves above the 40-dB open-loop gain level. Thus, the op-amp should be able to produce more than 40 dB of open-loop gain.

Supply Voltage. The positive and negative supply voltages should both be 15 V, since this is the nominal value listed. Most IC op-amp datasheets list certain characteristics as "maximum" (temperature range, total dissipation, maximum supply voltage, maximum input signal, and so on) and then list the remaining characteristics as "typical" with a given "nominal" supply voltage. In no event can the supply voltage exceed the 19 V maximum. Since the available supply voltage (15 V) is subject to a 20% variation, or 18 V maximum, the supply voltage is within the 19-V limits.

Decoupling or Bypass Capacitors. The values of C1 and C2 should be found on the datasheet. In the absence of a value, use 0.1 μF for any frequency up to 10 MHz. If this value produces a response problem at any frequency (high or low), try a value between 0.001 and 0.1 μF.

Closed-Loop Resistances.   The value of R2 should be 100 times the value of R1 to get the desired gain of 100. The value of R1 should be selected so that the voltage drop across R1 (with the nominal input-bias current) is comparable to the input signal (never larger than the input signal).

A 50-Ω value for R1 produces a 10-μV drop even with the maximum 200-nA input-bias current. Such a 10-μV drop is less than 10% of the 80-mV input signal. Thus, the fixed drop across R1 should have no appreciable effect on the input signal. With a 50-Ω value for R1, the value of R2 is 5 kΩ (50 × 100 gain = 5000).

Offset Minimizing Resistance.   The value of R3 can be found using the equation of Fig. 7–31, once the values of R1 and R2 are established. Note that the value of R3 works out to about 49 Ω, using the Fig. 7–31 equation. A simple trial value for R3 is always slightly less than the R1 value. The final value of R3 should be such that the no-signal voltages at each input are equal.

Comparison of Circuit Characteristics.   Once the values of the external circuit components are selected, the characteristics of the op-amp and the closed-loop circuit should be checked against the requirements of the design example. The following is a summary of the comparison.

Gain versus Phase Compensation.   The closed-loop gain (gain with feedback through R2) should always be less than the open-loop gain (gain with no feedback). As a guideline, the open-loop gain should be at least 20 dB greater than the closed-loop gain.

Figure 7–14 shows that open-loop gain up to about 66 dB is possible with a proper phase-compensation capacitor. Figure 7–14 also shows that a capacitance of 1000 pF produces an open-loop gain of slightly less than 60 dB at 300 kHz, whereas a 300-pF capacitor produces a 66-dB open-loop gain. To assure an open-loop gain of 60 dB, use a phase-compensation value of about 700 pF.

With a 60-dB open-loop gain and values of 50 and 5000 Ω, respectively, for R1 and R2, the closed-loop circuit should have a flat frequency response of 40 dB (gain of 100) from zero up to 300 kHz. A rolloff starts at frequencies above 300 kHz. Thus, the closed-loop gain is well within tolerance.

Input Voltage.   The peak input voltage must not exceed the rated maximum input voltage. In this case, the rated maximum is + 10 V and − 13 V, whereas the input signal is 80 mV rms, or about 112-mV peak (80 × 1.4). This is well below the + 10-V maximum limit.

When the rated maximum input signal is an uneven positive and negative value, always use the lowest value for total swing of the input signal. In this case, the input swings from + 112 mV to − 112 mV, far below + 10 V. An input signal that starts from zero can swing as much as + 10 V and − 10 V without damaging the op-amp. An input signal that starts from -2 V can swing as much as ±11 V.

Output Voltage.   The peak-to-peak output voltage must not exceed the rated maximum output voltage swing (with the required input signal and selected amount of gain). In this case, the rated output voltage swing is 23 V (peak to peak), whereas the actual output is about 22.4 V (80 mV rms input × a gain of 100 = 8000 mV output; 8000 mV × 2.8 = 22.4 V peak to peak). Thus, the anticipated maximum output voltage is within the rated capacity.

However, the actual output voltage depends on slew rate, which, in turn, depends on compensation capacitance. As shown in the characteristics, the slew rate is given as 33 V/μs when gain is 100. The datasheet does usually show a relationship between slew rate and compensation capacitance. However, since slew rate is always maximum with the lowest value of compensation capacitance, it can be assumed that the slew rate is near 33 V/μs with a 700-pF capacitance (which is near the lowest recommended value of 300 pF, Fig. 7–14).

Using the equation of Fig. 7–31, it is possible to calculate the output voltage capability of the op-amp. With a maximum operating frequency of 300 kHz and an assumed slew rate of 33 V/μs, the peak-output-voltage capability is 33,000,000 (6.28 × 300,000), or about 17 V. Thus, the peak-to-peak output voltage capability is 34 V, well above the anticipated 22.4 V.

Output Power.   The output power of an op-amp is usually computed on the basis of rms output power (rather than peak or peak to peak) and output load. In this example, the output voltage is 8 V rms (80 mV × 100 gain = 8000 mV = 8 V). The load resistance or impedance is 500 Ω, as stated in the design assumptions. Thus, the output power is $(8)^2/500 = 0.128$ W $= 128$ mW.

A 128-mW output is well below the 250-mW typical output power of the op-amp. Also, 128-mW output plus a device dissipation of 300 mW is 428 mW, well below the rated 750-mW total device dissipation. Thus, the op-amp should be capable of delivering full-power output to the load.

Note that power output ratings usually are at some given temperature, 25 °C in this case. Assume that the temperature is raised to 50 °C. The total device dissipation must be derated by 8 mW/ °C, or a total of 200 mW for the 25 °C increase in temperature. This derates or reduces the 750-mW total device dissipation to 550 mW. However, this 550 mW is still well above the anticipated 428 mW.

7-5   IC OP-AMP APPLICATIONS

The remainder of this chapter covers a few of the many applications for the basic IC op-amp. The reader will note that the power-supply and phase/frequency compensation connections are omitted from the schematics. In all applications it is assumed that the IC op-amp has been connected to a power source and that a suitable phase/frequency compensation has been selected for the IC, as described thus far in this chapter.

Before going into specific applications, this section describes some basic IC op-amp characteristics related to all applications. These characteristics include the virtual-ground concept, d-c stabilization and zero correction, and overload protection.

7-5.1   *Virtual-Ground Concept*

As shown in Fig. 7–32, when the positive input of an op-amp is grounded, a concept of virtual ground is applied to the negative input. Actually, the d-c level at the negative input of op-amp is *very close to ground*. When an input signal is applied, the signal tends to move the negative input away from ground. However, the negative feedback from the amplifier output resists this tendency.

The amount that the negative input voltage varies with a signal depends on the open-loop gain of the amplifier; the higher the gain is the less the negative input voltage varies. This is the same for any feedback amplifier (an increase in feedback decreases distortion).

With the high open-loop gain normally found in an op-amp, the negative input voltage varies only slightly under closed-loop conditions. It is convenient to assume that (for all practical purposes) the negative input voltage does not change with signal. Thus, it appears as though the negative input is grounded. The term *virtual ground* is used to indicate that, although the amplifier input appears grounded, the input is not truly grounded. (Many equations for the functions performed by an op-amp can be derived most easily by means of the virtual-ground concept.)

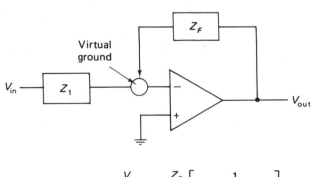

$$\text{Closed-loop gain} = \frac{V_{out}}{V_{in}} \approx -\frac{Z_F}{Z_1}\left[\frac{1}{1 - \frac{1}{A}\left(1 + \frac{Z_F}{Z_1}\right)}\right]$$

A = open-loop gain

FIGURE 7–32   Theoretical feedback circuit and gain equation for op-amp showing virtual ground concept.

It should be noted that since a virtual ground exists at the negative input the input impedance of the amplifier is essentially determined by the value of the Zi component. For example, when an op-amp is used in laboratory test equipment, the input component is typically a 50-Ω resistor. Thus, the source sees 50 Ω no matter what signal level is applied to the input.

7-5.2 *Direct-Current Stability and Zero Correction*

Since most op-amps are direct coupled, they are subject to various forms of drift. For example, a d-c amplifier cannot tell the difference between a change in supply voltage or a change in the signal voltage. Because of the instability problems, many op-amps have a zero-correction circuit (in addition to the chopper stabilization circuit discussed in Chapter 5). Generally, the zero-correction circuit is external to the op-amp, but can be made part of the internal circuit if desired.

In theory, the output voltage of an op-amp depends on the ratio of input and feedback resistors, as well as the input voltage. If the input voltage is zero, the output voltage is zero. In practice, there is a small, unwanted drift voltage, which appears at the amplifier output terminals, even when the input terminals are short circuits (no signal).

The undesired voltage (often called the *zero offset*) has both a random component (very low frequency noise) and a component that is temperature sensitive. The temperature-dependent component can be offset easily by injecting an extremely small offset voltage or current into the input circuit. This cancels the internally generated offset (or offset because of poor design). The zero-correction circuit of Fig. 7–33 is used with an op-amp that is part of a laboratory instrument.

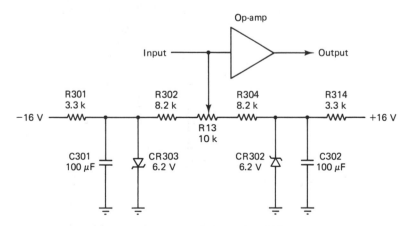

FIGURE 7-33 Zero-correction circuit for an op-amp.

7-5.3 Overload Protection for Op-Amps

Op-amps are often provided with overload-protection circuits. This prevents damage to the transistors, which can occur on overloads because of saturation. The protection also prevents the various a-c coupling capacitors (if any) from charging up. Such capacitors are usually large enough in value to ensure good low-frequency response. If these capacitors become fully charged, the transient recovery time of the amplifier is long. With proper overload protection, recovery time can be kept to about 20 μs for a typical IC op-amp.

Figure 7–34 is the overload-protection circuit for a typical op-amp. Input diodes CR210 and CR211 prevent the input voltage from ever exceeding ± 0.6 V, thus protecting the input circuits of the IC from being destroyed. Zener diodes CR301 and CR302 have nominal breakdown voltages of 10 V so that, whenever the input signal exceeds about ± 11.8 V, either CR301 or CR302 sets the level at 10 V.

Feedback current from the input terminal cannot flow back to the input unless one of CR314 or CR315 and one of CR201 or CR202 are turned on. Thus, when the output voltage reaches 11.8 V (10 + 0.6 + 0.6 + 0.6), large amounts of current from the output terminal are available to oppose the incoming overload current and keep the entire op-amp from going into saturation.

CR301 and CR302 cannot be used alone since the reverse leakage of these diodes feeds input currents of about 1 μA directly into the input. The remain-

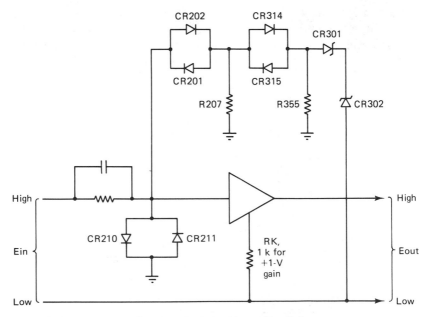

FIGURE 7-34 Overload protection circuit for an op-amp.

ing four diodes (CR201, CR202, CR314, and CR315) and resistors R207/R355 shunt the normal zener-diode reverse leakage current to ground.

7-6 OP-AMP SUMMING AMPLIFIER (ANALOG ADDER)

Figure 7-35 shows an IC op-amp used as a summing amplifier (also known as an analog adder). This circuit can sum a number of voltages (one of the original uses for op-amps in analog computers).

When the open-loop gain is high, the circuit functions with a minimum or error to sum a number of voltages. One circuit input is provided for each voltage to be summed. The single circuit output is the sum of the various input voltages (a total of four in this case) multiplied by any circuit gain. Generally, gain is set so that the output is at some given voltage value when all inputs are at their maximum value. In other cases, the resistance values are selected for unity gain.

The circuit values shown in Fig. 7-35 are based on the following conditions. Each of the four inputs varies from 2 to 50 mV rms. The output is a nominal 1 V rms with full input on all four channels, but must not exceed 2

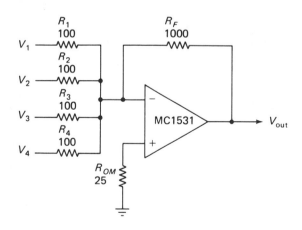

$$V_{out} = \frac{R_F}{R_1} V_1 + \frac{R_F}{R_2} V_2 + \frac{R_F}{R_3} V_3 + \frac{R_F}{R_4} V_4$$

When $R_1 = R_2 = R_3 = R_4 = R_F$

then $V_{out} = V_1 + V_2 + V_3 + V_4$

$$R_{OM} = \frac{1}{\dfrac{1}{R_1} + \dfrac{1}{R_2} + \dfrac{1}{R_3} + \dfrac{1}{R_4} + \dfrac{1}{R_F}}$$

FIGURE 7-35 Summing amplifier (analog adder).

V rms at any time. Resistors R1 through R4 are all the same value (to simplify design) and are selected so that the voltage drop (with nominal input-bias current) is comparable with the minimum input signal.

Assume a 5-μA input bias current for the IC. This results in 1.25 μA through each resistor. A 100-Ω value for R1 through R4 produces a 125-μA drop, which is less than 10% of the 2-mV minimum input signal.

The total (or maximum possible) signal voltage at the IC input is 200 mV (4 $\times$ 50 mV). Thus, the value of RF should be between 500 and 1000 Ω to get a nominal 1 V and a maximum 2 V (200 mV $\times$ 5 = 1 V; 200 mV $\times$ 10 = 2 V). The 1000-Ω value is selected for RF. The 25-Ω value of ROM is found using the equation of Fig. 7–35.

7-7 OP-AMP INTEGRATION AMPLIFIERS (INTEGRATORS)

Figures 7–36 and 7–37 show IC op-amps used as integration amplifiers (also called integrators). Integration of various signals (often square waves) can be accomplished using these circuits. The output voltage from the amplifier is inversely proportional to the time constant of the feedback network and directly proportional to the integral of the input voltage.

The circuit of Fig. 7–36 is best suited where the integrator must be used over a wide range of frequencies. Remember that the output amplitude depends on frequency once the values are selected. The circuit of Fig. 7–37 is best suited where only one frequency is involved.

The value of the RICF time constant should be about equal to the period of the signal to be integrated. The value of the RshuntCF time constant should be substantially longer than the period of the signal to be integrated (about 10 times longer). Therefore, Rshunt should be 10 times R1. Remember that Rshunt and CF form an impedance that is frequency sensitive (most noticed at low frequencies).

The circuit values used in Fig. 7–36 are based on the frequencies shown. Figure 7–38 shows a gain-versus-frequency curve for the classic Motorola MC1531 op-amp using the component values specified in the circuit of Fig. 7–36.

If the input-offset voltage of the op-amp is low, the noninverting input can be connected to ground through a fixed resistance R3. As a first trial value, use the same resistance value as R1.

If greater precision is required or if the op-amp offset is large, use the potentiometer network shown in Fig. 7–36. With this arrangement, potentiometer R is adjusted for no offset voltage under no-signal conditions.

In the circuit of Fig. 7–37, the value of R1 is chosen on the basis of input-bias current and voltage drop. Assuming a 5000-nA input-bias current, the 33-kΩ value for R1 shown produces an approximate 165-mV drop, which is 10% of 1.65 V. Thus, the 33-kΩ R1 is suitable for input signals 1.65 V or greater.

Assume that the circuit of Fig. 7–37 is to be an integrator for 1-kHz square

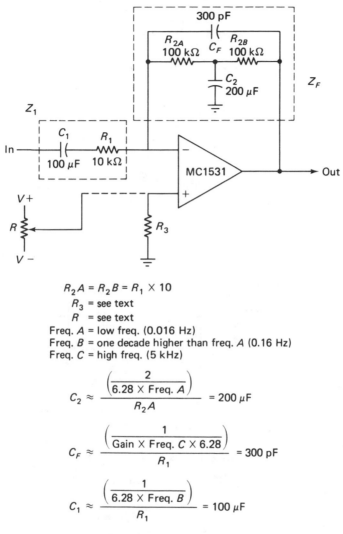

FIGURE 7-36 Integration amplifier (integrator) for wide-frequency-range applications.

waves. This requires a period of about 1 ms, or 0.001 s. Any combination of CF and R1 can be used, provided that the value of CF times R1 is about 0.001. Using the value of 33 Ω for R1, the value of CF is 0.03 μF. The value of Rshunt must then be at least 330 kΩ.

Note that the purpose of Rshunt is to provide feedback. This feedback is necessary so that an offset voltage cannot continuously charge CF (which can result in amplifier limiting). If the offset is small or can be minimized by ROM, it is possible to eliminate Rshunt.

Resistor Rshunt may have the effect of limiting gain at very low fre-

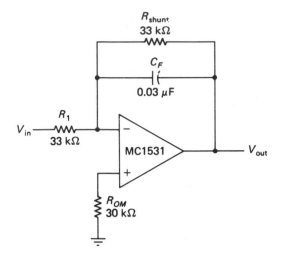

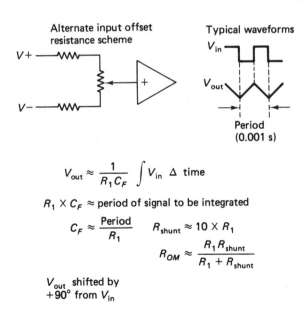

$$V_{out} \approx \frac{1}{R_1 C_F} \int V_{in} \, \Delta \text{ time}$$

$R_1 \times C_F \approx$ period of signal to be integrated

$$C_F \approx \frac{\text{Period}}{R_1} \qquad R_{shunt} \approx 10 \times R_1$$

$$R_{OM} \approx \frac{R_1 R_{shunt}}{R_1 + R_{shunt}}$$

V_{out} shifted by
$+90°$ from V_{in}

FIGURE 7-37 Integration amplifier (integrator) for fixed-frequency operation.

quencies. However, above about 15 Hz, the effect of Rshunt is negligible (because of CR in parallel). If greater precision is required, particularly at low frequencies, the input-offset resistance ROM can be replaced by the potentiometer networks shown in Fig. 7-37. With this arrangement, potentiometer R is adjusted for no offset-voltage under no-signal conditions.

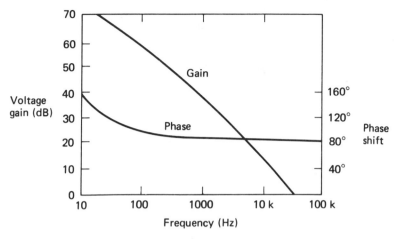

FIGURE 7–38 Gain versus frequency response for MC1531 op-amp used as an integration amplifier.

7-8 OP-AMP DIFFERENTIATION AMPLIFIER (DIFFERENTIATOR)

Figure 7–39 shows an op-amp used as a differentiation amplifier (or differentiator). Differentiation of various signals (usually square waves or sawtooth and sloping waves) can be accomplished using this circuit. The output voltage from the amplifier is inversely proportional to the *time rate of change* of the input voltage.

The value of the RFC1 time constant should be about equal to the period of the signal to be differentiated. In practical applications, the time constant is chosen on a trial-and-error basis to get a reasonable output level.

The main problems in design of differentiating amplifiers is that the gain increases with frequency. Therefore, differentiators are very susceptible to high-frequency noise. The classic remedy for this effect is to connect a small resistor (about 50 Ω) in series with the input capacitor so that the high-frequency gain is decreased. The addition of the resistor results in a more realistic model of the differentiator function because a resistance is always added in series with the input capacitor by the signal-source impedance.

In some applications, a differentiator is used to *detect the presence of distortion* or high-frequency noise in the signal. A differentiator can often detect hidden information that could not be detected in the original signal. This is because differentiation permits slight changes in input slope to produce very significant changes in the output.

An example of this is to determine the linearity of a sweep sawtooth waveform. Nonlinearity results from changes in waveform slope. So, if nonlinearity is present, the differentiated waveform indicates the points of nonlinearity quite

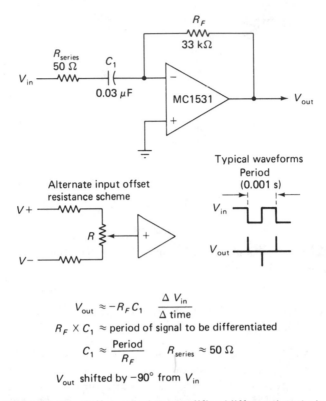

$$V_{out} \approx -R_F C_1 \frac{\Delta V_{in}}{\Delta \text{time}}$$

$R_F \times C_1 \approx$ period of signal to be differentiated

$$C_1 \approx \frac{\text{Period}}{R_F} \qquad R_{series} \approx 50 \, \Omega$$

V_{out} shifted by $-90°$ from V_{in}

FIGURE 7-39 Differentiation amplifier (differentiator) circuit.

clearly. (However, it should be pointed out that repetitive waveforms with a rise and fall of differing slopes can show erroneous waveforms.)

The circuit values shown in Fig. 7–39 are based on the assumption that 1-kHz waves must be differentiated, so the values used for the integrator of Sec. 7–7 can be used as initial trial values, even though the components are interchanged. Thus, C1 is 0.03 μF, with a 33-kΩ value for RF. The value of Rseries (if used) is an arbitrary 50 Ω. Remember that Rseries and C1 form an impedance that is frequency sensitive (that is, most noticed at high frequencies).

7-9 OP-AMP NARROW-BANDPASS AMPLIFIERS (TUNED PEAKING FILTER)

Figure 7–40 shows an op-amp used as a narrow-bandpass amplifier (also called a tuned peaking filter or amplifier). Circuit gain is determined by the ratio of R1 and RF in the usual manner. However, the frequency at which maximum gain occurs (or the narrowband peak) is the resonant frequency of the L1C1 circuit. Capacitor C1 and coil L1 form a parallel-resonant circuit that rejects the resonant frequency. Thus, there is minimum feedback (and maximum gain) at the resonant frequency.

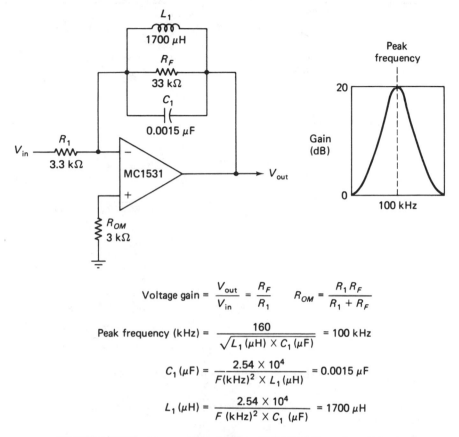

$$\text{Voltage gain} = \frac{V_{out}}{V_{in}} = \frac{R_F}{R_1} \qquad R_{OM} = \frac{R_1 R_F}{R_1 + R_F}$$

$$\text{Peak frequency (kHz)} = \frac{160}{\sqrt{L_1\,(\mu H) \times C_1\,(\mu F)}} = 100 \text{ kHz}$$

$$C_1\,(\mu F) = \frac{2.54 \times 10^4}{F(kHz)^2 \times L_1\,(\mu H)} = 0.0015 \text{ }\mu F$$

$$L_1\,(\mu H) = \frac{2.54 \times 10^4}{F\,(kHz)^2 \times C_1\,(\mu F)} = 1700 \text{ }\mu H$$

FIGURE 7-40 Narrow-bandpass amplifier (tuned peaking).

The circuit values shown in Fig. 7–40 are based on the gain (20 dB) and frequency (100 kHz) involved. Assume an arbitrary value of 3.3 kΩ for R1. The value of ROM is then the same, or slightly less. The value of RF must be 33 kΩ (or R1 × 10) for a 20-dB gain.

Any combination of L1 and C1 can be used, provided that the resonant frequency is 100 kHz. For frequencies below 1 MHz, the value of C1 should be between 0.001 and 0.01 μF. Assume an arbitrary 0.0015 μF for C1. Using the equations of Fig. 7–40, the value of L1 is about 1700 μH.

7-10 OP-AMP WIDE-BANDPASS AMPLIFIERS

Figure 7–41 shows an op-amp used as a wide-bandpass amplifier. Maximum circuit gain is determined by the ratio of RR and RF. That is, the gain of the passband or flat portion of the response curve is set by RF/RR. Minimum circuit gain is set by the ratio of R1 and RF in the usual manner. The frequencies at which rolloff starts and ends (at both high- and low-frequency limits) are

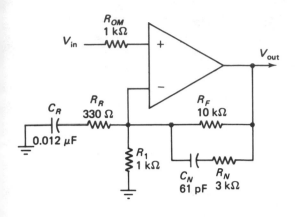

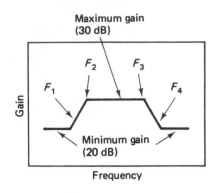

$$\text{Maximum gain} = \frac{R_F}{R_R} = 30 \text{ dB}$$

$$\text{Minimum gain} = \frac{R_F}{R_1} = 20 \text{ dB}$$

$$R_N \approx 30\% \text{ of } R_F = 3 \text{ k}\Omega$$

$$R_{OM} \approx R_1 = 1 \text{ k}\Omega$$

$$C_R = \frac{1}{6.28F_2 R_R} = 0.012 \text{ } \mu\text{F}$$

$$C_N = \frac{1}{6.28F_3 (R_N + R_F)} = 61 \text{ pF}$$

$$F_1 = \frac{10}{6.28C_R (R_F + 10R_R)} = 10 \text{ kHz}$$

$$F_2 = \frac{1}{6.28C_R R_R} = 40 \text{ kHz}$$

$$F_3 = \frac{1}{6.28C_N (R_N + R_F)} = 200 \text{ kHz}$$

$$F_4 = \frac{40}{6.28C_N (40R_N + R_F)} = 800 \text{ kHz}$$

$$R_F = R_1 \times \text{minimum gain} = 10 \text{ k}\Omega$$

$$R_R = R_F / \text{maximum gain} = 330$$

FIGURE 7-41 Wide-bandpass amplifier.

set by impedances of the various circuit combinations, as shown by the equations of Fig. 7–41.

The circuit values used in Fig. 7–41 are based on the gain and frequencies shown. Note that if phase compensation is required for the basic IC the compensation values should be based on the minimum gain of 20 dB, not on the passband gain of 30 dB.

The value of R1 is chosen on the basis of input-bias current and voltage drop in the usual manner. With a value of 1 kΩ for R1, a value of 10 kΩ is required for RF and a value of 330 Ω for RR. These relationships produce gains of 20 and 30 dB, respectively. In practice, it may be necessary to reduce both of these trial values to get the desired gain relationship.

7-11 OP-AMP UNITY-GAIN AMPLIFIERS (VOLTAGE FOLLOWER)

Figure 7–42 shows an op-amp used as a unity-gain amplifier (also known as a voltage follower). Feedback and input resistances are completely eliminated from the circuit. With this arrangement, the output voltage equals the input voltage (or may be slightly less than the input voltage). However, the input im-

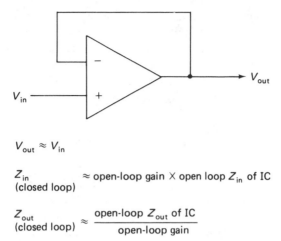

$$V_{out} \approx V_{in}$$

$$\begin{array}{l} Z_{in} \\ \text{(closed loop)} \end{array} \approx \text{open-loop gain} \times \text{open loop } Z_{in} \text{ of IC}$$

$$\begin{array}{l} Z_{out} \\ \text{(closed loop)} \end{array} \approx \dfrac{\text{open-loop } Z_{out} \text{ of IC}}{\text{open-loop gain}}$$

FIGURE 7-42 Basic IC op-amp unity-gain amplifier (voltage follower).

pedance is very high, while the output impedance is very low (as shown by the equations).

In effect, the op-amp input impedance is multiplied by the open-loop gain, whereas the output impedance is divided by the open-loop gain. For example, if the IC op-amp has an open-loop gain of 1000 (60 dB), an output impedance of 200 Ω, and an input impedance of 15 kΩ, the circuit of Fig. 7–42 produces an output impedance of 0.2 Ω (200/1000) and an input impedance of 15 MΩ (1000 × 15,000). Remember that open-loop gain varies with frequency, so closed-loop impedances are frequency-dependent.

Another consideration that is sometimes overlooked in this application is the need to supply input-bias current to the op-amp. In a conventional circuit, bias is supplied through the input resistances. In the circuit of Fig. 7–42, the input bias must be supplied through the signal source. This may alter input impedance.

One more problem with the circuit of Fig. 7–42 is that the total input voltage is a common-mode voltage. That is, a voltage equal to the total input voltage appears across the two inputs. Should the input signal consist of a large d-c value plus a large-signal variation, the common-mode input-voltage range may be exceeded.

One solution for this problem is to capacitively couple the input signal to the noninverting input. This eliminates any d-c voltage, and only the signal appears at the inputs. This solves the common-mode problem, but requires a resistor from the noninverting input to ground. The resistance provides a path for the input-bias current. However, the resistance also sets the input impedance of the op-amp. If the resistance is large, the input-bias current produces a large offset voltage (both input and output) across the resistance.

Some manufacturers recommend that resistances (of equal value) be used in the feedback loop (from output to inverting input) and in the noninverting input. This still produces unity gain, and the circuit performs as a voltage or source follower. However, the input and output impedances are then set (primarily) by the resistance values rather than by the op-amp characteristics, as is the case with the circuit of Fig. 7–42.

7-12 OP-AMP UNITY-GAIN AMPLIFIERS WITH FAST RESPONSE

One problem of a unity-gain amplifier is that the slew rate is very poor. That is, the response time is very slow and the power bandwidth is decreased. The reason for poor bandwidth and unity gain is that most op-amp datasheets recommend a large-value compensating capacitor for unity gain.

As an example, assume that the op-amp has the characteristics shown in Fig. 7–14 and that the desired operating frequency is 200 kHz (the unity-gain op-amp must have a full-power bandwidth up to 200 kHz). The recommended compensation for 60-dB gain is about 0.01 μF, whereas the unity-gain compensation is 1 μF.

Now assume that the op-amp also has the characteristics of Fig. 7–43 and that the load is 100 Ω. With a compensation of 0.001 μF, full power can be

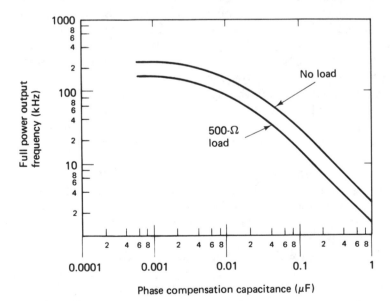

FIGURE 7-43 Frequency for full-power output as a function of phase-compensating capacitance.

delivered at frequencies well above 200 kHz. However, if the compensating capacitor is 1 μF, the maximum frequency at which full power can be delivered is below 4 kHz.

Several methods are used to provide fast response time (high slew rate) and a power bandwidth with unity gain. One such method is described next.

7-12.1 Using Op-Amp Datasheet Phase Compensation

The circuit of Fig. 7-44 shows a method of connecting an op-amp for unity gain but with high slew rate (fast response and good power bandwidth). The datasheet phase-compensation method is used in modified form. (Instead of using the unity-gain compensation, the datasheet compensation for a gain of 100 is used.) Then the values of R1 and R3 are selected to provide unity gain (R1 = R3).

As shown by the equations, the values of R1 and R3 must be about 100 times the value of R2. The values shown in Fig. 7-44 provide for an input/output signal of 8 V and an input-bias current of 200 nA.

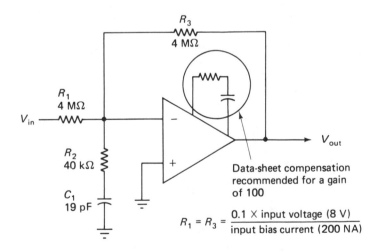

$$R_1 = R_3 = \frac{0.1 \times \text{input voltage (8 V)}}{\text{input bias current (200 NA)}}$$

$$R_2 = \frac{R_1}{100} \qquad C = \frac{1}{6.28 \times R_2 \times F}$$

Slew rate $\approx$ slew rate for gain of 100

FIGURE 7-44 Unity-gain op-amp with fast response (good slew rate) using datasheet phase compensation.

7-13   OP-AMP HIGH-INPUT-IMPEDANCE
AMPLIFIERS

Figure 7–45 shows an op-amp used as a high-input-impedance amplifier. The high-input-impedance feature of the unity-gain amplifier (Sec. 7–11) is combined with gain, as shown by the equations. Note that the circuit of Fig. 7–45 is similar to that of the basic op-amp, except that there is no input offset compensating resistance (in series with the noninverting input) for the high-input-impedance circuit. This results in a trade-off or higher input impedance, with some increase in output offset voltage.

In the basic op-amp, an offset compensating resistance is used to nullify the op-amp input offset voltage. This (theoretically) results in no offset at the output. The output of the basic op-amp is at zero volts in spite of the tremendous gain. In the unity-gain amplifier (Sec. 7–11), there is no offset compensating resistance; but since there is no gain, the output is at the same offset as the input. In a typical IC op-amp, the input offset is less than 10 mV. This figure should not be critical for the output of a typical unity-gain amplifier application.

In the circuit of Fig. 7–45, the offset-compensation resistance is omitted, and the output is offset by an amount equal to the op-amp input-offset voltage (multiplied by the closed-loop gain). However, since the circuit of Fig. 7–45 is to be used only for modest gains, modest output offset occurs.

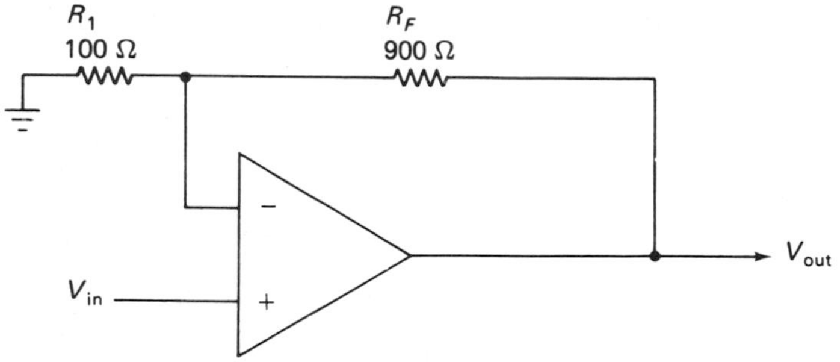

$$V_{out} = \frac{R_1 + R_F}{R_1} \times V_{in} = 10 \times V_{in}$$

$$\begin{array}{l} Z_{out} \\ \text{(closed loop)} \end{array} \approx \frac{\text{open-loop } Z_{out} \text{ of IC}}{1 + \text{open-loop gain} \left(\dfrac{R_1}{R_1 + R_F} \right)} \approx 2\ \Omega$$

$$R_F \approx (\text{gain} - 1) \times R_1 = 900$$

$$\begin{array}{l} Z_{in} \\ \text{(closed loop)} \end{array} \approx \text{open-loop } Z_{in} \text{ of IC} \times \text{closed-loop gain} \approx 150\ \text{k}\Omega$$

FIGURE 7–45   High-input-impedance amplifier.

The values shown in Fig. 7-45 are based on the assumption that the circuit is to provide a gain of 10, with high (150-kΩ) input impedance, and low (2-Ω) output impedance. The values assume an IC with an open-loop output impedance of 200 Ω, open-loop input impedance of 15 Ω, and a gain of 1000 (60 dB). The value of R1 is usually chosen on the basis of input-bias current and voltage drop and is arbitrarily set for 100 Ω.

7-14 OP-AMP DIFFERENCE AMPLIFIER (SUBTRACTOR)

Figure 7-46 shows an op-amp used as a difference amplifier and/or subtractor. One signal voltage is subtracted from another through simultaneous applications of signals to both inputs.

If the values of all resistors are the same, the output is equal to the voltage at the positive (noninverting) input, less the voltage at the negative (inverting) input. The output also represents the difference between the two input voltages.

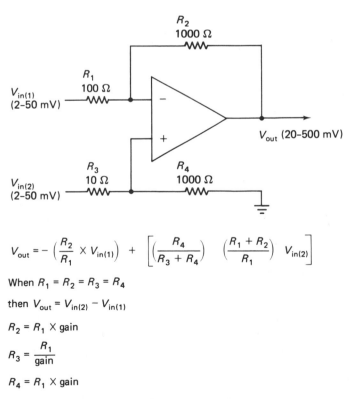

$$V_{out} = -\left(\frac{R_2}{R_1} \times V_{in(1)}\right) + \left[\left(\frac{R_4}{R_3 + R_4}\right)\left(\frac{R_1 + R_2}{R_1}\right)V_{in(2)}\right]$$

When $R_1 = R_2 = R_3 = R_4$

then $V_{out} = V_{in(2)} - V_{in(1)}$

$R_2 = R_1 \times$ gain

$R_3 = \dfrac{R_1}{\text{gain}}$

$R_4 = R_1 \times$ gain

FIGURE 7-46 Difference amplifier (subtractor).

Thus, the circuit can be used as a subtractor or difference amplifier, whichever is required.

If the values of all resistors are not the same, the output is the algebraic sum of the gains for the two input voltages, as shown by the equations of Fig. 7–46. Generally, it is simpler to use the same values for all resistors.

The values shown in Fig. 7–46 are based on the assumption that the circuit is to be used as a difference amplifier for two input voltages. Each of the voltage inputs varies from 2 to 50 mV rms. The output must vary between 20 and 500 mV rms. Since this example requires a gain of 10, all resistor values cannot be the same. To provide a gain of 10 for the negative inputs, R2 must be 10 times R1.

The value of R1 should be selected so that the voltage drop (with nominal input-bias current) is comparable with the minimum input signal, in the usual manner. Assume an arbitrary value of 100 Ω for R1. With a 100-Ω value for R1, the value of R2 must be 1000 Ω to provide a negative input gain of 10. With a 100-Ω value for R1, the values of R3 and R4 should be 10 and 1000 Ω, respectively.

7–15 OP-AMP VOLTAGE-TO-CURRENT CONVERTER (TRANSADMITTANCE AMPLIFIER)

Figure 7–47 shows an op-amp used as a voltage-to-current converter (also known as a transadmittance amplifier). This circuit can be used to supply a current to a load that is proportional to the voltage applied to the input of the amplifier. The current supplied to the load is relatively independent of the load characteristics. This circuit is essentially a *current-feedback amplifier*.

Current sampling resistor R provides the feedback to the positive input. When R1 through R4 are all the same value, the feedback maintains the voltage across R at the same value as the input voltage. If a constant input voltage is applied, the voltage across R also remains constant (within very close tolerances), regardless of the load.

If the voltage across R remains constant, the current through R must also remain constant. With R3 and R4 normally much higher than the load impedance, the current through the load must remain nearly constant, regardless of changes in impedance.

The values of R1 through R4 should normally be the same. The current-sampling resistor R is then selected for the desired load current. The value of R should be selected to limit the output power, $I^2 \times (R + \text{load})$, to a value within the capability of the IC. For example, if the IC is rated at 600-mW total dissipation, with 100-mW dissipation for the basic IC, the total output must be limited to 500 mW. As a guideline, the value of R should be about one-tenth of the load (Z).

The values shown in Fig. 7–47 are based on the assumption that the circuit is to be used as a voltage-to-current converter, the output load is 45 Ω

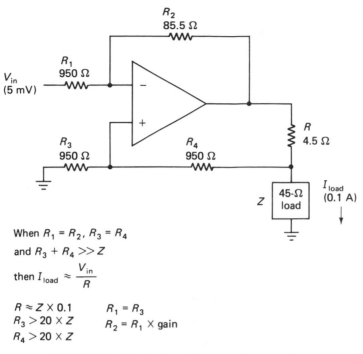

When $R_1 = R_2$, $R_3 = R_4$
and $R_3 + R_4 \gg Z$

then $I_{load} \approx \dfrac{V_{in}}{R}$

$R \approx Z \times 0.1$ $R_1 = R_3$
$R_3 > 20 \times Z$ $R_2 = R_1 \times gain$
$R_4 > 20 \times Z$

FIGURE 7-47 Voltage-to-current converter.

(nominal), the maximum power output of the IC is 500 mW, and the circuit is to maintain the maximum output current, regardless of variation in output load, with a constant input voltage of 5 mV.

With a 45-Ω load, the value of R should be about 4.5 Ω. The combined resistance or R and the load is then 49.5 Ω (rounded off to 50 Ω). With a total resistance of 50 Ω and a maximum power output of 0.5 W for the IC, the maximum possible output current is 0.1 A ($I = \sqrt{P/R} = \sqrt{0.5/50} = \sqrt{0.01} = 0.1$ A).

With a value of 4.5 Ω for R and 0.1 A through R, the drop across R is 450 mV. With 450 mV required at the output and 5 mV at the input, the amplifier gain must be 90. The values of R3 and R4 should be at least 950 Ω each, with a nominal load of 50 Ω. The value of R1 is the same as R3, or 950 Ω. With a value of 950 Ω for R1 and a gain of 90, the value of R2 should be 85.5 kΩ.

7-16 OP-AMP VOLTAGE-TO-VOLTAGE CONVERTER (VOLTAGE-GAIN AMPLIFIER)

Figure 7–48 shows an op-amp used as a voltage-to-voltage converter (also known as a voltage-gain amplifier). This circuit is similar to the voltage-to-current converter (Sec. 7–15), except that the load and current-sensing resistors are trans-

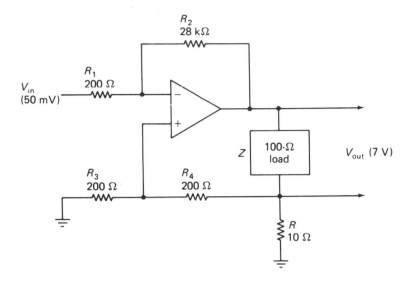

When $R_1 = R_2$, $R_3 = R_4$
then $V_{out} = V_{in}$

$R \approx Z \times 0.1$ $R_1 = R_3$

$R_3 > 20 \times R$ $R_2 = R_1 \times$ gain

$R_4 > 20 \times R$

FIGURE 7-48   Voltage-to-voltage converter or amplifier.

posed. The voltage across the load is relatively independent of the load characteristics.

The values of R1 through R4 should normally be the same. The current-sampling resistor R is then selected for the desired load current. The value of R is selected to limit the output power, $I^2 \times (R + load)$, to a value within the capability of the IC (Sec. 7–15).

The values shown in Fig. 7–48 assume that the circuit is to be used as a voltage-to-voltage converter, the output load is 100 Ω (nominal), the maximum power output of the IC is 500 mW, and the circuit is to maintain the maximum output voltage across the load, regardless of variation in load (within the current capabilities of the IC), with a constant input voltage of 50 mW.

With a 100-Ω load, the value of R should be about 10 Ω. The combined resistance of R and the nominal load is 110 Ω. With a total resistance of 110 Ω and a maximum power output of 0.5 W for the IC, the maximum possible output current is about 0.07 A ($I = \sqrt{P/R} = \sqrt{0.5/110} = \sqrt{0.0045} = 0.07A$).

With a nominal value of 100 Ω for the load and 0.07 A through the load, the maximum drop across the load is 7 V. This value may be used, provided that the IC is capable of a 7-V output. If not, use a voltage output that is within the IC capabilities.

With 7 V required at the output and 50 mV at the input, the amplifier gain must be 140. The values of R3 and R4 should be at least 200 Ω each, with a value of 10 Ω for R. The value of R1 should be the same as R3, or 200 Ω. With a value of 200 Ω for R1 and a gain of 140, the value of R2 should be 28 kΩ.

7-17 OP-AMP LOW-FREQUENCY SINE-WAVE GENERATOR

Figure 7–49 shows an op-amp used as a low-frequency sine-wave generator. (This circuit is a parallel-T oscillator.) Feedback to the negative input becomes positive at the frequency indicated in the equations, while positive feedback is provided at all times. The amount of positive feedback (set by the ratio of

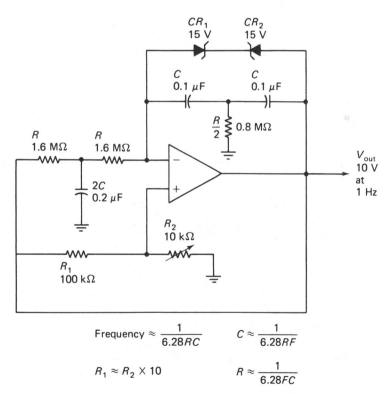

$$\text{Frequency} \approx \frac{1}{6.28RC} \qquad C \approx \frac{1}{6.28RF}$$

$$R_1 \approx R_2 \times 10 \qquad R \approx \frac{1}{6.28FC}$$

Zener point of CR_1 and $CR_2 \approx 1.5 \times V_{out}$ (peak to peak)

$R < 2$ MΩ

FIGURE 7-49 Low-frequency sine-wave generator using an IC op-amp.

R1 to R2) is sufficient to cause the IC amplifier to oscillate. In combination with the feedback to the negative input, feedback to the positive input can be used to stabilize the oscillation amplitude.

The value of R1 should be about 10 times the value of R2. The ratio of R1 and R2, as set by the adjustment of R2, controls the amount of positive feedback. Thus, the setting of R2 determines the stability of oscillation.

The *oscillation amplitude* is set by the peak-to-peak output capability of the IC and the values of the zener diodes CR1/CR2. As shown by the equations, the zener voltage should be about 1.5 times the desired peak-to-peak output voltage. The nonlinear resistance of the back-to-back zener diodes is used to limit the output amplitude and maintain good linearity.

The *oscillation frequency* is set by the values of C and R. The upper-frequency limit is about equal to the bandwidth of the basic IC. That is, if the open-loop gain drops 3 dB at 100 kHz, the circuit should provide full-voltage output up to 100 kHz.

The values shown in Fig. 7–49 assume that the circuit is to provide 10-V sine-wave signals at 1 Hz. The value of R2 is chosen on the basis of bias current in the usual manner. Assume a value of 10 kΩ for R2. With a value of 10 kΩ for R2, the value of R1 is 100 kΩ. With a required 10-V peak-to-peak output, the values (zener voltage) of CR1/CR2 are 15 V. It is assumed that the IC is capable of 10 V of peak-to-peak output.

The values of R and C are related to the desired frequency of 1 Hz. Any combination of R and C can be used, provided that the combination works out to a frequency of 1 Hz. However, for a practical design, the value of R should not exceed about 2 MΩ (for a typical IC). Using a value of 1.6 MΩ for R, C is 0.01 μF.

7–18   OP-AMPS WITH DIFFERENTIAL INPUT/DIFFERENTIAL OUTPUT

Figure 7–50 shows two op-amps used to provide differential-input/differential-output amplification. (Note that some op-amps have a differential output.) The circuit of Fig. 7–50 can be used at the input of a system in which there is considerable noise pickup at the input leads, but the input differential is small. Since the noise signals are common mode (appearing on both input leads simultaneously at the same amplitude and polarity), the noise is not amplified. However, the differential input is amplified and appears as an amplified differential output voltage.

The circuit of Fig. 7–50 can be formed with two identical op-amps or, preferably, a dual-channel IC op-amp. The dual-channel IC is preferable since both channels have identical characteristics (gain, bias, and so on), because both channels are fabricated on the same chip. However, the circuit works satis-

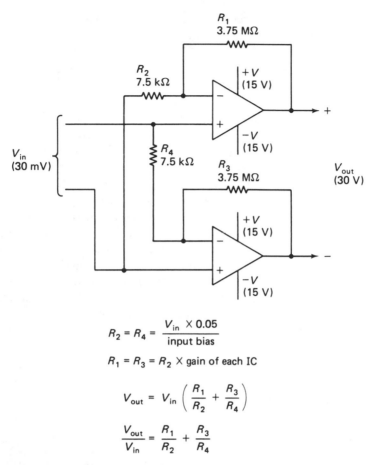

$$R_2 = R_4 = \frac{V_{in} \times 0.05}{\text{input bias}}$$

$$R_1 = R_3 = R_2 \times \text{gain of each IC}$$

$$V_{out} = V_{in} \left(\frac{R_1}{R_2} + \frac{R_3}{R_4} \right)$$

$$\frac{V_{out}}{V_{in}} = \frac{R_1}{R_2} + \frac{R_3}{R_4}$$

FIGURE 7-50 Differential-input/differential-output op-amp circuit.

factorily if the two op-amps are carefully matched as to characteristics (particularly input-bias and input-offset voltage).

In addition to common-mode rejection, the circuit of Fig. 7–50 has certain other advantages. For example, each op-amp provides one-half of the total differential output gain. Thus, if each op-amp provides a gain of 10, the differential output is 20 times the differential input. Also, the maximum output swing of a single-ended op-amp is slightly less than the IC supply voltage. Using the circuit of Fig. 7–50, the maximum differential output voltage swing is twice that of the supply voltage.

The values shown in Fig. 7–50 assume that the circuit is to provide a differential output with a swing of about 30 V, the available differential input is 30 mV, and the input-bias current is 200 nA. With an input of 30 mV and a desired output of 30 V, the circuit must provide a total gain of 1000 (30 V/0.03 V). Each op-amp must provide a gain of 500.

7-19 OP-AMP TEMPERATURE SENSORS

Figure 7-51 shows an op-amp used as the major active element in a temperature-sensor circuit. Temperature sensing with thermistors is popular because thermistors are inexpensive and easy to use. However, thermistors are nonlinear and can supply only very small output signals (generally a few microwatts).

The circuit of Fig. 7-51 overcomes these limitations. The output voltage is relatively linear over the temperature range of the op-amp. The output is exactly linear near the temperature at which the thermistor resistance equals the resistance of R1. Thus, R1 should be equal to RT (the thermistor resistance) at the center of the desired temperature range.

The reference voltage Vref is taken from zener CR1 and the voltage divider. The upper limit is a value determined by the power rating (PT) of the thermistor, as shown by the equations. With the values shown, R2 is adjusted so that Vref is −0.067 V. Under these conditions, the maximum power dissipated by the thermistor is about 2.5 μW, well under the 5-μW rating.

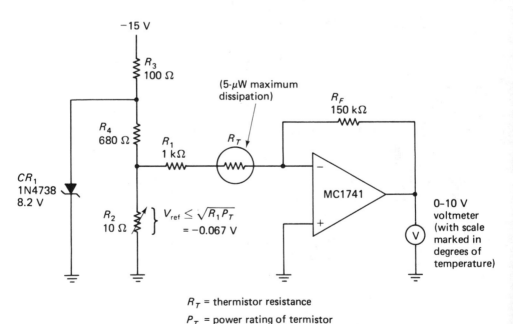

R_T = thermistor resistance

P_T = power rating of termistor

R_1 = R_T at center of temperature range

FIGURE 7-51 Temperature sensor using an IC op-amp.

7-20   OP-AMP ANGLE GENERATORS

Figure 7–52 shows two op-amps connected as an angle generator. As shown by the equations, the output of op-amp 1 is proportional to the sine of the input phase angle, and the output of op-amp 2 is proportional to the cosine of the input phase angle. The op-amps are connected as a Scott-T transformer into a three-wire synchro line.

When all resistor values are the same, the output of op-amp 2 is equal to twice the input voltage, multiplied by the cosine of the phase angle. For example, if the phase angle is 33 ° and the input voltage (line-to-line, three-phase input) is 1 V, the op-amp 2 output is 2 × 1 = 2; 2 × 0.8387 (the cosine of 33°) = 1.6774 V.

With the same conditions (phase angle 33 °, input 1 V), the op-amp 1 output is 1.732 × 0.5446 (the sine of 33°) = about 0.94 V.

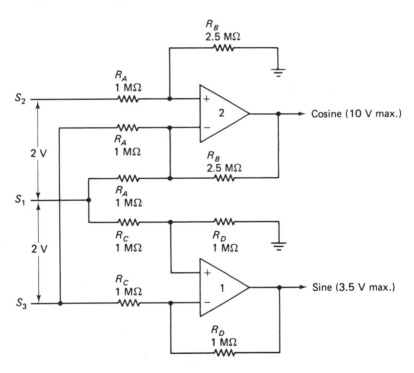

$$\text{Op amp 2} = 2 \frac{R_B}{R_A} E_{max} \text{ cosine of phase angle}$$

$$\text{Op amp 1} = 1.732 \frac{R_D}{R_C} E_{max} \text{ sine of phase angle}$$

$$E_{max} = \text{maximum line-to-line output}$$

FIGURE 7-52   Angle generator using an IC op-amp.

The accuracy of the Fig. 7–52 circuit depends on matching of the op-amps, as well as matching of the resistors. A dual-channel IC op-amp is ideal for the circuit because both channels are fabricated on the same chip. However, two separate op-amps with closely matched characteristics produce satisfactory results. The resistors should have a tolerance of 1% (or better).

The circuit of Fig. 7–52 is most effective when the three-phase voltages are about 1 V, or a fraction of 1 V (such as some analog computer servo systems), and the output readings are to be in the 5- to 10-V range (typical of digital logic systems).

The values shown in Fig. 7–50 assume that the circuit is to provide output for a three-phase system (S1, S2, S3) with a maximum line-to-line voltage of 2 V, the op-amp has an input-bias current of 200 nA, the sine output should not exceed 3.5 V, and the cosine output should be 10 V or less.

With no gain (RC = RD), the output from op-amp 1 goes from 0 V (when the phase angle is 0°) to about 3.464 V (when the phase angle is 90°). Since the cosine output is not to exceed 10 V, find the maximum output from op-amp 2 with no gain (4 V, or twice the input of 2 V, in this case); then divide the maximum output limit by the no-gain output, or 10/4 = 2.5. Thus, a gain of 2.5 is required for op-amp 2, and the values of both RB resistors should be 2.5 MΩ.

The circuit of Fig. 7–52 has several advantages over direct measurement of the phase angle by meters. First, op-amps present far less loading than a meter. (Of course, this is more important with low-voltage systems than with three-phase systems in the 120-V range.) Second, the circuit is independent of frequency (unlike most phase-angle meters, which are for one frequency only).

Probably of the greatest advantage for analog computer applications, the output can be "weighted" or "scaled." For example, some design problems may require that the sine output be multiplied by 5 (or any other number), with the cosine multiplied by 10 (or another number). This is done by setting the gain of the individual op-amps to different levels (by different ratios of RA/RB and RC/RD).

7-21 OP-AMP PEAK DETECTORS

Figure 7–53 shows an op-amp used as the major active element in a peak detector system. Where accuracy is required for peak detection, the conventional diode-capacitor detector is often inadequate because of changes in the diode forward-voltage drop (because of variations in the charging current and temperature). Ideally, if the forward drop of the diode can be made negligible, the peak value is the absolute value of the peak input and is not diminished by the diode-voltage drop (typically 0.5 to 0.7 V for a silicon diode).

The circuit of Fig. 7–53 uses the base–collector junction of a transistor as the detecting diode. The transistor (acting as a diode) is contained within

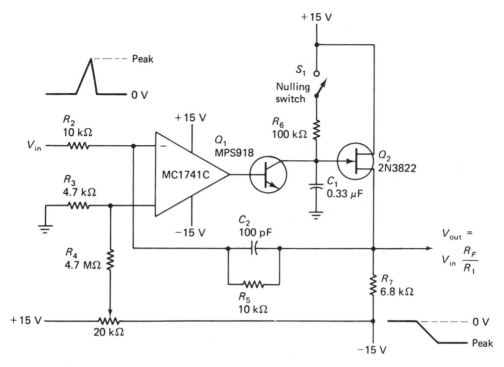

FIGURE 7-53 Peak detector using IC op-amp.

the feedback loop (between the op-amp output and the circuit output). This reduces the effective forward voltage drop of the diode by an amount equal to the loop gain (or ratio of open-loop gain to closed-loop gain).

For example, assume that the transistor has a base–emitter drop of 0.5 V, that the open-loop gain of the op-amp is 1000, and that the closed-loop gain is 1 (unity). Under these conditions, the loop gain is 1000, and the effective forward voltage drop is 0.0005 V. Any transistor can be used provided that the leakage is low (preferably 10 nA or less, with +15 V at the collector).

The storage time of the Fig. 7–53 circuit depends on leakage of the diode (transistor) and the FET, as well as the value of C1. A larger value of C1 increases storage time.

Note that the value shown in Fig. 7–53 apply to a typical op-amp that requires a V+ and V− supply of 15 V. These values can be used as a starting point for design. Also note that the output is adjusted to zero offset (zero volts output with no signal input) by closing the nulling switch S1 and adjusting offset potentiometer R1.

8

OPERATIONAL TRANSCONDUCTANCE AMPLIFIERS

This chapter is devoted entirely to operational transconductance amplifiers or OTAs. An OTA is similar to the op-amps described in Chapter 7. However, OTAs and op-amps are not always interchangeable. For that reason, an explanation of unique characteristics found in OTAs is in order.

The OTA not only includes the usual differential inputs of an op-amp, but also contains an *additional control input* in the form of an amplifier bias current or IABC. This control increases the OTAs flexibility for use in a wide range of applications.

The characteristics of an ideal OTA are similar to those of an ideal op-amp except that the OTA has an extremely *high output impedance*. Because of this difference, the output signal of an OTA is best described in terms of current that is proportional to the difference between the voltages at the two inputs (inverting and noninverting).

The OTA transfer characteristics (or input/output relationship) are best defined in terms of *transconductance* rather than voltage gain. Transconductance, or gm, is the ratio between the difference in current output (Iout) for a given difference in voltage input (Ein). Except for the high output impedance and the definition of input/output relationships, OTA characteristics are similar to those of a typical op-amp.

With an OTA, the user can select the optimum circuit conditions for a specific application simply by varying the bias IABC. For example, if low power consumption, low bias, and low offset current, or high input impedance are desired, then low IABC is selected. On the other hand, if operation into a moderate load impedance is the main consideration, the higher levels of IABC are used.

8-1 BASIC OTA CIRCUITS

Figure 8-1 shows a simplified circuit diagram of an OTA. The output signal is a "current" that is proportional to the OTA transconductance (established by the IABC) and the differential input voltage. The OTA can either source or sink current at the output, depending on the polarity of the input signal.

The circuit shown in Fig. 8-1 is for an RCA OTA available in IC form. Note that the IC incorporates a unique zener diode regulator system (D4, D5, Q10) that permits current regulation supply voltages below those normally associated with similar ICs.

8-1.1 Definition of OTA Terms

The following terms apply to all types of OTA circuits. However, the terms were first applied to IC OTA devices developed by RCA.

Amplifier bias current or IABC is the current supplied to the amplifier bias terminal to establish the operating point (such as the IABC current at the base of Q3 in Fig. 8-1).

Amplifier supply current or IA is the current drawn by the amplifier from the positive supply source. *The total supply current* (which includes the sum

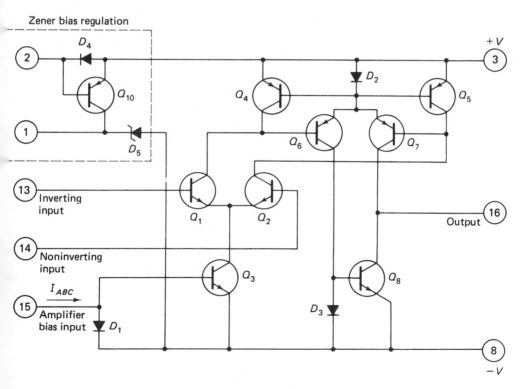

FIGURE 8-1 Simplified diagram of OTA with bias regulator.

of the amplifier supply current, the amplifier bias current, and the regulator bias current) is not to be mistaken for the amplifier supply current.

Bias regulator current is the current flowing from the zener bias regulator (such as at terminal 2 of Fig. 8–1) set by an external source, which establishes the operating conditions of the bias regulator.

Bias terminal voltage or VABC is the voltage existing between the amplifier bias terminal and the negative supply voltage terminal (such as between the IABC terminal and terminal 8 of Fig. 8–1).

Peak output current or IOM is the maximum current drawn from a short circuit on the amplifier output (positive IO) or the maximum current delivered into a short-circuit load (negative IO). Peak-to-peak current swing is twice the peak output current.

Peak output voltage or VOM is the maximum positive voltage swing (VOM +) or the maximum negative voltage swing (VOM −) for a specific supply voltage and amplifier bias.

Power consumption or P is the product of the sum of the supply voltage and the supply current, or (V + plus V −) × IA. This is not the total power consumed by an operating circuit. The power in the regulator must also be included for total power consumed.

Zener regulator voltage or VZ is the regulator voltage (such as across terminals 1 and 8 or Fig. 8–1), measured with current flowing in the bias regulator.

Unlike conventional op-amps, the characteristics of OTAs can be altered by adjusting IABC. In effect, many of the OTA characteristics can be programmed to meet specific design problems. The following is a summary of the effects of IABC on typical OTAs. Note that the characteristics listed here for OTAs are the same as for op-amps (Chapter 7).

Input offset current is directly affected by IABC, and increases almost in direct proportion with increases in IABC. The same is essentially true for *input bias current, amplifier supply current, device dissipation, transconductance,* and *peak output current.*

Input offset voltage is not drastically affected by variations in IABC. A possible exception is when the OTA is operated at high temperatures. The same is essentially true of *peak output voltage*, which is set (primarily) by supply voltage, as is the case with a conventional op-amp.

Input and output capacitances, as well as *amplifier bias voltage*, increase with IABC. However, these characteristics do not increase in direct proportion to IABC. That is, a large increase in IABC produces a small increase in input/output capacitance and amplifier bias voltage.

Input and output resistances both decrease with increases in IABC.

8–2 BASIC OTA SYSTEM CIRCUITS

Figure 8–2 shows the basic OTA system, complete with external circuit components. The following paragraphs provide a specific design example for the circuit. Note that the OTA used in Fig. 8–2 is shown in Fig. 8–1.

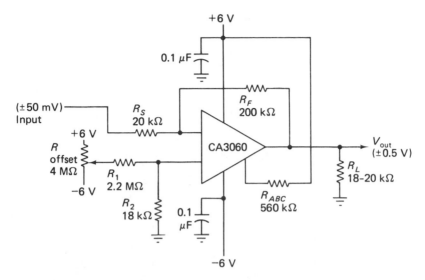

FIGURE 8-2 Basic OTA system.

8-2.1 Design Example for Basic OTA

Assume that the circuit is to provide a closed-loop gain of 10 (20 dB), the input offset voltage must be adjustable to zero, current drain must be as low as possible, the supply voltage is ±6 V, maximum input voltage is ±50 mV, input resistance is 20 kΩ, and load resistance is 20 kΩ.

Transconductance.   As in the case of a conventional op-amp, closed-loop gain is set by the ratio of feedback resistance RF to input resistance RS. With RS specified as 20 kΩ and a desired closed-loop gain of 10, the value of RF is 10 × 20, or 200 kΩ.

The next step is to calculate the required transconductance or gm (also known as g21 on some OTA datasheets) to provide a suitable open-loop gain. Assume that the open-loop gain AOL must be at least 10 times the closed-loop gain. With a closed-loop gain of 10, the open loop gains must be 10 × 10, or 100.

Open-loop gain is related directly to load resistance RL and transconductance gm. However, the actual load resistance is the parallel combination of RL and RF, or about 18 kΩ. With an AOL of 100 and an actual load of 18 kΩ, the gm should be 100/18,000, or about 5.5 millimhos (mmho).

The transconductance gm is set by IABC. With a datasheet curve similar to that of Fig. 8-3, select an IABC from the minimum-value curve to assure that the OTA provides sufficient gain. As shown in Fig. 8-3, for a gm of 5.5 mmho, the required IABC is 20 μA.

Output Swing.   Check that the calculated IABC of 20 μA produces the desired output-swing capability. With an input of ±50 mV and a gain of 10, the output swing is ±0.5 V, and the 0.5 V appears across the output load (of

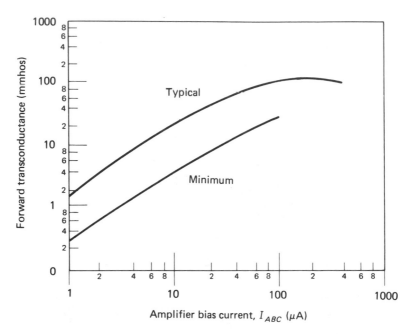

FIGURE 8-3   Forward transconductance versus amplifier bias current for OTA.

about 18 kΩ). With a 0.5-V swing and approximate load of 18 kΩ, the total amplifier current output is about 0.5 V/18 kΩ, or 27.7 μA. With a datasheet curve similar to that of Fig. 8–4, use the minimum-value curve to check that an IABC of 20 μA produces an IOM of at least 27.7 μA. As shown by Fig. 8–4, an IABC of 20 μA produces an IOM of about 40 μA, well above the required 27.7.

Calculating RABC.   As shown by Fig. 8–2, RABC is connected to the + 6-V supply. With this arrangement, RABC and diode D1 are then in series between the + V and − V supplies, and there is a total of 12 V across the series components (as shown in Fig. 8–1).

The drop across D1, which is VABC, can be found by reference to a curve similar to Fig. 8–5, which shows that an IABC of 20 μA produces a VABC of about 0.63 V. The drop across RABC is 12 − 0.63 V, or 11.37 V. For a drop of 11.37 V and an IABC of 20 μA, the value of RABC is 11.37/20 μA, or about 568 kΩ. Use the next lowest standard resistor of 560 kΩ, as shown in Fig. 8–2, to assure that a minimum IABC of 20 μA flows.

Input Offset Circuit.   The final step is to calculate values for the input offset adjustment circuit Roffset, R1, and R2 shown in Fig. 8–2. To reduce the loading effect of the offset adjustment circuit on the power supply, the values should be selected in a manner similar to that of a conventional op-amp. For

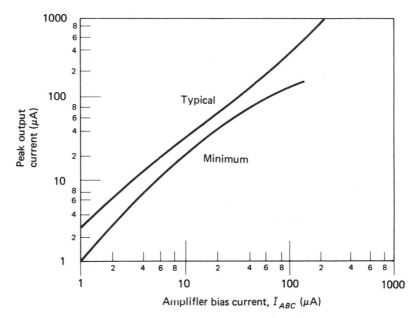

FIGURE 8-4   Peak output current versus amplifier bias current for OTA.

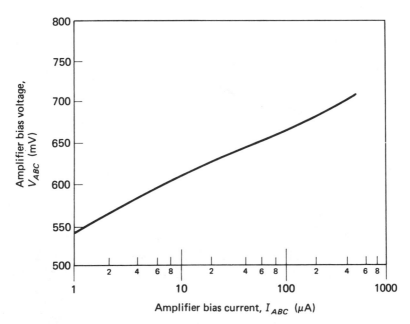

FIGURE 8-5   Amplifier bias voltage versus amplifier bias current for OTA.

example, the value of R2 should be about equal to the parallel combination of RS and RF, or about 18 kΩ.

With a datasheet curve similar to that of Fig. 8–6, find the input offset current. As shown in Fig. 8–6, for an IABC of 20 μA, the input offset current should be a maximum of 200 nA. With 200 nA flowing through R2, the voltage across R2 is 3.6 mV. This 3.6 mV must be added to the maximum input offset voltage possible for the OTA. The datasheet for the time-honored CA3060 shows a maximum input offset voltage of 5 mV. Thus the maximum voltage required at the noninverting input is 5 mV + 3.6 mV, or 8.5 mV.

The current necessary to provide a possible offset voltage of 8.6 mV across an 18-kΩ resistance is about 0.48 μA. This current must flow through R1. A possible ±6 V is available from Roffset to R1. However, for a more stable circuit, assume that ±1 V is available to R1. With 1 V available and a required current of 0.48 μA, the value of R1 is about 2 MΩ. Use the next larger standard value of 2.2 MΩ. The value of Roffset is not critical. A larger value of Roffset draws less current from the supplies. As a guideline, the maximum value of Roffset should be less than twice the value of R1, or less than 4.4 MΩ in this case. Use a standard value of 4 MΩ.

This completes the final step in the design of the basic OTA circuit. Other circuit design considerations are essentially the same as for conventional op-amps. For example, the CA3060 datasheet recommends open-loop compensa-

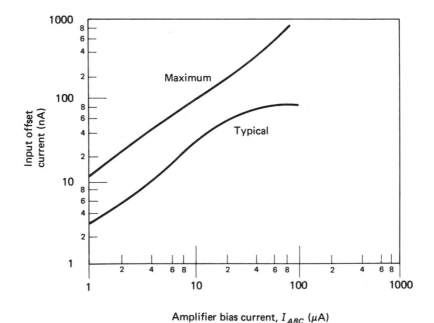

FIGURE 8-6   Input-offset current versus amplifier bias current for OTA.

tion by means of a resistor and capacitor across the differential input terminals. As always, the user should follow all datasheet recommendations.

8-2.2 *Typical OTA Applications*

The remaining sections of this chapter describe how OTAs can be used in practical circuits. Remember that an OTA can be used as a substitute for a conventional op-amp in any application provided that the OTA has comparable characteristics (frequency response, power output, and so on). The OTA can also substitute for many conventional amplifier applications.

The area in which an OTA cannot be substituted for an op-amp is one in which low output impedance is required. Even when OTA output impedance is reduced to the minimum by adjustment of IABC, the impedance is generally higher than for a comparable op-amp.

8-3 OTA MULTIPLEXER

Because an OTA has the variable bias (IABC) feature, the OTA can be gated for multiplex applications. That is, the OTA can be gated for full on or full off by pulses applied to the IABC input. In a multiplex circuit, two or more OTA units are connected so that their outputs are summed together, with the input receiving signals from separate sources. The resultant output is the combination of the input signals.

8-3.1 *Three-Channel Multiplexer Circuit*

Figure 8–7 shows a three-channel multiplex system. The CA3060 is an IC package that contains three identical OTA units. The 3N138 is a classic MOSFET used as a buffer/amplifier.

Each OTA is connected as a high-input-impedance voltage follower (similar to the op-amp voltage followers described in Chapter 7). A gating or *strobe* pulse is applied to each OTA unit at some specific time interval. When the strobe pulse is applied, each OTA is activated, and the OTA output swings to the approximate level of the input (at that OTA unit). All three OTAs can be strobed simultaneously, if required. The resultant output is then the sum of the three OTA inputs.

The 3N128 provides about 100 dB of voltage gain at the output. Part of the output is fed back to each OTA. Without this feedback, gain is unstabilized and the system is not accurate. Generally, multiplex systems are designed so that the output is identical to the sum of the inputs or is at a level equal to the sum of the inputs, multiplied by a fixed amount of gain (such as the sum of inputs times 100 dB).

The values shown in Fig. 8–7 are for operation with supply voltage of ± 15 V. The system can be operated at ± 6 V, with modifications. The 300-kΩ resistors connected between the stobe pulses and the IABC inputs must be re-

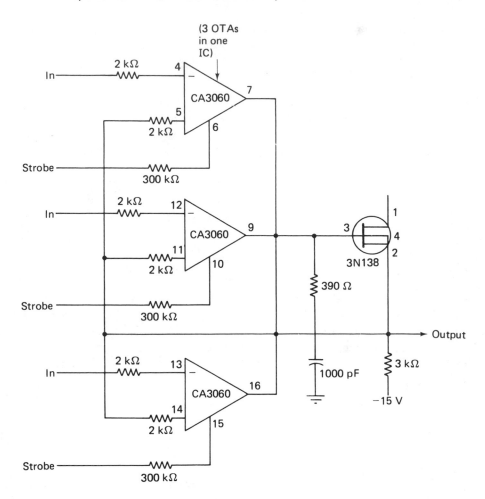

FIGURE 8-7 Three-channel multiplexer circuit using three OTAs in one IC.

duced to about 100 kΩ. Similarly, the 3-kΩ drain resistance of the 3N128 must be reduced to about 1 kΩ. When reduced (±6 V) supply voltages are used, it is possible to use other MOSFET amplifiers, such as the RCA 40811.

The only phase compensation required consists of the single 390-Ω resistor and a 1000-pF capacitor, connected at the interface between the OTA outputs and the MOSFET input. With these values, the bandwidth of the system is about 1.5 MHz and the slew rate is 0.3 V/ms.

8-4   OTA SAMPLE-AND-HOLD CIRCUITS

The multiplex system described in Sec. 8-3 can be modified to produce sample-and-hold functions, using the strobe or gating characteristics of an OTA. That is, the IABC input of the OTA is pulsed to switch the OTA on and off. The OTA output then represents a *sample* of the OTA signal input (taken during the "on" period). The *hold* function is provided by a capacitor at the OTA output.

Figure 8-8 shows the basic sample-and-hold system. In this circuit, the OTA functions as a simple voltage follower, with the phase-compensation capacitor C serving the additional function of sampled-signal storage. When the IABC is at 0 V (SAMPLE), the OTA is on. Under these conditions, the OTA output is at the same level as the signal input, and the 300-pF phase-compensation capacitor C charges to the level of the OTA output. When the IABC input is at − 15 V (HOLD), the OTA is off. However, capacitor C remains charged.

The main problem with any sample-and-hold system using a charging capacitor is that the capacitor may discharge through leakage during the OFF or HOLD cycle. Such leakage can occur through the amplifier output circuit or the 3N138 input circuit. However, since an OTA has a very high output

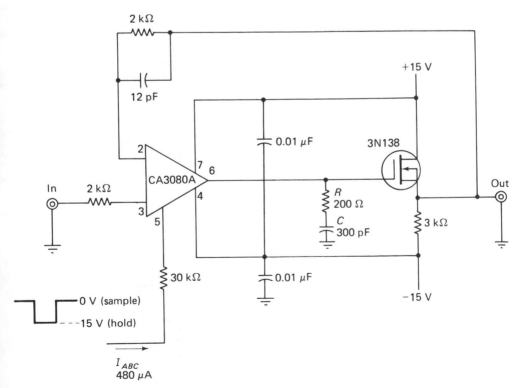

FIGURE 8-8   Sample-and-hold circuit using an OTA.

resistance, the leakage path through the OTA is practically nil. (The CA3080 has an output resistance in excess of 1000 MΩ under cutoff conditions.) Similarly, the gate leakage of the 3N138 is very low (typically 10 pA) since the transistor is a MOSFET (with insulated gate).

The open-loop voltage gain of the system is about 100 dB. The open-loop output impedance of the 3N138 is about 220 Ω (with a gm of about 4600 µmho at an operating current of 5 mA). The system closed-loop output impedance is about equal to the open-loop impedance of the 3N138, divided by open-loop voltage gain, or 220 Ω/100 dB = 220 Ω/10^5 = about 0.0022 Ω. This output impedance is comparable to the closed-loop output impedance of an op-amp.

Tilt.   Variation in the stored signal level during the hold period (or tilt) is of concern in any sample-and-hold system using the charging-capacitor method. The variation or tilt is primarily a function of the cutoff leakage current in the OTA output (a maximum limit of 5 nA), the leakage of the storage element (capacitor C), and other extraneous paths (such as gate leakage of the 3N138).

The leakage currents may be either positive or negative. Consequently, the stored signal may rise or fall (tilt) during the hold period. Figure 8–9 shows the expected pulse tilt (in microvolts) as a function of time for various values

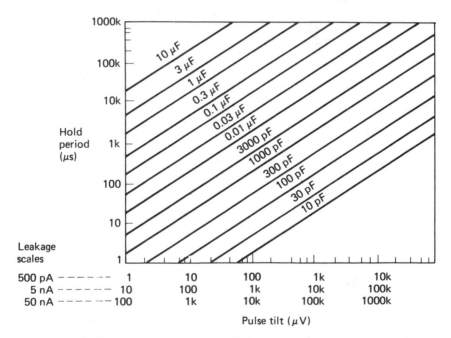

FIGURE 8-9   Chart showing tilt in sample-and-hold potentials as a function of hold time with load capacitance as a parameter.

of the compensation/storage capacitor C. The horizontal axis shows three scales representing typical leakage currents of 50 nA, 5 nA, and 500 pA.

As an example of how Fig. 8–9 can be used, assume that the hold period is 20 μs and that the capacitor is 100 pF. If leakage can be limited to 500 pA, the pulse tilt is 100 μV. That is, the system output can shift by 100 μV during the hold period. If the leakage current increases to 5 nA, the pulse tilt increases to 1000 μV. The effects of level shift (or pulse tilt) are cumulative. For example, if 10 hold periods are required for a system signal input, the total level shift is 10 times that for each hold period.

8–5 OTA GYRATOR (SYNTHETIC INDUCTANCE)

One problem in designing filters used at very low frequencies is the need for a very large inductance. This problem can be overcome by means of active-filter gyrator circuits. A gyrator is a circuit that appears as a variable inductance of high value (typically in the kilohenry range) but does not contain any inductive components.

Figure 8–10 shows a gyrator circuit composed of two OTAs. The circuit appears as an inductance of up to 10 kH across terminals A and B. The circuit can be used to exact inductance values by means of R1.

The setting of R1 varies the IABC current of both OTAs simultaneously. When one IABC current is increased, the opposite IABC current is decreased. This action varies the transconductance of both OTAs simultaneously and serves to "tune" the circuit by changing the resistance of the OTAs.

The circuit of Fig. 8–10 makes use of the high-impedance output available in OTAs. The Q (Chapter 9) of the 10-kH variable inductance is about 13, using the circuit values shown. The 20-kΩ to 2-MΩ attenuators in this circuit extend the dynamic range of each OTA by a factor of 100.

8–6 OTA GAIN-CONTROL AND MODULATION CIRCUITS

An obvious function of an OTA is that of a gain-control element. The gain of signals passing through an OTA can be controlled by variation of the IABC. This is because transconductance of the OTA (and thus gain of the circuit) is directly proportional to IABC.

In the simplest form, an OTA can be connected as a conventional amplifier, but with the IABC input connected to a voltage source through a variable resistance (which acts as the gain control). For a specified value of IABC (as set by the variable resistance), the output current of the OTA is equal to the product of the transconductance and the input-signal magnitude. The output voltage swing is the product of the output current and the load resistance.

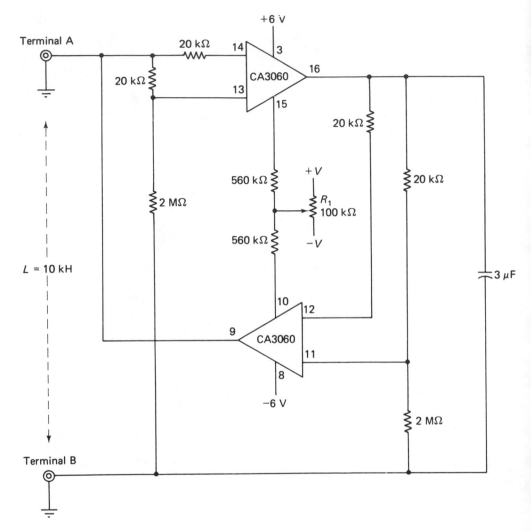

FIGURE 8-10 Gyrator (synthetic inductance) for an active filter circuit using two OTAs.

8-6.1 Amplitude-Modulation Circuit

The basic gain-control function can be applied to amplitude modulation of signals. Figure 8–11 shows a basic amplitude-modulation circuit using an OTA as the modulator. In this circuit, the signal input is a voltage VX at some carrier frequency, and the IABC input is a voltage VM at some modulating frequency. The output signal current IO is equal to gm times VX. The sign of the output signal is negative because the input signal is applied to the inverting input of the OTA.

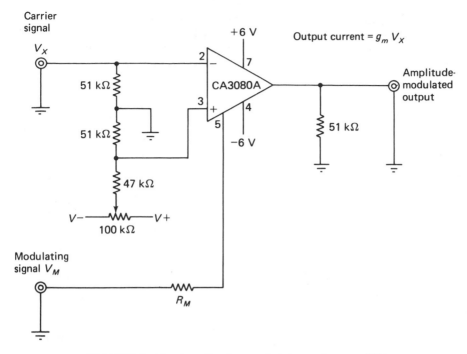

FIGURE 8-11 Amplitude modulator using an OTA.

The gm of the OTA is controlled by adjustment of IABC, as usual. However, in this circuit, the level of the unmodulated carrier output is established by a particular IABC through resistor RM. Amplitude modulation of the carrier frequency occurs because variations of the voltage VM force a change in the IABC supplied via RM. When VM goes in the negative direction (toward the IABC terminal potential), the IABC decreases and reduces gm of the OTA. When VM goes positive, the IABC increases, resulting in an increase of the gm.

For the particular OTA shown, the gm is about equal to $19.2 \times \text{IABC}$, where gm is millimhos and IABC is in milliamperes. In this case, IABC is about equal to

$$\frac{\text{VM} - \text{V}^-}{\text{RM}} = \text{IABC}$$

$$\text{IO} = -\text{gm VX}$$

$$\text{gm VX} = (19.2)(\text{IABC})(\text{VX})$$

$$\text{IO} = \frac{-19.2(\text{VM} - \text{V}^-)\text{VX}}{\text{RM}}$$

$$\text{IO} = \frac{19.2(\text{VX})(\text{V}^-)}{\text{RM}} = \frac{19.2(\text{VX})(\text{VM})}{\text{RM}} = \text{modulation equation}$$

Note that there are two terms in the modulation equation. The first term represents the fixed carrier input, independent of VM. The second term represents the modulation, which either adds to or subtracts from the first term. When VM is equal to the V⁻ term, the output is reduced to zero.

In the preceding modulation equations, the term

$$(19.2)(VX)\frac{VABC}{RM}$$

involving the amplifier bias-current terminal voltage VABC is neglected. This term is assumed to be small because VABC is small compared with V⁻ in the equation.

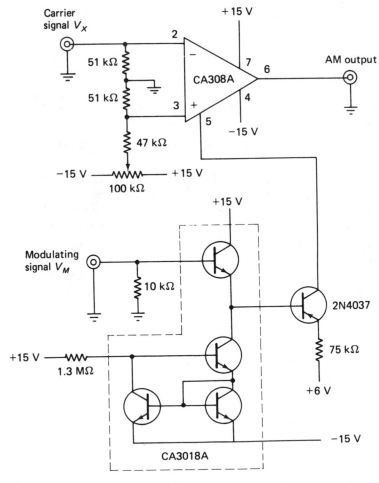

FIGURE 8-12 Amplitude modulator using an OTA controlled by PNP and NPN transistors.

If the VABC terminal is driven by a current source (such as from the collector of a transistor), the effect of VABC variation is eliminated. Instead, any variation depends on the transistor base–emitter junction characteristics.

Figure 8–12 shows an amplitude-modulation configuration using one transistor in an array as an emitter follower and the three remaining transistors of the array connected as a current source for the emitter follower. (Note that the 100-kΩ potentiometer shown in Figs. 8–11 and 8–12 is used to set the output voltage symmetrically about zero.) The OTA modulation circuit of Fig. 8–12 permits a range exceeding 1000 : 1 in gain, and thus provides modulation of the carrier signal input in excess of 99%.

8-7   OTA TWO-QUADRANT MULTIPLIER

Figure 8–13 shows an OTA used as a two-quadrant multiplier. Note that the circuit of Fig. 8–13 is essentially the same as for the modulator circuits described in Sec. 8–6. That is, when modulation is applied to the IABC input and the

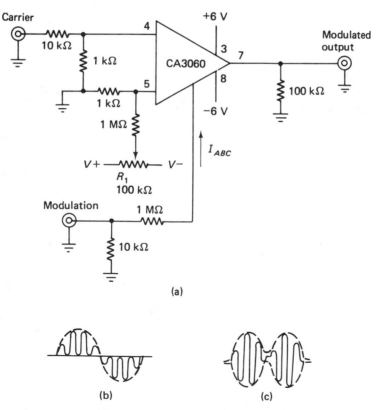

(a)

(b)   (c)

FIGURE 8-13   Two-quadrant multiplier using an OTA.

carrier voltage is applied to the differential input, the waveform shown in Fig. 8–13b is obtained.

The input offset control R1 is adjusted to balance the circuit so that no modulation can occur at the output without a carrier input. The linearity of the modulator is indicated by the solid trace of the superimposed modulating frequency, as shown in Fig. 8–13b. The maximum depth of modulation (or percentage of modulation) is determined by the ratio of the peak input modulating voltage to V^-.

The two-quadrant multiplier characteristic of the circuit is seen if modulation and carrier are reversed (modulation to differential input, carrier to IABC),

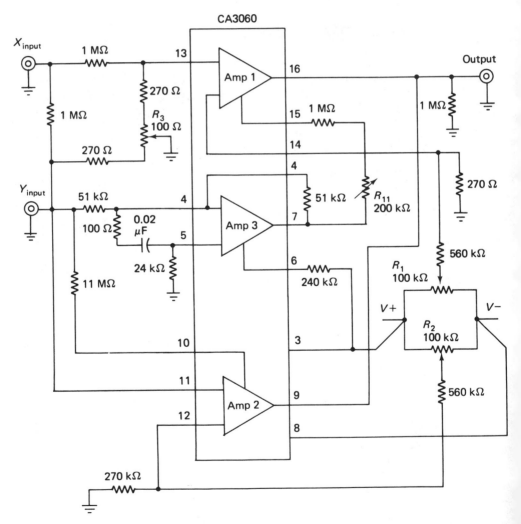

FIGURE 8–14 Typical four-quadrant multiplier using an OTA.

as shown in Fig. 8–13c. The polarity of the output must follow that of the differential input. Thus, the output is positive only during the first (or positive) half-cycle of the modulation and is negative only during the second half-cycle. Note that both input and output signals are reference to ground. The output signal is zero when either the differential input or IABC is zero.

8-8   OTA FOUR-QUADRANT MULTIPLIERS

OTA units can be used as four-quadrant multipliers. Many circuit configurations are possible. Figure 8–14 shows a typical four-quadrant multiplier, including all the adjustment controls associated with differential input and an adjustment for equalizing the gains of amplifiers 1 and 2.

Adjustment of the circuit is quite simple. With both the X and Y voltages at zero, connect terminal 10 to terminal 8. This disables amplifier 2. Adjust the offset voltage of amplifier 1 to zero by means of R1. Remove the short between terminals 8 and 10. Connect terminal 15 to terminal 8. This disables amplifier 1. Adjust the offset voltage of amplifier 2 to zero by means of R2. With a-c signals on both the X and Y input, adjust R3 and R11 for symmetrical output signals.

Figure 8–15 shows the output waveform with the circuit adjusted. Figure 8–15a shows suppressed-carrier modulation of a 14-kHz carrier with a triangular

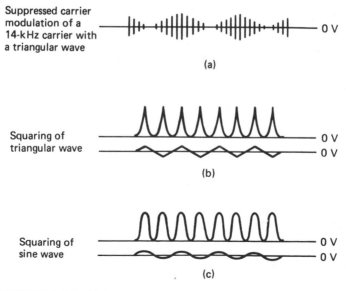

FIGURE 8-15   Voltage waveforms of four-quadrant multiplier circuit.

wave. Figures 8–15b and c, respectively, show the squaring of a triangular wave and a sine wave. Note that in both cases the outputs are always positive and return to zero after each cycle.

8-9 OTA WITH HIGH-CURRENT OUTPUT STAGES

An OTA can be combined with other devices to produce high-gain, high-current circuits. For example, the sample-and-hold circuit of Sec. 8–4 combines an OTA with a MOSFET. The resultant voltage gain is about 100 dB. The actual gain of the overall circuit is equal to the product of the OTA gain and output resistance (which is typically 142,000 or 103 dB). Thus, the overall gain is set by the OTA characteristics. However, the output voltage swing and circuit swing are set by the MOSFET (and source-terminal load).

Figure 8–16 shows an OTA combined with a MOS inverter/amplifier to form a simple open-loop amplifier circuit. Each of the three inverter/amplifiers in the CD4007A has a typical voltage gain of 30 dB. This gain, combined with the typical 100-dB gain of the OTA, results in a total voltage gain of about 130 dB. (Note that OTA gain is affected by IABC which, in turn, is set by RABC.)

Several circuit configurations are available when an OTA is combined with

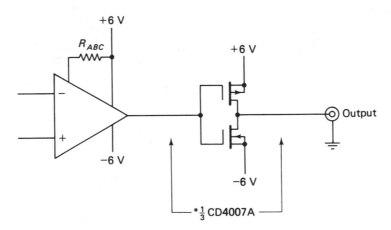

*Additional current output can be obtained when remaining two amplifiers of the CD4007A are connected in parallel with the single stage shown.

FIGURE 8-16 OTA combined with MOS to form open-loop amplifier circuit with high current gain.

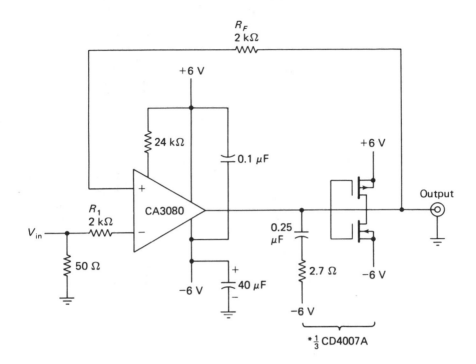

*Additional current output can be obtained when remaining two amplifiers of the CD4007A are connected in parallel with the single stage shown.

FIGURE 8-17 OTA combined with MOS to form unity-gain amplifier with increased current capacity.

MOS inverter/amplifiers. Some of these configurations provide additional gain, while others provide additional current capacity.

The circuit shown in Fig. 8-17 provides unity gain (with the same resistance values for RF and R1) and increased current capacity. The MOS inverter/ amplifier can source or sink a current of about 6 mA.

The circuit of Fig. 8-18 provides a voltage gain of about 160 dB and a source or sink current capacity of about 12 mA. The voltage gain results from open-loop operation. The OTA provides about 100 dB, inverter/amplifier A provides about 30 dB, and the parallel-connected inverter/amplifiers B and C provide the remaining 30 dB. Since the outputs of B and C are connected in parallel, the normal source/sink capacity of 6 mA is doubled to 12 mA.

The circuit of Fig. 8-19 provides unity gain (with the same resistance values for RF and R1) but with increased current capacity. The parallel-connected B and C outputs provide about 12 mA of sink or source current capacity.

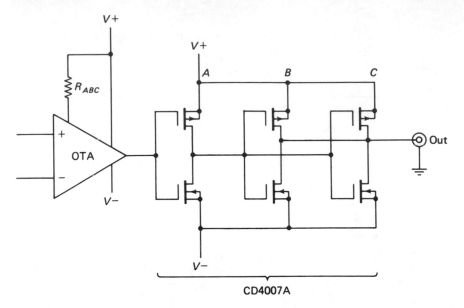

FIGURE 8-18 OTA combined with three MOS stages to form open-loop amplifier circuit with high current output and increased gain.

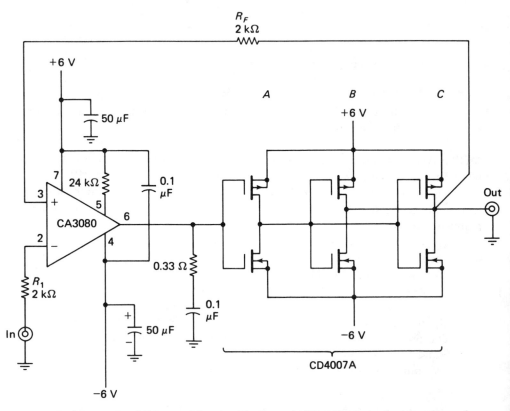

FIGURE 8-19 OTA combined with three MOS stages to form unity gain amplifier with increased current capacity.

8-10   OTA MULTISTABLE CIRCUITS

An OTA can be set to draw very little standby current by proper adjustment of IABC. Similarly, MOS inverter/amplifiers draw very little current when used in switching applications. (A MOS inverter/amplifier switch draws current when changing states, but very little current when the output signal voltage swings either positive or negative. The low standby-power capabilities of the OTA, when combined with the MOS inverter/amplifier characteristics, are ideally suited for use in precision multistable circuits.

8-10.1   OTA Astable (Free-Running) Multivibrator

Figure 8–20 shows an OTA combined with a MOS inverter/amplifier to form an astable or free-running multivibrator. As shown by the equations, the frequency of operation is set by feedback resistance R and capacitor C, as well as resistors R1 and R2. Since resistors R1/R2 also set output impedance of the circuit, a trade-off may be necessary in selecting values of R and C.

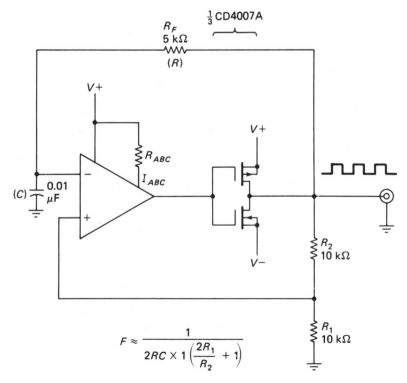

$$F \approx \cfrac{1}{2RC \times 1\left(\cfrac{2R_1}{R_2} + 1\right)}$$

FIGURE 8-20   OTA combined with a MOS to form an astable (free-running) multivibrator.

The value of RABC is set by the IABC requirements as usual (Sec. 8–2). For a multistable circuit, IABC is usually set so that the OTA draws minimum power, but enough that the MOS inverter/amplifier is properly driven to provide the desired output-voltage swing. Using the values shown, the output frequency is about 7.7 kHz, with an output-voltage swing equal to V + and V − (about 12 V).

8-10.2 OTA Monostable (One-Shot) Multivibrator

Figure 8–21 shows an OTA combined with a MOS inverter/amplifier to form a monostable (one-shot) multivibrator. Generally, monostable multivibrators are used to produce output pulses of some specific time duration (T), regardless of the trigger input pulse duration and frequency. As shown by the equations, the time duration T of output pulses is set by feedback capacitance C and resistance R, as well as resistors R1 and R2.

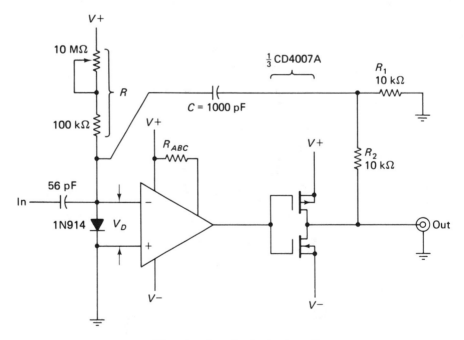

Time duration of output pulse = T

$$T = RC \times 1\left[\frac{\dfrac{R_1}{R_1 + R_2}(V^+ - V^-) + (V^+ - V_D)}{V+}\right]$$

FIGURE 8-21   OTA combined with MOS to form a monostable (one-shot) multivibrator.

The relationship of V+, V−, and VD also affects output pulse duration T. Note that VD, the voltage across the input diode, is typically 0.5 V. Again, RABC is adjusted so that the OTA provides just enough output to drive the MOS inverter/amplifier for the desired output-voltage swing. Also, it may be necessary to trade off the values of R1, R2, R, and C, since R1 and R2 set the circuit output impedance.

8-10.3 OTA Threshold Detector

Figure 8-22 shows an OTA combined with a MOS inverter/amplifier to form a threshold detector. The threshold point is set by R1 and R2, as well as the supply voltage. Standby power consumption is set by IABC, which, in turn, is set by RABC in the usual manner (Sec. 8-2).

The standby power consumption of the circuits shown in Figs. 8-20 through 8-22 is typically 6 mW. However, the standby power can be made to operate in the micropower region by changes in the value of RABC. Also, for greater current output from any of the circuits shown in Figs. 8-20 through 8-22, the remaining MOS inverter/amplifiers in the IC can be connected in parallel with the single stage. Each of the three elements in the IC sink or source

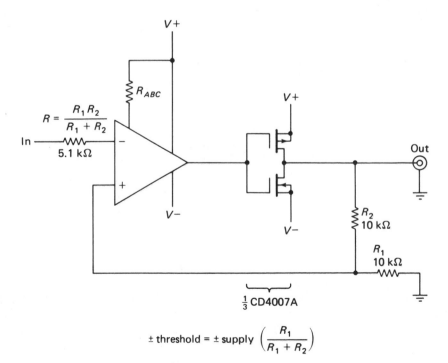

$$\pm\,\text{threshold} = \pm\,\text{supply}\left(\frac{R_1}{R_1 + R_2}\right)$$

FIGURE 8-22 OTA combined with MOS to form a threshold detector.

about 6 mA. Thus, with all three elements in parallel, the circuit should be able to sink or source about 18 mA.

8-11   OTA MICROPOWER COMPARATOR

Figure 8–23 shows an OTA combined with two MOS inverter/amplifiers to form a micropower comparator. Circuit output is proportional to the differential signal at the OTA inputs. If both inputs are at the same level, there is no output.

If desired, either of the inputs (inverting or noninverting) can be adjusted to some reference level by means of the voltage-divider network shown in Fig.

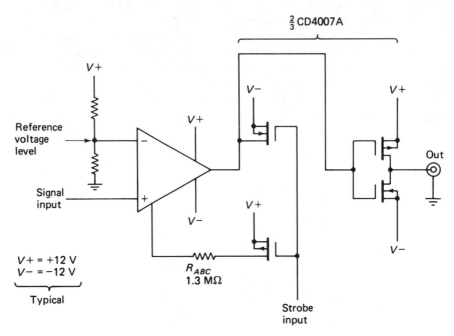

FIGURE 8-23   OTA combined with two MOS stages to form a micropower comparator.

8–23. Under these conditions, the output is proportional to the difference between the signal input level and the reference level.

The circuit is "on" only when a strobe or gate pulse is applied. The standby power consumption of the circuit is about 10 μW. When the circuit is strobed, the OTA consumes about 420 μW. Under these conditions, the circuit responds to a differential input-voltage signal of about 6 μs. By decreasing the value of RABC, the circuit response time can be decreased to about 150 ns. However, the standby power consumption rises to about 20 μW.

The differential input common-mode range of the circuit is about -1 V to $+10$ V. The voltage gain of the circuit is about 130 dB.

9

AMPLIFIER TEST PROCEDURES

This chapter is devoted entirely to test procedures for amplifiers. These procedures can be applied to a complete amplifier (such as a stereo system) or to specific circuits (such as the audio and RF circuits of a radio transmitter or receiver).

The procedures can also be applied to amplifier circuits at any time during design or experimentation. As a minimum, the tests should be made when the circuit is first completed in experimental form. If the test results are not as desired, the component values can be changed as necessary to obtain the desired results. The circuit can then be retested in final form (with all components and/or ICs soldered in place). This shows if there is any change in circuit performance because of physical relocation of components.

The chapter includes a series of notes regarding changes in component values on test results. This information is summarized in Sec. 9–2 and is of particular interest to hobbyists and experimenters.

It is assumed that the reader is already familiar with the use of common electronic test equipment. Such subjects are covered thoroughly in the author's many books on test equipment and procedures.

9–1 BASIC AUDIO-AMPLIFIER TESTS

The following paragraphs do not describe every possible test to which audio circuits can be subjected. However, the paragraphs do include all basic tests necessary for typical audio-amplifier circuit operation.

9-1.1 Frequency Response

The frequency response of an audio amplifier, or filter, can be measured with an audio signal generator and a meter or scope. When a meter is used, the signal generator is tuned to various frequencies, and the resultant circuit output response is measured at each frequency. The results are then plotted in the form of a graph or response curve, such as shown in Fig. 9–1. The procedure is essentially the same when a scope is used to measure audio-circuit response. However, the scope gives an added benefit of visual distortion analysis, as discussed in Secs. 9–1.10 and 9–1.11.

9-1.2 Basic Frequency-Response Tests

The basic frequency-response test or measurement procedure (with either meter or scope) is to apply a *constant-amplitude signal* to the circuit input while monitoring the circuit output. The input signal is varied in frequency (but not amplitude) across the entire operating range of the circuit.

Any well-designed audio circuit should have a constant response from about 20 Hz to 20 kHz. With direct-coupled amplifiers and many ICs, the response can be extended from a few hertz up to 100 kHz (and higher). The voltage output at the various frequencies across the range is plotted on a graph similar to that shown in Fig. 9–1.

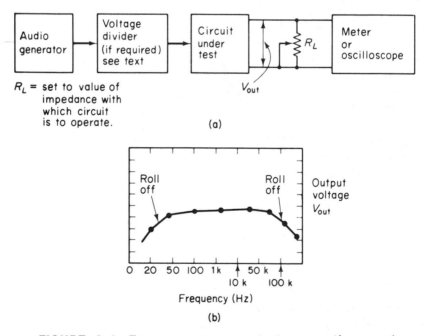

FIGURE 9-1 Frequency-response test connections and typical response curve.

The basic frequency-response test procedure is as follows:

1. Connect the equipment as shown in Fig. 9-1. Set the generator, meter, or scope controls as necessary. It is assumed that the audio generator is provided with controls to vary the output in both frequency and amplitude.

If the audio generator cannot be adjusted in amplitude, the output can be applied to the test-circuit input through a *precision voltage divider*, such as shown in Fig. 9-2. This divider provides test voltage in precision steps (1 mV, 10mV, 100 mV, and 1 V).

In use, the audio generator is adjusted for 10-V output, and each of the voltage-divider outputs or steps is checked with a precision voltmeter. Accuracy of the step voltage depends on accuracy of the resistance values. Precision resistors with 1% accuracy (or better) should be used. It is also assumed that the audio generator is provided with an output amplitude meter. If not, the generator output can be monitored with an external meter.

2. Initially, set the generator output frequency to the low end of the frequency range. Then set the generator output amplitude to the desired input level. For example, most audio amplifiers are rated as to some specific input voltage for a given output (1-V input for full output, 1-V input for 10-W output, and so on).

3. In the absence of a realistic test input voltage, set the generator output to an arbitrary value. A simple method of finding a satisfactory input level is to monitor the circuit output (with the meter or scope) and increase the generator output at the circuit center frequency (or at 1 kHz) until the circuit is overdriven. This point is indicated when further increases in generator output do

Basic Design Rules

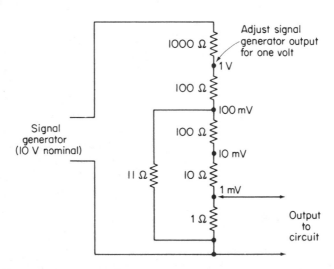

FIGURE 9-2 Voltage divider with low output impedance.

not cause further increases in meter reading (or the output waveform peaks begin to flatten on the scope display). Set the generator output *just below* this point. Then return the meter or scope to monitor the generator voltage (at the test circuit input) and measure the voltage. Keep the generator at this voltage level throughout the test.

4. If the circuit is provided with any operating or adjustment controls (volume, loudness, gain, treble, bass, and so on), set these controls to some arbitrary point when making the initial frequency-response measurement. The response measurements can then be repeated at different control settings if desired. (Use the control setting recommended in service literature, if available.)

5. Record the circuit output voltage on the graph. *Without changing the generator output amplitude*, increase the generator frequency by some fixed amount, and record the new circuit output voltage. The amount of frequency increase between each measurement is an arbitrary matter. Use an increase of 10 Hz at the low end and high end (where rolloff occurs) and an increase of 100 Hz at the middle frequencies.

6. Repeat the process, checking and recording the circuit output voltage at each checkpoint to produce a frequency-response curve. For a typical audio circuit, the curve resembles that of Fig. 9–1, with a flat portion across the middle frequencies and a rolloff at each end. A bandpass filter has a similar response curve. High- and low-pass filters produce curves with rolloff at one end only. (High-pass has a rolloff at the low end, and vice versa.)

7. After the initial frequency-response check, the effect of operating or adjustment controls should be checked. Volume, loudness, and gain controls should have the same effect all across the frequency range. Treble and bass controls may also have some effect at all frequencies. However, a treble control should have the greatest effect at the high end, whereas the bass control affects the low end most.

8. Note that generator output may vary with changes in frequency, a fact often overlooked in making a frequency-response test of any circuit (not just audio circuits). Even precision laboratory generators can vary in amplitude output with changes in frequency, thus resulting in considerable error.

It is recommended that the generator output be monitored after *each change* in frequency (some audio generators have a built-in output meter). Then, if necessary, the generator output amplitude can be reset to the correct value. Within extremes, it is more important that the generator output amplitude *remain constant* rather than at some specific value when making a frequency-response check.

9–1.3 *Voltage-Gain Tests*

Voltage-gain tests of an audio circuit are made in the same way as for frequency response. The ratio of output voltage to input voltage (at any given frequency or across the entire frequency range) is the voltage gain. Since the input voltage (generator output) must be held constant for a frequency-response test, a voltage-gain curve should be identical to a frequency-response curve.

9-1.4   Power Output and Gain Tests

The *power output* of an audio circuit is found by noting the output voltage Vout across load resistance RL (Fig. 9-1) at any frequency or across the entire frequency range. Power output is found by $(Vout)^2/RL$.

To find the *power gain* of an audio circuit, it is necessary to find both the input and output power. Input power is found in the same way as output power, except that the *impedance at the input* must be known (or calculated). This is not always practical in some audio circuits, especially in designs where input impedance depends on transistor or IC gain. (The procedure for finding input impedance is given in Sec. 9-1.8.) With input power known (or calculated), the power gain is the ratio of output power to input power.

An *input sensitivity* specification is often used in place of power gain for some audio circuits and amplifiers. Input sensitivity specifications require a minimum power output with a given voltage input (such as 100-W output with 1-V input).

9-1.5   Power-Bandwidth Tests

Many audio-circuit or amplifier specifications include a power-bandwidth factor. Such specifications require that the audio circuit deliver a given power output across a given frequency range. For example, a certain audio amplifier circuit produces full power output up to 200 kHz (as shown in Fig. 9-3.), even though the frequency response is flat up to 100 kHz. That is, voltage (without a load) remains constant up to 100 kHz, while power output (across a normal load) remains constant up to 20 kHz.

9-1.6   Load-Sensitivity Tests

An audio circuit is sensitive to changes in load. This is especially true of audio power amplifiers, but can also be the case with voltage amplifiers. An amplifier produces maximum power (when the amplifier output impedance is the same

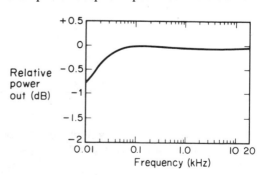

FIGURE 9-3 Typical power-bandwidth graph.

as the load impedance. This is shown by the curve of Fig. 9–4 (the load sensitivity for a typical audio-amplifier circuit).

If the load is twice the output-circuit impedance (ratio of 2.0), the output is reduced to about 50%. If the load is 40% of the output impedance (ratio of 0.4), the output power is reduced to about 25%. Generally, a power-amplifier circuit should be checked for load sensitivity during some stage of design. Such a test often shows defects in design that are not easily found with the usual frequency-response and power-output tests.

The circuit for a load-sensitivity test is the same as for frequency response (Fig. 9–1), except that the load resistance RL must be variable. (Never use a wirewound load resistance. The reactance can result in considerable error. If a non-wirewound variable resistance of sufficient wattage rating is not available, use several fixed carbon or composition resistances arranged to produce the desired resistance value.)

Measure the power output at various load-impedance/output-impedance ratios. To make a comprehensive test of an audio circuit under design, repeat the load-sensitivity test across the entire frequency range.

9–1.7 Dynamic Output Impedance Tests

The load-sensitivity test described in Sec. 9–1.6 can be reversed to find the dynamic output impedance of an audio circuit. The connections and procedure (Fig. 9–1) are the same, except that the load resistance RL is varied until maximum output power is found. Power is removed and RL is disconnected from

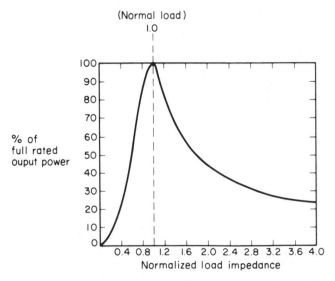

FIGURE 9–4 Typical output-power versus load-impedance (load-sensitivity) graph.

the circuit. The d-c resistance of RL (measured with an ohmmeter) is equal to the dynamic output impedance. Of course, the value applies only at the frequency of measurement. The test should be repeated across the entire frequency range of the circuit.

9-1.8 Dynamic Input Impedance Tests

Use the circuit of Fig. 9–5 to find the dynamic input impedance of an audio circuit. The test conditions should be identical to those for frequency response, power output, and so on. That is, the same audio generator, operating load, meter or scope, and frequencies should be used.

Adjust the signal source to the frequency (or frequencies) at which the circuit is operated. Move switch S back and forth between position A and B, while adjusting resistance R until the voltage reading is the same in both positions of the switch. Disconnect resistance R from the circuit, and measure the d-c resistance of R with an ohmmeter. The d-c resistance of R is then equal to the dynamic impedance at the circuit input.

Accuracy of the impedance measurement depends on the accuracy with which the d-c resistance is measured. A noninductive resistance must be used. The impedance found by this method applies only to the frequency used during the test.

9-1.9 Audio-Circuit Signal-Tracing Tests

A scope is the most logical instrument for tracing and testing signals throughout audio circuits, whether the circuits are complete audio-amplifier systems or a single stage. The scope duplicates every function of an electronic voltmeter or DVM in troubleshooting, signal tracing, and performance testing. In addition, the scope offers the advantage of a visual display for such common audio-circuit conditions as distortion, hum, noise, ripple, and oscillation.

A scope is used in a manner similar to that of a voltmeter when signal-tracing audio circuits. A signal is introduced into the input by the signal generator. The amplitude and waveform of the input signal are measured on the scope. The scope probe is then moved to the input and output of each stage,

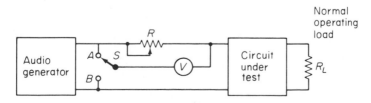

FIGURE 9-5 Test circuit for measurement of dynamic input impedance.

in turn, until the final output is reached. The gain of each stage is measured as a voltage on the scope. In addition, it is possible to observe any change in waveform from that applied to the input. Thus, stage gain and distortion (if any) are established quickly with a scope.

9-1.10 Checking Distortion by Sine-Wave Analysis Tests

The connections for audio-circuit signal-tracing tests with sine waves are shown in Fig. 9–6. The procedure for checking audio-circuit distortion with sine waves is essentially the same as that described in Sec. 9–1.9. The primary concern, however, is *deviation* of the amplifier (or stage) output waveform from the input waveform. If there is no change (except in amplitude), there is no distortion. If there is a change in the waveform, the nature of the change often reveals the cause of distortion. For example, the presence of second or third harmonics distorts the fundamental signal (Sec. 9–1.12).

In practice, analyzing sine waves to pinpoint audio-circuit problems that produce distortion is a difficult job, requiring considerable experience. Unless

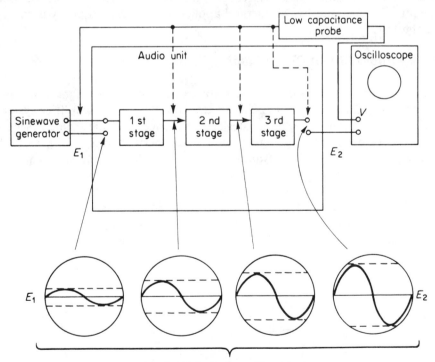

Amplitude increases with each stage
waveform remains substantially the same

FIGURE 9-6   Basic audio signal tracing with an oscilloscope.

distortion is severe, the distortion may pass unnoticed. Sine waves are best used where harmonic-distortion or intermodulation-distortion meters are combined with scopes for distortion analysis. If a scope is to be used alone, square waves provide the best basis for distortion analysis. (The reverse is true for frequency-response and power measurements.)

9-1.11 Checking Distortion by Square-Wave Analysis Tests

The procedure for checking distortion with square waves is essentially the same as for sine waves. Distortion analysis is more effective with square waves because of their high odd-harmonic content, and because it is easier to see deviation from a straight line with sharp corners than from a curving line.

As in the case of sine-wave-distortion testing, square waves are introduced into the circuit input, while the output is monitored on a scope, as shown in Fig. 9–7. The primary concern is deviation of the amplifier (or stage) output waveform from the input waveform (which is also monitored on the scope). If the scope has a dual-trace feature, the input and output can be monitored simultaneously.

If there is a change in waveform, the nature of the change often reveals the cause of distortion. For example, a comparison of the square-wave response shown in Fig. 9–8 against the typical patterns of Fig. 9–7 shows possible low-frequency phase shift. In any event, such a comparison shows some problem in the low-frequency response.

The third, fifth, seventh, and ninth harmonics of a clean square wave are emphasized. Thus, if an audio circuit passes a signal of some given frequency and produces a clean square-wave output, it is safe to assume that the frequency response is good up to at least nine times the fundamental frequency. For example, if the response at 10 kHz is as shown in Fig. 9–8, it is fair to assume that the response is the same at 90 to 100 kHz. This is convenient since not all audio generators provide an output up to 100 kHz (required by many audio-circuit design specifications).

9-1.12 Harmonic Distortion Tests

No matter what audio circuit is used or how well the circuit is designed, there is always the possibility of odd or even harmonics being present with the fundamental. These harmonics combine with the fundamental and produce distortion, as is the case when any two signals are combined. The effects of second and third harmonic distortion are shown in Fig. 9–9. As shown, when harmonic signals swing negative simultaneously with a positive swing of the fundamental, or vice versa, the fundamental signal is distorted by the combination.

Commercial harmonic-distortion meters operate on the *fundamental suppression* principle. As shown in Fig. 9–10, a sine wave is applied to the circuit

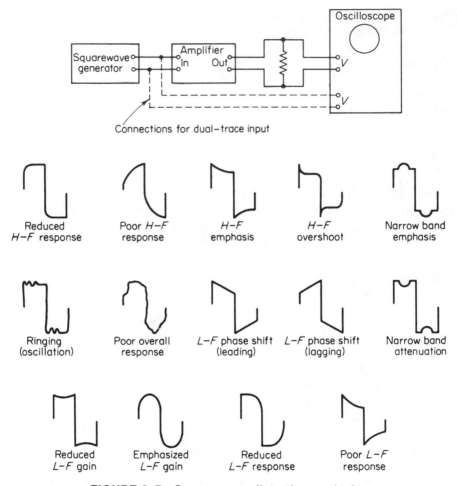

FIGURE 9-7 Square-wave distortion analysis.

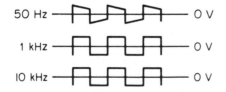

FIGURE 9-8 Square-wave response of audio amplifier.

input, and the circuit output is measured on the scope. The output is then applied through a filter that suppresses the fundamental frequency. Any output from the filter is then the result of harmonics.

This output can also be displayed on the scope. (Some commercial harmonic distortion meters use a built-in meter instead of, or in addition to, an

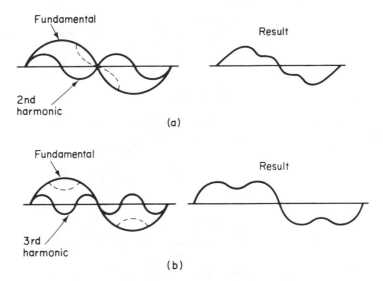

FIGURE 9-9 Effects of second and third harmonic distortion.

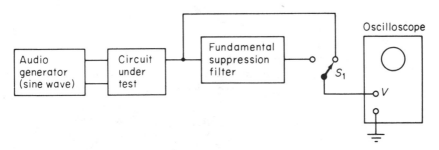

FIGURE 9-10 Basic harmonic distortion meter circuit using fundamental-suppression principle.

external scope.) When the scope is used, the filter-output signal frequency is checked to determine harmonic content. For example, if the input is 1 kHz and the output (after filtering) is 3 kHz, third-harmonic distortion is indicated.

The percentage of harmonic distortion can also be determined by this method. For example, if the output without filter is 100 mV and with filter is 3 mV, a 3% harmonic distortion is indicated.

In some commercial harmonic-distortion meters, the filter is tunable so that the amplifier can be tested over a wide range of fundamental frequencies. In other harmonic distortion meters, the filter is fixed in frequency, but can be detuned slightly to produce a sharp null.

When a design circuit is tested over a wide range of frequencies for harmonic distortion and the results plotted on a graph similar that of Fig. 9–11,

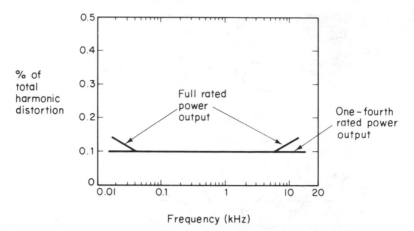

FIGURE 9-11   Typical total harmonic distortion (THD) graph.

the percentage is known as the *total harmonic distortion* or THD. Note that the THD shown in Fig. 9-11 is less than 0.2%. Also note that harmonic distortion can vary with frequency and power output.

9-1.13   Intermodulation Distortion Tests

When two signals of different frequency are mixed in any circuit, there is a possibility of the lower-frequency signal amplitude-modulating the higher-frequency signal. This produces a form of distortion known as *intermodulation distortion*.

Commercial intermodulation-distortion meters consist of a signal generator and high-pass filter as shown in Fig. 9-12. The signal-generator portion of the meter produces a high-frequency signal (usually about 7 kHz) that is modulated by a low-frequency signal (usually 60 Hz). The mixed signals are applied to the circuit input.

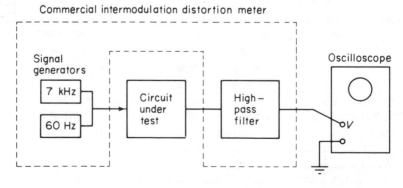

FIGURE 9-12   Basic intermodulation distortion meter circuit.

The circuit output is connected through a high-pass filter to the scope vertical channel. The high-pass filter removes the low-frequency (60-Hz) signal. Thus, the only signal appearing on the scope vertical channel should be the high-frequency (7-kHz) signal. If any 60-Hz signal is present on the display, the signal is being passed through as modulation on the 7-kHz signal.

Figure 9–13 shows an intermodulation test circuit that can be fabricated in the shop or laboratory. Note that the high-pass filter is designed to pass signals above 200 Hz. The purpose of the 39- and 10-kΩ resistors is to set the 60-Hz signals at four times the 7-kHz signal. Many audio generators provide for a line-frequency output (60 Hz) that can be used as the low-frequency modulation source.

If the circuit of Fig. 9–13 is used instead of a commercial meter, set the generator line-frequency (60-Hz) output to some fixed value (1 V, 2 V, and so on). Then set the generator output (7 kHz) to the same value. If the line-frequency output is not adjustable, measure the actual value of the line-frequency output and then set the generator audio output to the same value.

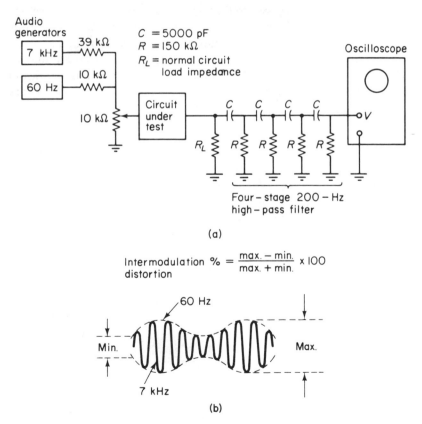

(a)

$$\text{Intermodulation \% distortion} = \frac{\text{max.} - \text{min.}}{\text{max.} + \text{min.}} \times 100$$

(b)

FIGURE 9-13 Test circuit for measurement of intermodulation distortion percentage.

The percentage of intermodulation distortion can be calculated using the equation of Fig. 9–13.

9-1.14 Background Noise Tests

If the vertical channel of a scope is sufficiently sensitive, a scope can be used to test and measure the background noise level of an audio circuit, as well as to test for the presence of hum, oscillation, and so on. The scope vertical channel should be capable of a measurable deflection with about 1 mV (or less), since this is the background noise level of many audio amplifiers.

The basic procedure consists of measuring circuit output with the volume or gain control (if any) at maximum, but without an input signal. The scope is superior to a voltmeter for noise-level measurement since the frequency and nature of the noise (or other signal) are displayed visually.

The basic connections for background-noise-level tests are shown in Fig. 9–14. The scope gain or sensitivity control is increased until there is a noise or "hash" indication. It is possible that a noise indication could be caused by pickup in the test connections. If in doubt, disconnect the test leads from the circuit, but not from the scope. If the noise indications are removed when the test leads are disconnected, the noise is being produced in the circuit under test.

If you suspect that there is 60-Hz line hum present in the circuit output (picked up from the power supply or any other source), set the scope sync control to line. If a stationary pattern appears, there is line hum.

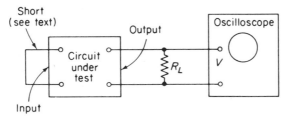

R_L = normal circuit load impedance

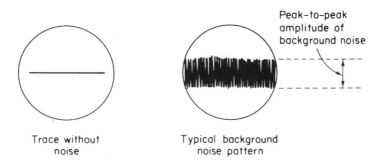

Trace without noise

Typical background noise pattern

FIGURE 9-14 Measuring circuit background noise and hum.

If a signal appears that is not at the line frequency, the signal may be caused by oscillation in the circuit or by stray pickup. Short the circuit input terminals. If the signal remains on the scope display, the circuit is probably oscillating.

9-1.15 Phase-Shift and Feedback Measurements

In any amplifier there is some phase shift between input and output signals. This is usually not critical for audio-amplifier circuits. One exception is in op-amps where feedback from output to input is used to control gain, as discussed in Chapter 7. For that reason, the procedures for measurement of phase-shift and feedback levels in amplifier circuits are described in Sec. 9-4.

9-2 ANALYZING AMPLIFIER OPERATION WITH TEST RESULTS

The following notes apply primarily to solving *design problems* (poor frequency response, lack of gain, and the like) in amplifiers, but can also be applied to analyzing circuits of existing commercial amplifiers.

When design circuits of any kind of fail to perform properly (or as hoped), a planned procedure for isolating the problem is very helpful. Remember that circuit troubleshooting (either discrete component or IC) is difficult at best. This is especially true when the circuit involves more than one stage, since the stages are interdependent, or when more than one IC is involved.

9-2.1 Failure in Design Circuits

A special problem arises in analyzing the failure of design circuits. The first requirement in logical troubleshooting is a thorough knowledge of the circuit's performance when *operating normally*. However, a failure in a trial circuit, just designed, can be the result of component failure or improper trial values for components. For example, an existing amplifier may show low gain based on past performance. A newly designed amplifier circuit may show the same identical results simply because that is the best gain possible with the selected trial components.

To minimize the problem in newly designed amplifiers, try to isolate problems on a stage-by-stage basis. For example, if the circuit has two or more stages and gain is low for the overall circuit, measure the gain for each stage. With trouble isolated to a particular stage, try to determine which half of the stage is at fault.

9-2.2 Stage Input/Output Relationship

Any stage (discrete component or IC) has two halves, input and output. Generally, for a transistor stage, the input is base–emitter (or gate–source for a FET), with the emitter–collector (source–drain) acting as the output. Remember that a defect in one-half affects the other half. An obvious example of this is where low input current (base) produces low output current (collector). Sometimes less obvious is the case where output affects input. For example, in a stage with an emitter resistor (for feedback stabilization), an open collector can appear to reduce the input impedance.

9-2.3 Loop-Feedback Problems

Circuits with loop-feedback or overall-feedback present a particular problem. A closed feedback loop causes all stages to respond as a unit, making it difficult to know which stage is at fault. Of course, with an IC, it makes no difference which stage is faulty, since the IC is replaced as a unit.

The loop-feedback problem can be solved by opening the loop. To do so, however, creates another problem, since the operating Q point of one (or possibly all) stages is disturbed. The subject of loop-feedback troubleshooting is discussed further in Chapter 10.

9-2.4 Saturated and/or Cutoff Transistors

Look for any transistor that is full-on (collector or drain voltage very low) or full-off (collector or drain voltage very high, probably near the supply voltage) during test procedures. Either of these conditions in a linear amplifier of any type is the result of component failure or improper design. In effect, the transistor is operating as a switch, rather than an amplifier.

9-2.5 Bias Problems

Design faults that appear under no-signal conditions (such as improper Q point) are generally the result of improper bias relationships. Faults that appear only when a signal is applied can be caused by poor bias, but can also result from wrong component values.

A high-gain circuit is generally more difficult to bias than a low-gain circuit. Use the following procedure when it is difficult to find a good bias point (or Q point) for a high-gain amplifier.

Increase the signal input until the output waveform appears as a square wave (Fig. 9-15). That is, *overdrive* the amplifier circuit.

Keep reducing the input signal, while adjusting the bias until both positive and negative peaks are clipped by the same amount. If it is impossible to find any bias point where the signal peaks can be clipped symmetrically, a defect in design or components can be suspected.

Unsymmetrical clipping can be caused by operating the transistor on a nonlinear portion of the transfer curve or load line (improper bias) or by the fact that the transistor does not have a linear curve (or a very short linear curve). Try a different transistor. If the results are the same, change the circuit trial values (particularly collector, emitter, and base resistance).

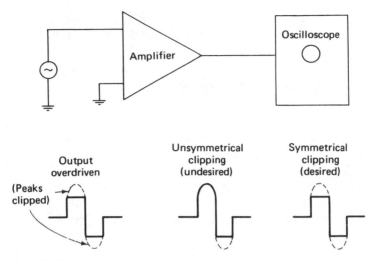

FIGURE 9-15 Determining proper bias (Q point) by means of oscilloscope display.

9-2.6 *Oscillation and Feedback Problems*

If it is impossible to get any Q point except very near full-on or full-off, this can be caused by *excess positive feedback*. While this condition is desirable in a multivibrator or oscillator, the condition must be avoided in linear amplifiers.

High-gain amplifiers should be carefully checked to find any tendency to oscillate or to exhibit abnormal noise. High-gain amplifiers may also be very sensitive to supply-voltage changes. When an amplifier design circuit is complete, it is often helpful to repeat all the basic test procedures with various supply voltages. Unless otherwise specified by design requirements, any amplifier should perform equally well with a ±10% supply-voltage variation.

If oscillation occurs in any amplifier circuit (high or low gain), try moving the input and output leads. Even a well-designed amplifier may oscillate simply because input and output leads are close together. It may be necessary to physically relocate parts or to shield parts of an amplifier to prevent feedback that results in oscillation.

Low-frequency oscillation is often the result of poor supply-voltage filtering or too many stages connected to the same supply-voltage point. Try isolating the stages with separate supply-voltage filter capacitors.

9-2.7 *Poor Low-Frequency Response*

The most common cause of poor low-frequency response is *low capacitor values*. The design procedures of Chapters 1 and 3 provide for an approximate 1- or 3-dB loss at the low-frequency limit. If a greater loss can be tolerated, a lower capacitor value can be used. (This applies to coupling capacitors, emitter-bypass capacitors, or source-bypass capacitors, but not to power-supply or decoupling capacitors.) If better low-frequency response is desired, all other factors being equal, *increase capacitor values.*

9-2.8 Poor High-Frequency Response

The most common cause of poor high-frequency response is the *input capacitance of transistors.* As frequency increases, transistor input capacitance decreases, changing the input impedance. This change in input impedance usually results in decreased gain, all other factors remaining equal. Generally, poor high-frequency response is not a problem over the audio range (up to about 20 kHz), but can be a problem beyond about 200 kHz. The only practical solutions are to accept reduced stage gain or change transistors.

9-3 BASIC RADIO-FREQUENCY AMPLIFIER TESTS

This section describes test procedures for RF amplifiers. The first paragraphs are devoted to test and measurement procedures for the resonant circuits used at radio frequencies (resonant-frequency measurements, Q measurements, and the like). The remaining sections cover test procedures for complete RF amplifiers in such typical applications as a radio transmitter.

Experimental RF circuits should always be retested in final form (with all components soldered in place). This shows if there is any change in circuit characteristics because of the physical relation of components. Such tests are especially important at the higher radio frequencies. Often, there is capacitance or inductance between components, from components to wiring, and between wires. These stray "components" can add to the reactance and impedance of circuit components. When the physical locations of parts and wiring are changed, the stray reactances change and alter circuit performance.

9-3.1 Basic RF Voltage Measurement

When the voltages to be measured are at radio frequencies and are beyond the frequency capabilities of the meter or scope, an *RF probe* is required. Such probes rectify the RF signals into a d-c output that is almost equal to the peak RF voltage. The d-c output of the probe is then applied to the meter or scope input and is displayed as a voltage readout in the normal manner.

If a probe is available as an accessory for a particular meter, that probe should be used in favor of any experimental or homemade probe. The manufacturer's probe is matched to the meter in calibration, frequency compensation, and so on. If a probe is not available for a particular meter or scope, the following notes discuss the fabrication of probes suitable for measurement and testing of RF circuits.

Half-Wave RF Probe.   The half-wave probe of Fig. 9–16 provides an output to the meter (or scope) that is about equal to the peak value of the voltage being measured. Since most meters are calibrated to read in rms values, the probe output must be reduced to 0.707 of the peak value by R1.

The value of R1 can be found by calculation. For practical purposes, a variable resistor should be used during calibration and then be replaced by a fixed resistor of the correct value. The following steps describe the calibration procedure:

1. Connect the experimental probe circuit to a signal generator and meter.

2. Set the meter to measure d-c voltage. Either a VOM or digital meter can be used, but best results are found with a high-input-impedance meter.

3. Adjust the signal-generator voltage amplitude to some precise value, such as 10 V (rms), as measured on the generator's output meter.

4. Adjust the calibrating resistor R1 until the meter indicates the same value (10 V rms).

5. As an alternative procedure, adjust the signal generator for a 10-V peak output; then readjust R1 for a reading of 7.07 on the meter being calibrated.

6. Remove the power, disconnect the circuit, measure the value of R1, and replace the variable resistor with a fixed resistor of the same value.

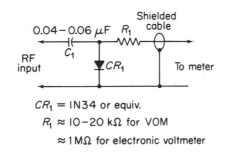

CR_1 = IN34 or equiv.

$R_1 \approx$ 10–20 kΩ for VOM

$\approx$ 1 MΩ for electronic voltmeter

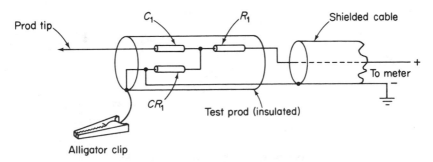

FIGURE 9-16   Half-wave RF probe.

7. Repeat the test with the fixed resistance in place. If the reading is correct, assemble the circuit in a suitable package, such as within a test probe. Repeat the test with the circuit in final form. Also repeat the test over the entire frequency range of the probe. Generally, the probe provides satisfactory response up to about 250 MHz.

8. Remember that the meter must be set to *measure direct current*, since the probe output is dc.

Demodulator Probe.   When RF signals contain modulation, a demodulator probe (Fig. 9–17) is most effective for testing RF circuits during experimentation (or at any time, actually). For example, an RF signal modulated by a fixed audio tone can be applied to an RF amplifier being tested. The demodulator measures the RF amplifier response in terms of both RF signal and audio signal.

The demodulator probe is similar to the half-wave probe except for the low capacitance of C1 and parallel resistor R2. These two components act as a filter. The demodulator probe produces both an a-c and a d-c output. The RF signal is converted into a d-c voltage about equal to the peak value. The low-frequency modulation voltage on the RF signal appears as ac at the probe output.

In use, the meter is set to dc and the RF signal is measured. Then the meter is set to ac and the modulating voltage is measured. The calibrating resistor R1 is adjusted so that the d-c scale reads the rms value. The procedure for calibra-

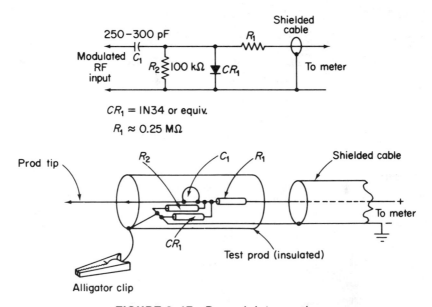

FIGURE 9-17 Demodulator probe.

tion and fabrication of the demodulator probe is the same as for the half-wave probe, except that the schematic of Fig. 9–17 should be used. Also remember that R1 should be adjusted on the basis of the RF signal (not the modulating signal) with the meter set to dc.

9-3.2 Measuring the Resonant Frequency of LC Circuits

RF equipment is based on the use of resonant circuits (called tank circuits), consisting of a capacitor and a coil (inductance) connected in series or parallel, as shown in Fig. 9–18.

At the resonant frequency, the inductive and capacitive reactances are equal, and the LC circuit acts as a high impedance (in a parallel circuit) or a low impedance (in a series circuit). In either case, any combination of capacitance and inductance has some resonant frequency.

Either (or both) the capacitance or inductance can be variable to permit tuning of the resonant circuit over a given frequency range. When the inductance is variable, tuning is usually done by means of a metal slug inside the coil. The metal slug is screwdriver-adjusted to change the inductance (and thus the inductive reactance) as required. Typical RF circuits used in receivers (AM, FM, communications, and so on) often include two resonant circuits in the form of a transformer (RF or IF transformer, and so on). Either the capacitance or inductance can be variable.

Figure 9–18 contains the equations that show the relationships among capacitance, inductance, reactance, and frequency as they relate to resonant circuits. Note that there are two sets of equations. One set of equations includes reactance (inductive and capacitive). The other set omits reactance.

A meter can be used in conjunction with an RF signal generator to find the resonant frequency of either series or parallel LC circuits. The generator must be capable of producing a signal at the resonant frequency of the circuit, and the meter must be capable of measuring the frequency. If the resonant frequency is beyond the normal range of the meter, an RF probe must be used. The following steps describe the measurement procedure:

1. Connect the equipment as shown in Fig. 9–19. Use the connections of Fig. 9–19a for parallel-resonant LC circuits or the connections of Fig. 9–19b for series-resonant LC circuits.

2. Adjust the generator output until a convenient midscale indication is obtained on the meter. Use an unmodulated signal output from the generator.

3. Starting at a frequency well below the lowest possible frequency of the circuit under test, slowly increase the generator output frequency. If there is no way to judge the approximate resonant frequency, use the lowest generator frequency.

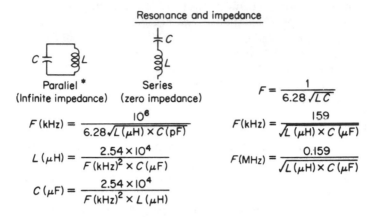

Resonance and impedance

Paraliel * (Infinite impedance)　　**Series** (zero impedance)

$$F = \frac{1}{6.28\sqrt{LC}}$$

$$F\,(\text{kHz}) = \frac{10^6}{6.28\sqrt{L\,(\mu H) \times C\,(\text{pF})}}$$

$$F\,(\text{kHz}) = \frac{159}{\sqrt{L\,(\mu H) \times C\,(\mu F)}}$$

$$L\,(\mu H) = \frac{2.54 \times 10^4}{F\,(\text{kHz})^2 \times C\,(\mu F)}$$

$$F\,(\text{MHz}) = \frac{0.159}{\sqrt{L\,(\mu H) \times C\,(\mu F)}}$$

$$C\,(\mu F) = \frac{2.54 \times 10^4}{F\,(\text{kHz})^2 \times L\,(\mu H)}$$

*Approximate; accurate when circuit Q is 10 or higher

Inductive reactance

$$Z = \sqrt{R^2 + X_L^2} \qquad Q = \frac{X_L}{R} \qquad L = \frac{X_L}{6.28\,F}$$

Series

$$Z = \frac{RX_L}{\sqrt{R^2 + X_L^2}} \qquad Q = \frac{R}{X_L} \qquad F = \frac{X_L}{6.28\,L}$$

Parallel

$$X_L = 6.28 \times F\,(\text{Hz}) \times L\,(\text{H})$$
$$X_L = 6.28 \times F\,(\text{kHz}) \times L\,(\text{mH})$$
$$X_L = 6.28 \times F\,(\text{MHz}) \times L\,(\mu H)$$

Capacitive reactance

$$Z = \sqrt{R^2 + X_C^2} \qquad Q = \frac{X_C}{R} \qquad F = \frac{1}{6.28\,CX_C}$$

Series

$$Z = \frac{RX_C}{\sqrt{R^2 + X_C^2}} \qquad Q = \frac{R}{X_C} \qquad C = \frac{1}{6.28\,FX_C}$$

Parallel

$$X_C = \frac{1}{6.28 \times F\,(\text{Hz}) \times C\,(\text{F})}$$
$$X_C = \frac{159}{F\,(\text{kHz}) \times C\,(\mu F)}$$

FIGURE 9-18　Resonant circuit equations.

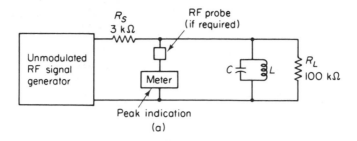

Peak indication

(a)

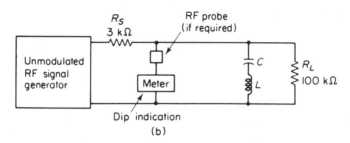

Dip indication

(b)

FIGURE 9-19 Measuring resonant frequency of LC circuits.

4. If the circuit being tested is parallel resonant, watch the meter for a maximum, or peak, indication.

5. If the circuit being tested is series resonant, watch the meter for a minimum, or dip, indication.

6. The resonant frequency of the circuit under test is the one at which there is a maximum (for parallel) or minimum (for series) indication on the meter.

7. There may be peak or dip indications at harmonics of the resonant frequency. Therefore, the test is most efficient when the approximate resonant frequency is known.

8. The value of load resistor RL is not critical. The load is shunted across the LC circuit to flatten or broaden the resonant response (to lower the circuit Q, as discussed in Sec. 9-3.5). Thus, the voltage maximum or minimum is approached more slowly. A suitable trial value for RL is 100 kΩ. A lower value of RL sharpens the resonant response, and a higher value flattens the curve.

9-3.3 *Measuring the Inductance of a Coil*

Once a coil has been designed and wound, using the theoretical values such as described in Chapter 4, it is often convenient to measure the actual inductance. The same is true of a commercially wound coil where the inductance is unknown.

A meter can be used in conjunction with an RF signal generator and a fixed capacitor of known value and accuracy to find the inductance of a coil. The generator must be capable of producing a signal at the resonant frequency of the test circuit, and the meter must be capable of measuring the frequency. If the resonant frequency is beyond the normal range of the meter, an RF probe must be used. The following steps describe the measurement procedure.

1. Connect the equipment as shown in Fig. 9–20. Use a capacitive value such as 10-μF, 100 pF, or some other even-number value to simplify the calculation.

2. Adjust the generator output until a convenient midscale indication is obtained on the meter. Use an unmodulated signal output from the generator.

3. Starting at a frequency well below the lowest possible resonant frequency of the inductance–capacitance combination under test, slowly increase the generator frequency. If there is no way to judge the approximate resonant frequency, use the lowest generator frequency.

4. Watch the meter for maximum or peak indication. Note the frequency at which the peak indication occurs. This is the resonant frequency of the circuit.

5. Using this resonant frequency and the known capacitance value, calculate the unknown inductance using the equation of Fig. 9–20.

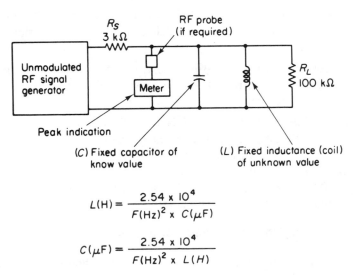

$$L(H) = \frac{2.54 \times 10^4}{F(Hz)^2 \times C(\mu F)}$$

$$C(\mu F) = \frac{2.54 \times 10^4}{F(Hz)^2 \times L(H)}$$

FIGURE 9-20   Measuring inductance and capacitance in LC circuits.

6. Note that the procedure can be reversed to find an unknown capacitance value when a known inductance value is available.

9-3.4 *Measuring the Self-Resonance and Distributed Capacitance of a Coil*

No matter what design or winding method is used, there is some distributed capacitance in any coil. When the distributed capacitance combines with the coil's inductance, a resonant circuit is formed. The resonant frequency is usually quite high in relation to the frequency at which the coil is used. However, since self-resonance may be at or near a harmonic of the frequency to be used, the self-resonance effect may limit the coil's usefulness in LC circuits. Some coils, particularly RF chokes, may have more than one self-resonant frequency.

A meter can be used in conjunction with an RF signal generator to find both the self-resonant frequency and distributed capacitance of a coil. The generator must be capable of producing a signal at the resonant frequency of the circuit, and the meter must be capable of measuring voltages at that frequency. Use an RF probe if required. The following steps describe the measurement procedure:

1. Connect the equipment as shown in Fig. 9-21.

2. Adjust the generator output until a convenient midscale indication is obtained on the meter. Use an unmodulated signal output from the generator.

3. Tune the signal generator over the entire frequency range, starting at the lowest generator frequency. Watch the meter for either peak or dip indications. Either a peak or dip indicates that the inductance is at a self-resonant point. The generator output frequency at that point is the self-resonant frequency. Make certain that peak or dip indica-

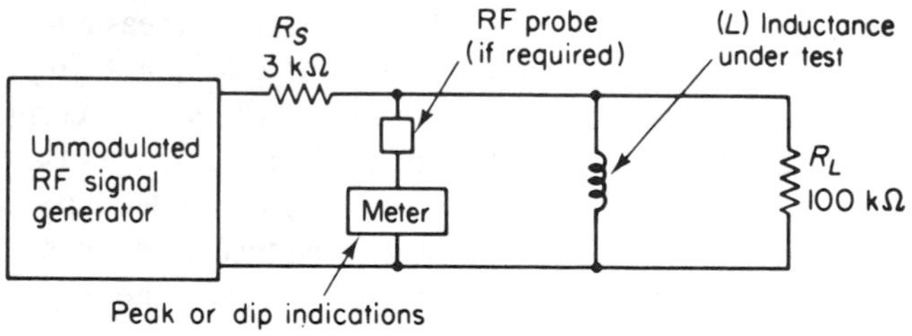

$$C(\mu F) = \frac{2.54 \times 10^4}{F(Hz)^2 \times L(\mu H)}$$

FIGURE 9-21 Measuring self-resonance and distributed capacitance of coils.

tions are not the result of changes in generator output level. Even the best laboratory generators may not produce a flat (constant-level) output over the entire frequency range.

4. Since there may be more than one self-resonant point, tune through the entire signal-generator frequency range. Try to cover a frequency range up to at least the *third harmonic of the highest frequency involved* in a resonant-circuit design.

5. Once the resonant frequency is found, calculate distributed capacitance using the equation of Fig. 9–21. For example, assume that a coil with an inductance of 7 μH is found to be self-resonant at 50 MHz:

$$C \text{ (distributed capacitance)} = \frac{2.54 \times 10^4}{(50)^2 \times 7} = 1.45 \text{ pF}$$

9–3.5   *Measuring the Q of Resonant Circuits*

A resonant circuit has a Q, or quality, factor. The circuit Q depends on the ratio of reactance to resistance. If a resonant circuit has pure reactance, the Q is high (actually infinite). However, this is not practical. For example, any coil has some d-c resistance, as do the leads of the capacitor. Also, as frequency increases, the a-c resistance presented by the leads increases. The sum total of these resistances is usually lumped together and is considered as a resistor in series or parallel with the circuit. The total resistance, usually termed the *effective resistance*, is not to be confused with the reactance.

The resonant circuit Q depends on the individual Q factors of inductance and capacitance used in the circuit. For example, if both the inductance and capacitance have a high Q, the circuit has a high Q, provided that a minimum of resistance is produced when the inductance and capacitance are connected to form a resonant circuit.

From a practical test standpoint, a resonant circuit with a high Q produces a sharp resonance curve (narrow bandwidth), whereas a low Q produces a broad resonance curve (wide bandwidth). For example, a high-Q resonant circuit provides good harmonic rejection and efficiency, in comparison with a low-Q circuit, all other factors being equal. The *selectivity* of a resonant circuit is thus related directly to Q. A very high Q (or high selectivity) is not always desired. Sometimes it is necessary to add resistance to a resonant circuit to broaden the response (increase the bandwidth, decrease the selectivity), as discussed in Chapter 4 (Fig. 4–8).

Usually, resonant-circuit Q is measured at points on either side of the resonant frequency where the signal amplitude is down 0.707 of the peak resonant value, as shown in Fig. 9–22. Note that Q must be increased for increases in resonant frequency if the same bandwidth is to be maintained.

As an example, if the resonant frequency is 10 MHz, with a bandwidth of 2 MHz, the required circuit Q is 5. If the resonant frequency is increased

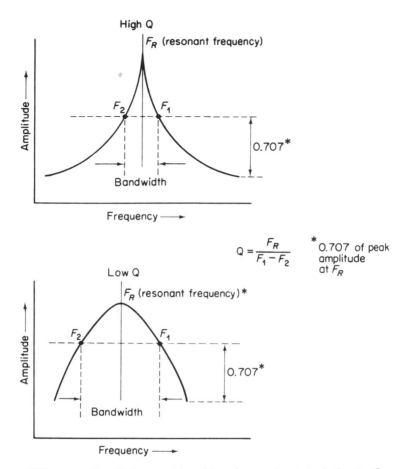

$$Q = \frac{F_R}{F_1 - F_2}$$

* 0.707 of peak amplitude at F_R

FIGURE 9-22 Relationship of bandpass characteristics to Q of resonant circuits.

to 50 MHz, with the same 2-MHz bandwidth, the required Q is 25. Also note that Q must be decreased for increases in bandwidth if the same resonant frequency is to be maintained. For example, if the resonant frequency is 30 kHz, with a bandwidth of 2 kHz, the required Q is 15. If the bandwidth is increased to 10 kHz, with the same 30-kHz resonant frequency, the required Q is 3.

The Q of a circuit can be measured using a signal generator and a meter with an RF probe. A high-impedance digital meter generally provides the least loading effect on the circuit and thus provides the most accurate indication. Figure 9-23a shows the test circuit in which the signal generator is connected directly to the input of a complete stage, and Fig. 9-23b shows the indirect method of connecting the signal generator to the input.

When a stage or circuit has sufficient gain to provide a good reading on the meter with a nominal output from the generator, the indirect method (with isolating resistor) is preferred. Any signal generator has some output impedance

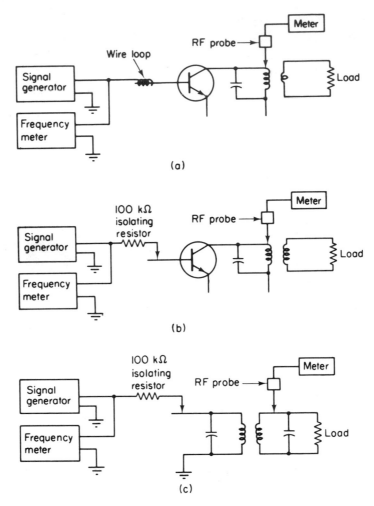

FIGURE 9-23 Measuring Q of resonant circuits.

(typically 50Ω). When this resistance is connected directly to the tuned circuit, the Q is lowered and the response becomes broader. (In some cases, the generator output impedance seriously detunes the circuit.)

Figure 9-23c shows the test circuit for a single component (such as an IF transformer). The value of the isolating resistance is not critical and is typically in the range of 100 kΩ. The procedure for determining Q using any of the circuits in Fig. 9-23 is as follows:

1. Connect the equipment as shown in Fig. 9-23a, b, or c, as applicable. Note that a load is shown in Fig. 9-23c. When a circuit is normally used with a load, the most realistic Q measurement is made with the circuit terminated in that load value. A fixed resistance can be used

to simulate the load. The Q of a resonant circuit often depends on the load value.

2. Tune the signal generator to the circuit resonance frequency. Operate the generator to produce an unmodulated output.

3. Tune the generator frequency for maximum reading on the meter. Note the generator frequency.

4. Tune the generator below resonance until the meter reading is 0.707 of the maximum reading. Note the generator frequency. To make the calculation more convenient, adjust the generator output level so that the meter reading is some even value, such as 1 V or 10 V, after the generator is tuned for maximum. This makes it easy to find the 0.707 mark.

5. Tune the generator above resonance until the meter reading is 0.707 of the maximum reading. Note the generator frequency.

6. Calculate the circuit Q using the equation of Fig. 9–22. For example, assume that the maximum meter indication occurs at 455 kHz (FR), the below-resonance indication is 453 kHz (F2), and the above-resonance indication is 458 kHz (F1). Then $458 - 453 = 5$, and $455/5 =$ a Q of 91.

9–3.6 Measuring the Impedance of Resonant Circuits

Any resonant circuit has some impedance at the resonant frequency. The impedance changes as frequency changes. This includes transformers (tuned and untuned), tank circuits, and so on. In theory, a series-resonant circuit has zero impedance, while a parallel-resonant circuit has infinite impedance at the resonant frequency. In practical circuits, this is impossible since there is always some resistance in the circuit.

It is often convenient to find the impedance of an experimental resonant circuit at a given frequency. Also, it may be necessary to find the impedance of a component in an experimental circuit so that other circuit values can be designed around the impedance. For example, an IF transformer presents an impedance at both the primary and secondary windings. These values may not be specified in the transformer datasheet.

The impedance of a resonant circuit or component can be measured using a signal generator and a meter with an RF probe. A high-impedance digital meter provides the least loading effect on the circuit and therefore provides the most accurate indication.

The procedure for impedance measurement at radio frequencies is the same as for audio frequencies, as discussed in Sec. 9–1.8, except as follows. An RF signal generator must be used as the signal source. The meter must be provided with an RF probe. If the circuit or component under measurement has both an input and output (such as a transformer), the opposite side or winding must be terminated in the normal load, as shown in Fig. 9–24. A fixed, noninductive, resistance can be used to simulate the load.

If the impedance of a tuned circuit is to be measured, tune the circuit to peak or dip and then measure the impedance at resonance. Once the resonant impedance is found, the signal generator can be tuned to other frequencies to find the corresponding impedance.

The RF signal generator is adjusted to the frequency (or frequencies) at which impedance is to be measured. Switch S is moved back and forth between positions A and B, while resistance R is adjusted until the voltage reading is the same in both positions of the switch. Resistor R is then disconnected from the circuit, and the d-c resistance of R is measured with an ohmmeter. The d-c resistance of R is then equal to the dynamic impedance at the circuit input.

Accuracy of the impedance measurement depends on the accuracy with which the d-c resistance is measured. A noninductive resistance must be used. The impedance found by this method applies only to the frequency used during the test.

9–3.7   *Testing Transmitter RF Circuits*

It is possible to test and adjust transmitter RF circuits using a meter and an RF probe. If an IF probe is not available (or as an alternative), it is possible to use a circuit such as shown in Fig. 9–25. This circuit is essentially a pickup coil that is placed near the RF circuit inductance and a rectifier that converts the RF into a d-c voltage for measurement on a meter. The basic procedures are as follows:

1. Connect the equipment as shown in Fig. 9–26. If the circuit being measured is an amplifier, without an oscillator, a drive signal must be supplied by means of a signal generator. Use an unmodulated signal at the correct operating frequency.

2. In turn, connect the meter (through an RF probe or the special circuit of Fig. 9–25) to each stage of the RF amplifier. Start with the first stage (this is the oscillator if the circuit under test is a complete transmitter) and work toward the final (or output) amplifier stage.

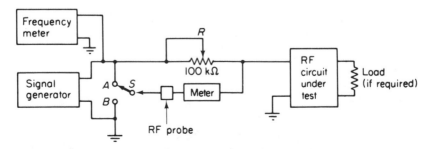

FIGURE 9–24   Measuring impedance of RF circuits.

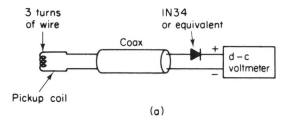

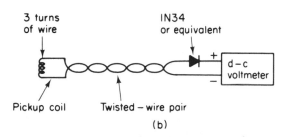

FIGURE 9-25 Test circuits for pickup and measurement of RF signals.

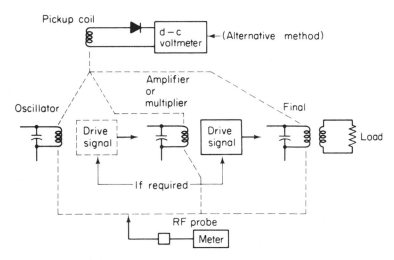

FIGURE 9-26 Testing RF transmitter circuits.

3. A voltage indication should be obtained at each stage. Usually, the voltage indication increases with each amplifier stage, as you proceed from oscillator to the final amplifier. However, some stages may be frequency multipliers and provide no voltage amplification (Chapter 4).

4. If a particular stage is to be tuned, adjust the tuning control for a maximum reading on the meter. If the stage is to be operated with a load (such as the final amplifier into an antenna), the load should be con-

nected or a simulated load should be used. A fixed, noninductive resistance provides a good simulated load at frequencies up to about 250 MHz.

5. It should be noted that this tuning method or measurement technique does not guarantee each stage is at the desired operating frequency. It is possible to get maximum readings on harmonics. However, it is conventional to design RF transmitter circuits so that the circuits do not tune to both the desired operating frequency and a harmonic. Generally, RF amplifier tank circuits tune on either side of the desired frequency, but not to a harmonic (unless the circuit is seriously de-tuned or the design calculations are hopelessly inaccurate).

9-4 OP-AMP AND OTA TEST PROCEDURES

This section describes test procedures for op-amps and OTAs. The basic procedures are essentially the same as for audio amplifiers (Sec. 9–1). The following paragraphs describe the *test differences* required for op-amp/OTAs, as well as any special test procedures unique to op-amp/OTAs.

As a minimum, the following tests should be made on the basic IC operating in an open-loop circuit. This will confirm (or deny) the IC characteristics found on the datasheet. The procedures can also be used to establish a set of characteristics for an IC where the datasheet is missing or inadequate. In the case of an OTA, it is assumed that IABC is set to some specific value (as recommended on the datasheet) and held at that value for all the following tests.

9-4.1 Frequency Response

Frequency response for an op-amp/OTA is tested in essentially the same way as for an audio amplifier (Secs. 9–1.1 and 9–1.2) except that the frequency is extended (usually) beyond the audio range. Generally, a scope is the better instrument for response measurement at higher frequencies. However, a voltmeter can be used. Both open- and closed-loop frequency response should be measured with the *same load*.

9-4.2 Voltage Gain

Voltage-gain measurement for an op-amp/OTA is essentially the same as for an audio amplifier (Sec. 9–1.3). Remember that the basic IC has a maximum input and output voltage limit, neither of which can be exceeded without possible damage to the IC and/or clipping of the waveform.

In general, the maximum rated input should be applied and the actual output measured. Check the output for clipping at the maximum level. If clipping occurs, decrease the input until clipping just stops and note the input voltage. Record these values as a basis for design.

Note the frequency at which the open-loop voltage gain drops 3 dB from the low-frequency value. This is the open-loop bandwidth. Keep in mind that the open-loop voltage gain and bandwidth are characteristics of the basic IC. Closed-loop gain is (or should be) dependent on the ratio of feedback and input resistances, while closed-loop bandwidth depends (mostly) on phase-compensation values.

Closed-loop characteristics are generally lower than open-loop characteristics (voltage gain is lower, frequency response is narrower, and so on). However, closed-loop characteristics are modifications of open-loop characteristics.

9-4.3 Power Output, Gain, and Bandwidth

Most IC op-amp/OTAs are not designed as power amplifiers. However, the power output, gain, and bandwidth can be measured in the same way as audio amplifiers (Secs. 9-1.4 and 9-1.5).

Remember that an IC has a power dissipation of its own, which must be subtracted from the *total device dissipation* to find the available power output.

9-4.4 Load Sensitivity

Since an IC op-amp/OTA is generally not used as a power amplifier, load sensitivity is not critical. However, if it should be necessary to measure the load sensitivity, use the procedure of Sec. 9-1.6.

9-4.5 Input and Output Impedance

Dynamic input and output impedances of an IC op-amp/OTA can be found using the procedures of Secs. 9-1.7 and 9-1.8. Remember that closed-loop impedances differ from open-loop impedances.

9-4.6 Distortion

Distortion measurements for IC op-amp/OTAs are the same as for audio amplifiers (Secs. 9-1.10 through 9-1.13). However, distortion requirements are usually not as critical.

9-4.7 Background Noise

Background noise measurements for IC op-amp/OTAs are the same as for audio amplifiers (Sec. 9-1.14). Generally, background noise should be measured under open-loop conditions. Some datasheets specify that both input and output voltages be measured. When input voltage is to be measured, a fixed resistance (usually 50 Ω) is connected between the input terminals.

9–4.8 Feedback Measurement

Since op-amp/OTA characteristics are usually based on the use of feedback signal, it is convenient to measure feedback voltage at a given frequency with given operating conditions.

The basic feedback-measurement connections are shown in Fig. 9–27. While it is possible to measure the feedback voltage as shown in Fig. 9–27a, a more accurate measurement is made when the feedback lead is terminated in the normal operating impedance.

If an input resistance is used in the normal circuit, and this resistance is considerably lower than the IC input impedance, use the resistance value.

If in doubt, measure the input impedance of the IC (Sec. 9–1.5); then terminate the feedback lead in that value to measure feedback voltage.

9–4.9 Input Bias Current

Input bias current can be measured using the circuit of Fig. 9–28. Any resistance value for R1 and R2 can be used, provided that the value produces a measurable voltage drop. A value of 1 kΩ is realistic for both R1 and R2 in a typical op-amp/OTA.

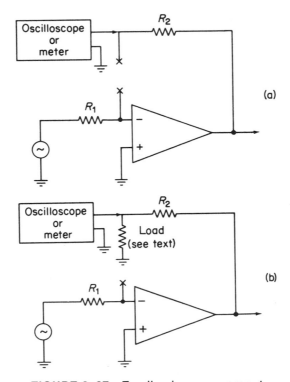

FIGURE 9-27   Feedback measurement.

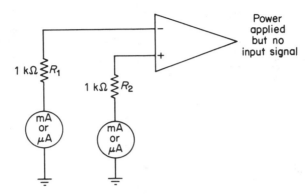

FIGURE 9-28 Input-bias current measurement.

If it is not practical to connect a meter in series with both inputs as shown in Fig. 9–28, measure the voltage drop across R1 and R2. Once the voltage drop is found, the input bias current can be calculated. For example, if the voltage drop is 3 mV across 1 kΩ, the input bias current is 3 μA.

In theory, the input bias current should be the same for both inputs. In practice, the bias currents should be *almost equal*. Any great difference in input bias is the result of unbalance in the input differential amplifier of the IC and can serious affect design (and usually indicates a defective IC).

9-4.10 Input-Offset Voltage and Current

Input-offset voltage and current can be measured using the circuit of Fig. 9–29. As shown, the output is alternately measured with R3 shorted and with R3 in the circuit. The two output voltages are recorded as E1 (S1 closed, R3 shorted), and E2 (S1 open, R3 in the circuit).

With the two output voltages recorded, the input-offset voltage and input-offset current can be calculated using the equations of Fig. 9–29. For example, assume that R1 = 51 Ω, R2 = 5.1 kΩ, R3 = 100 kΩ, E1 = 83 mV, and E2 = 363 mV (all typical values).

$$\text{input-offset voltage} = \frac{83 \text{ mV}}{100} = 0.83 \text{ mV}$$

$$\text{input-offset current} = \frac{280 \text{ mV}}{100 \text{ k}\Omega \ (1 \ + \ 100)} = 0.0277 \ \mu A$$

9-4.11 Common-Mode Rejection

Common-mode rejection or CMR can be measured using the circuit of Fig. 9–30. First find the open-loop gain under identical conditions of frequency, input, and so on, as described in Secs. 9–4.1 and 9–4.2.

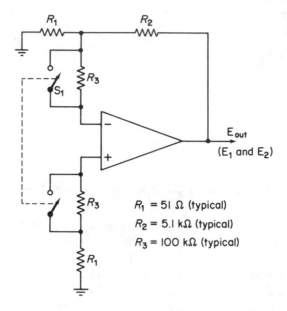

$E_1 = V_{out}$ with S_1 closed (R_3 shorted)

$E_2 = V_{out}$ with S_1 open (R_3 in circuit)

$$\text{Input offset voltage} = \frac{E_1}{\left(\dfrac{R_2}{R_1}\right)}$$

$$\text{Input offset current} = \frac{(E_2 - E_1)}{R_3\left(1 + \dfrac{R_2}{R_1}\right)}$$

FIGURE 9-29   Input-offset voltage and current measurements.

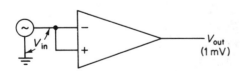

$$\frac{V_{out} \ (1\,mV)}{\text{open loop gain}} = \begin{array}{l}\text{equivalent}\\ \text{differential}\\ \text{input signal}\end{array}$$

$$\begin{array}{l}\text{Common}\\ \text{mode}\\ \text{rejection}\end{array} = \frac{V_{in}}{\begin{array}{c}\text{equivalent differential}\\ \text{input signal}\end{array}}$$

FIGURE 9-30   Common-mode rejection measurement.

Next, connect the IC in the common-mode circuit of Fig. 9–30. Increase the common-mode voltage Vin until a measurable output Vout is obtained. Be careful not to exceed the maximum specified input common-mode voltage swing. If no such value is specified, do not exceed the normal input voltage of the IC.

To simplify calculation, increase the input voltage until the output is 1 mV. With an open-loop gain of 100, this provides a differential input signal of 0.00001 V. Then measure the input voltage. Move the input voltage decimal point over five places to find the CMR.

9–4.12 Slew Rate

An easy way to observe and measure the slew rate of an op-amp/OTA is to measure the slope of the output waveform of a square-wave input signal, as shown in Fig. 9–31. The input square wave must have a *rise time that exceeds the slew-rate* capability of the amplifier. Therefore, the output does not appear as a square wave, but as an integrated wave. In the example shown, the output voltage rises (and falls) about 40 V in 1 μs.

Note that slew rate is usually measured in the closed-loop condition. Also note that the slew rate increases with higher gain.

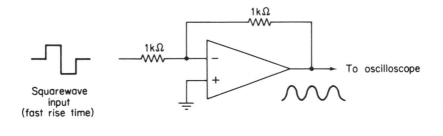

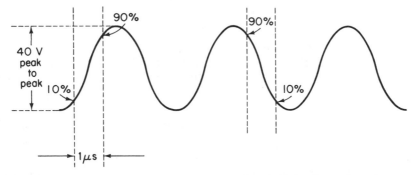

Example shows a slew rate of approximately 40 (40 Volts /μs) at unity gain.

FIGURE 9–31 Slew-rate measurement.

9-4.13 Power-Supply Sensitivity

Power-supply sensitivity can be measured using the circuit of Fig. 9–29 (the same test circuit as for input-offset voltage, Sec. 9–4.10). The procedure is the same as for measurement of input-offset voltage, except that one supply voltage is changed (in 1-V steps) while the other supply voltage is held constant. The amount of change in input-offset voltage for a 1-V change in one power supply is the power-supply sensitivity (or the input-offset voltage sensitivity, as it may be called).

For example, assume that the normal positive and negative supplies are 10 V and the input offset voltage is 7 mV. With the positive supply held constant, the negative supply is reduced to 9 V. Under these conditions, assume that the input-offset voltage is 5 mV. This means that the negative power-supply sensitivity is 2 mV/V. With the negative power supply held constant at 10 V, the positive supply is reduced to 9 V. Now assume that the input-offset voltage drops to 4 mV. This means that the positive power-supply sensitivity is 3 mV/V.

The test can be repeated over a wide range of power-supply voltages (in 1-V steps), if the IC is to be operated under conditions where the power supply may vary by a large amount.

9-4.14 Phase Shift

The phase shift between input and output of an op-amp/OTA is far more critical than with an audio amplifier. This is because an op-amp/OTA uses the principle of feeding back output signals to the input. All the phase-compensation schemes are based on this principle. Under ideal open-loop conditions, the output should be $180°$ out of phase with the negative input and in phase with the positive input.

The following paragraphs describe two procedures for the measurement of phase shift between the input and output of an op-amp/OTA. The same procedures can be used for any amplifier, provided that the signals are of a frequency that can be measured on a scope.

The scope is the ideal tool for phase measurement. The most convenient method requires a dual-trace scope (or a switching unit to produce a dual trace). If neither of these is available, it is still possible to provide accurate phase measurements up to about 100 kHz using the single-trace (or X-Y) method.

Dual-trace Phase Measurement. The dual-trace method of phase measurement provides a high degree of accuracy at all frequencies, but is especially useful at frequencies above 100 kHz, where X-Y phase measurements may prove inaccurate (because of inherent internal phase shift of the scope).

The dual-trace method also has the advantage of measuring phase differences between signals of different amplitude and waveshape, as is usually the case with input and output signals of an amplifier. The dual-trace method is best applied to those scopes having a built-in dual-trace feature.

If the scope is not dual trace, an electronic switch or "chopper" can be used, but the built-in dual-trace feature provides the most accuracy. No matter what scope is used, the procedure is essentially one of displaying both input and output signals on the scope screen simultaneously, measuring the distance (in screen scale divisions) between related points on the two traces, then converting the distance into phase.

The test connections for dual-trace phase measurement are shown in Fig. 9–32. For the most accurate results, the cables connecting the input and output signals should be of the same length and characteristics. At higher frequencies, a difference in cable length of characteristics can introduce a phase shift.

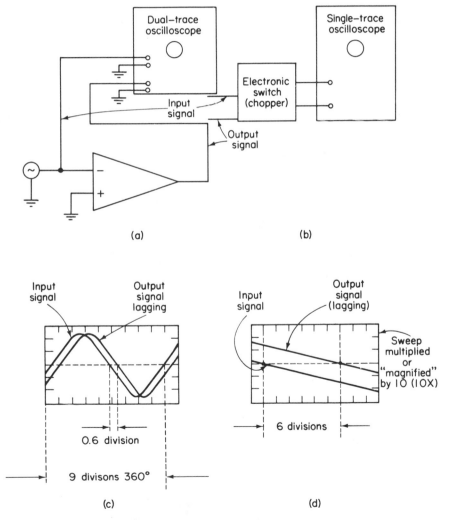

FIGURE 9-32 Dual-trace phase measurement.

The scope controls are adjusted until one cycle of the input signal occupies exactly nine divisions (typically 9 cm horizontally) of the screen. Then the phase factor of the input signal is found. For example, if 9 cm represents one complete cycle, or 360°, 1 cm represents 40° (360/9 = 40).

With the phase factor established, the horizontal distance between corresponding points on the two waveforms (input and output signals) is measured. The measured distance is then multiplied by the phase factor of 40°/cm to find the exact amount of phase difference. For example, assume a horizontal difference of 0.6 cm with a phase factor of 40° as shown in Fig. 9–32. Multiply the horizontal difference by the phase factor to find the phase difference (0.6 × 40 = 24° phase shift between input and output signals).

More accurate phase measurements can be made if the scope is provided with sweep magnification controls where the sweep rate is increased by some fixed amount (5×, 10×, and so on) and only a portion of one cycle is displayed. In this case, the phase factor and the approximate phase difference are found as described. Without changing any other controls, the sweep rate is increased (by the sweep magnification control or the sweep rate control), and a new horizontal-distance measurement is made, as shown in Fig. 9–32d.

For example, if the sweep rate is increased 10 times, the adjusted phase factor is 40°/10 = 4°/cm. Figure 9–32d shows the same signal as used in Fig. 9–32c, but with the sweep rate set by 10×. With horizontal difference of 6 cm, the phase difference is 6 × 4° = 24°.

Single-Trace (X-Y) Phase Measurement.   The single-trace (or X-Y) phase-measurement method can be used to measure the phase difference between input and output of an amplifier at frequencies up to about 100 kHz. Above this frequency, the inherent phase shift (or difference) between the horizontal and vertical systems of the scope makes accurate phase measurements difficult.

In the X-Y method, one of the signals (usually the input) provides horizontal deflection (X) and the other signal provides the vertical deflection (Y). The phase angle between the two signals can be determined from the resulting pattern.

The test connections for single-trace phase measurement are shown in Fig. 9–33, which shows the test connections necessary to find the inherent phase shift (if any) between the horizontal and vertical deflection systems of the scope. Such phase shift should be checked and recorded. If there is excessive phase shift (excessive in relation to the signals being measured), the scope should not be used. A possible exception is when the signals are of sufficient amplitude to be applied directly to the scope deflection plates, bypassing the horizontal and vertical amplifiers.

The scope controls are adjusted until the pattern is centered on the screen as shown in Fig. 9–33c. With the amplifier output connected to the vertical input, it is usually necessary to reduce vertical channel gain (to compensate for the increased gain through the amplifier). With the display centered in relation to the vertical line, distances A and B are measured, as shown in Fig. 9–33c.

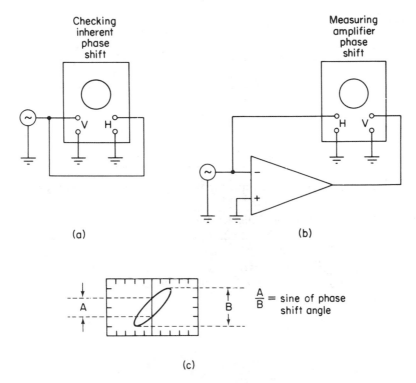

(a)

(b)

(c)

FIGURE 9-33 Single-trace (X–Y) phase measurement.

Distance A is the vertical measurement between two points where the traces cross the vertical center line. Distance B is the maximum vertical height of the display. Divide A by B to find the *sine of the phase angle*. This same procedure can be used to find inherent phase shift (Fig. 9–33a) or phase angle (Fig. 9–33b).

If the display appears as a diagonal straight line, the two signals are either in phase (tilted from the upper right to the lower left, positive slope) or 180° out-of-phase (tilted from the upper left to lower right, negative slope). If the display is a circle, the signals are 90° out of phase. Figure 9–34 shows the displays produced between 0° and 360°.

Notice that above a phase shift of 180° the resultant display is the same as at some lower frequency. Therefore, it may be difficult to tell whether the signal is leading or lagging. One way to find correct phase polarity (leading or lagging) is to introduce a small, known phase shift into one of the signals. The proper angle may then be found by noting the direction in which the pattern changes.

Once the scope's inherent phase shift has been established (Fig. 9–33a) and the amplifier phase shift measured (Fig. 9–33b), subtract the inherent phase difference from the phase angle to find the true phase difference.

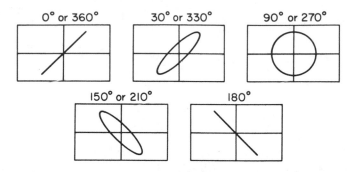

FIGURE 9-34   Approximate phase of typical X-Y displays.

For example, assume an inherent phase difference of 2° and a display as shown in Fig. 9-33c, where A is 2 cm and B is 4 cm. First find the sine of the phase angle, or $A/B = 2/4 = 0.5$. Then, from a *table of sines*, find the phase angle; $0.5 = 30°$ phase angle. To adjust for the phase difference between X and Y scope channels, subtract the inherent phase factor ($30° - 2° = 28°$, true phase difference).

─────── 10 ───────

AMPLIFIER TROUBLESHOOTING

This chapter describes the troubleshooting procedures for a cross section of amplifier circuits. It is assumed that you are familiar with the basics of trouble-shooting, including the use of test equipment. If not, you had better read the author's best-selling *Practical Electronic Troubleshooting* (Prentice-Hall, 1990). Before we get into circuit troubleshooting, let us discuss some basic problems common to all amplifiers.

10-1 BASIC FEEDBACK AMPLIFIER TROUBLESHOOTING

Troubleshooting amplifiers without feedback is a relatively simple procedure. The input and output waveforms of each stage can be monitored on a scope. Any stage showing an abnormal waveform (in amplitude, waveshape, and the like) or the absence of an output waveform, with a known, good input signal, points to a defect in that stage. Voltage measurements on all transistor or IC elements then pinpoint the problem.

Troubleshooting an amplifier with feedback is a more difficult task. Such problems as measurement of gain can be a particular concern. For example, if you try opening the loop to make a gain measurement, you usually find so much gain that the amplifier saturates, and the measurements are meaningless. On the other hand, if you start making waveform measurements on a closed-loop system, you often find the input and output signals are normal (or near normal), while inside the loop many of the waveforms are distorted. For this reason, feedback loops (especially internal-stage feedback loops) require special attention.

10-1.1   *Typical Feedback-Amplifier Circuit*

Figure 10–1 is the schematic of a basic feedback amplifier. Note the various waveforms around the circuit. These waveforms are similar to those that appear if the amplifier is used with sine waves. Note that there is an approximate 15% distortion *inside the feedback loop* (between Q1 and Q2), but only a 0.5% distortion at the output. This is only slightly greater distortion than at the input (0.3%). Open-loop gain for this circuit is about 4300; closed-loop gain is about 1000. The gain ratio (open-loop to closed-loop) of 4 to 1 is typical for feedback amplifiers used in laboratory work.

10-1.2   *Amplification of Signals*

Transistors in feedback amplifiers behave just like transistors in any other circuit. That is, the transistors respond to all the same rules for gain and input/output impedance. Specifically, each transistor amplifies the signal appearing *between emitter and base*. It is here that the greatest difference occurs between gain stages in feedback amplifiers and gain stages in nonfeedback (open-loop) amplifiers.

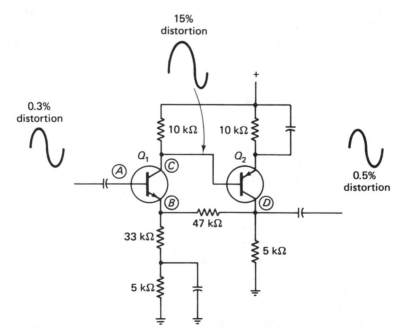

FIGURE 10-1   Basic feedback amplifier.

10-1.3 Difference in Open-Loop and Closed-Loop Gain

Transistor Q1 in Fig. 10-1 has a varying signal on *both the emitter and base* rather than on one element. In a nonfeedback amplifier, the signal usually varies at one element, either the emitter or base. Since most feedback systems use negative feedback, the signals at both the base and emitter are in phase. The resultant gain is much less than when one of these elements is fixed (no feedback, open loop). This accounts for the great amplifier gain increase when the loop is opened. Either the base or the emitter of the transistor stops moving, and the base–emitter control elements see a much larger effective input signal.

Assume that a "perfect" input signal is applied to the input (point A of Fig. 10-1). If the amplifier is "perfect" (produces no distortion), the signal returning to B is also undistorted. Since the system uses negative feedback, the signal that travels around the loop a second time is undistorted as well. If the amplifier is not perfect (assume an extreme case of clipping distortion), the returning signal shows the effect of distortion, as in Fig. 10-2.

10-1.4 Typical Clipping Problem

To simplify the explanation, assume that clipping is introduced in Q1 (Fig. 10-1) and that Q2 is perfect. The signals applied to the base and emitter of Q1 are not identical. The resultant signal at the control point of Q1 is quite distorted. In effect, the distortion is a mirror image of the distortion introduced by Q1. Transistor Q1 then amplifies this distortion and adds in its own counterdistortion.

After many trips around the loop, there is still distortion inside the loop, but the distortion is counterbalanced by the feedback. The final output from

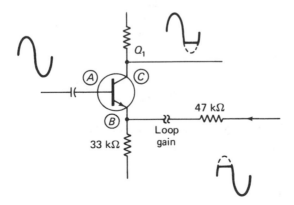

FIGURE 10-2 Amplifier-induced distortion in signal returning to point B.

Q2 is undistorted or is relatively free of amplifier-induced distortion. The higher the amplification and the greater the feedback are the more effective this cancellation becomes, and the lower the output distortion becomes.

10-1.5 Causes of Distortion

The previous facts mark the basic difference in troubleshooting a feedback amplifier. In any amplifier there are three basic causes of distortion: overdriving, operating the transistor at the wrong bias point, and the inherent nonlinearity of any solid-state device.

Overdriving.   Overdriving can be the result of many causes (too much input signal, too much gain in the previous stage, and the like). However, the net result is that the output signal is clipped on one peak because of the transistor being driven into saturation and on the other peak by driving the transistor below cutoff.

Wrong Bias Point.   Operating at the wrong bias point can also produce clipping, but of only one peak. For example, if the input signal is 1 V and the transistor is biased at 1 V, the input swings from 0.5 to 1.5 V. Assume that the transistor saturates at any point above 1.6 V and is cut off at any point below 0.4 V. No problems occur with the correct bias (1 V).

Now assume that the bias point is shifted (because of component aging, transistor leakage, or the like) to 1.3 V. The input now swings from 0.8 to 1.8 V, and the transistor saturates when one peak goes from 1.6 to 1.8 V. If, on the other hand, the bias point is shifted down to 0.7 V, the input swings from 0.2 to 1.2 V, and the opposite peak is clipped as the transistor goes into cutoff.

Nonlinear Operation.   Even if the transistor is not overdriven, it is still possible to operate a transistor on a nonlinear portion of the curve because of wrong bias. All transistors have some portion of the input/output curve that is more linear than other portions. That is, the output increases (or decreases) directly in proportion to input. An increase of 10% at the input produces an increase of 10% at the output. Ideally, transistors are operated at the center of this linear curve. If the bias point is changed, the transistor can operate on a portion of the curve that is less linear than the desired point.

Inherent Nonlinearity.   The inherent nonlinearity of any solid-state device (diode, transistor, IC, and so on) can produce distortion even if a stage is not overdriven and is properly biased. That is, the output never increases (or decreases) directly in proportion to the input. For example, an increase of 10% at the input can produce an increase of 13% (or any other percent) at the output. This is one of the main reasons for feedback in amplifiers where low distortion is required.

Negative Feedback versus Distortion.   In summary, a negative feedback loop operates to minimize distortion, in addition to stabilizing gain. The feedback takeoff point has the minimum distortion of any point within the loop. From a practical troubleshooting standpoint, if the *final output* (last stage or IC output) distortion and the overall gain are within limits, all the stages or ICs within the loop can be considered as operating properly. Even if there is some abnormal gain in one or more stages, the overall feedback system has compensated for the problem. Of course, if the overall gain and/or distortion is not within limits, the individual stages must be checked.

10-1.6   Feedback-Amplifier Troubleshooting Notes

Most feedback-amplifier problems can be pinpointed by waveform and voltage measurements. The following notes should be given special attention when troubleshooting any feedback amplifier circuit.

Opening the Loop.   Some troubleshooting literature recommends that the loop be opened and the circuits checked under no-feedback conditions. In some cases, this can cause circuit damage. Even if there is no damage, the technique is rarely effective. Open-loop gain is usually so high that some stage blocks or distorts badly. If the technique is used, as it must be for some circuits (typically op-amp/OTA), remember that distortion is increased. That is, a normally closed-loop amplifier can show considerable distortion when operated as an open loop, even though the amplifier is good.

Measuring Stage Gain.   Care should be taken when measuring the gain of amplifier stages in a discrete-component feedback amplifier. For example, in Fig. 10-1, if you measure the signal at the base of Q1, the base-to-ground voltage is not the same as the input voltage.

To get the correct value, connect the low side of the measuring device (meter or scope) to the emitter and the other lead to the base, as shown in Fig. 10-3. In effect, measure the signal across the base–emitter junction. This includes the effect of the feedback signal.

As a general safety precaution, never connect the ground lead of a voltmeter or scope to the base of a transistor, unless that lead connects back to an isolated inner chassis. The reason is because large a-c ground-loop currents (Sec. 6–3) can flow through the base–emitter junction (and then to ground) and easily blow out the transistor.

Low-Gain Problems.   As discussed, low gain in a feedback amplifier can also result in distortion. That is, if gain is normal in a feedback amplifier, some distortion can be overcome. With low gain, the feedback may not be able to bring the distortion within limits. Of course, low gain by itself is sufficient cause to troubleshoot an amplifier (feedback or not).

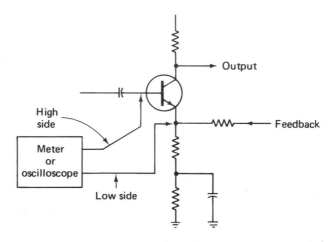

FIGURE 10-3 Measuring input signal voltage or waveforms.

Take the classic failure pattern of a discrete-component feedback amplifier that was working properly, but now the output is low by about 10%. This indicates a general deterioration of performance, rather than a major breakdown.

Remember that most feedback amplifiers have a very high open-loop gain that is set to some specific value by the ratio of resistors (feedback resistor to load resistor). If the closed-loop gain is low, it usually means that the open-loop gain has fallen far enough so that the resistors no longer set the gain. For example, if the beta of Q2 in Fig. 10–1 is lowered, the open-loop gain is lowered. Also, the lower beta lowers the input impedance of Q2, which, in turn, reduces the effective value of the load resistor for Q1. This also has the effect of lowering overall gain.

In troubleshooting such a situation, if waveforms show low gain, but element voltages are normal, try replacing the transistors. Of course, never overlook the possibility of open or badly leaking emitter-bypass capacitors. If the capacitors are open or leaking (acting as a resistance in parallel with the emitter resistor), there is considerable negative feedback and little gain. Of course, a completely shorted emitter-bypass capacitor produces an abnormal d-c voltage indication at the transistor emitter.

Distortion Problems. As discussed, distortion can be caused by improper bias, overdriving (too much gain), or underdriving (too little gain, preventing the feedback signal from countering the distortion). One problem often overlooked in a feedback amplifier with a distortion failure pattern is overdriving because of transistor leakage. (The problem of transistor leakage is discussed further in Sec. 10–2.)

Generally, it is assumed that the collector–base leakage reduces gain, since the leakage is in opposition to the signal-current flow. While this is true in the case of a single stage, it may not be true where more than one feedback stage is involved.

Whenever there is collector–base leakage, the base assumes a voltage nearly equal to that of the collector (nearer than is the case without leakage). This increases transistor forward bias and increases transistor current flow. An increase in the transistor current causes lower input resistances which, in turn, causes the stage gain to go up. At the same time, a reduction in input resistance causes a reduction in common-emitter input resistance, which may or may not cause a gain reduction (depending on where the transistor is located in the amplifier).

If the feedback amplifier is direct coupled (Chapter 5), the effects of feedback are increased. This is because the operating point (bias) of the following stage is changed, possibly resulting in distortion. For example, the collector of Q1 is connected directly to the base of Q2. If Q1 starts to leak (or the collector–base leakage increases with age), the base of Q2 (as well as the collector of Q1) shifts the Q point (no-signal voltage level).

10-2 EFFECTS OF LEAKAGE ON AMPLIFIER GAIN

When there is considerable leakage in a solid-state amplifier, the gain is reduced to zero and/or the signal waveform is drastically distorted. These indications make the problem easy, or relatively easy, to locate. The real difficulty occurs when there is just enough leakage to reduce amplifier gain, but not enough leakage to seriously distort the waveform or produce transistor voltages that are way off.

Collector–base leakage is the most common form of transistor leakage and produces a classic condition of low gain (in a single stage). Where there is any collector–base leakage, the transistor is forward biased, or the forward bias is increased. This condition is shown in Fig. 10-4.

Collector–base leakage has the same effect as a resistance between the collector and base. The base assumes the same polarity as the collector (although at a lower value), and the transistor is forward biased. If leakage is sufficient, the forward bias can be enough to drive the transistor into or near saturation. When a transistor is operated at or near the saturation point, the gain is reduced for a single stage. This is shown in the curve of Fig. 10-5.

If the normal transistor element voltages are known, excessive transistor leakage can be spotted easily, since all the transistor voltages are off. For example, in Fig. 10-4, the base and emitter are high and the collector is low when measured in reference to ground. However, if the normal operating voltages are not known, the transistor can appear to be good, since *all the voltage relationships are normal*. That is, the collector–base junction is reverse biased (collector more positive than base for an NPN), and the emitter–base junction is forward biased (emitter less positive than base for NPN).

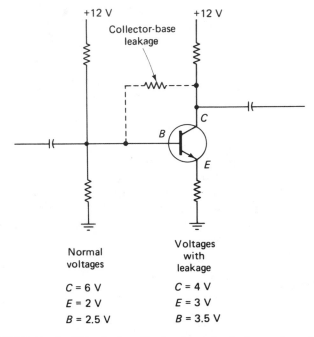

FIGURE 10-4 Effect of collector-base leakage on transistor element voltages.

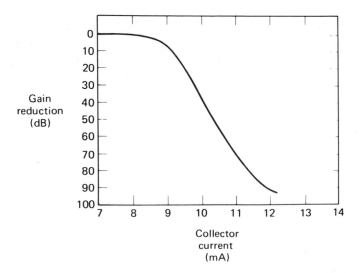

FIGURE 10-5 Relative gain of solid-state amplifier at various average collector-current levels.

10-2.1   Checking Transistor Leakage

Figure 10-6 shows a simple way to check transistor leakage. Measure the collector voltage to ground. Then short the base to the emitter and remeasure the collector voltage. If the transistor is not leaking, the base–emitter short turns the transistor off, and the collector voltage rises to the same value as the supply. If there is any leakage, a current path remains (through the emitter resistor, emitter–base short, collector–base leakage path, and collector resistor). There is some voltage drop across the collector resistor, and the collector has a voltage at some value lower than the supply.

Note that most meters draw current, and this current passes through the collector resistor. This can lead to some confusion, particularly if the meter draws heavy current (has a low ohms-per-volt rating). To eliminate any doubt, connect the meter to the supply through a resistor with the same value as the collector resistor. The drop, if any, should be the same as when the transistor collector is measured to ground. If the drop is much lower when the collector is measured, the transistor is leaking.

Example of Transistor Leakage Check.   Assume that in the circuit of Fig. 10-6 the supply is 12 V, the collector resistance is 2 kΩ, and the collector measures 4 V with respect to ground. This means that there is an 8-V drop across the collector resistor and a collector current of 4 mA (8/2000 = 4 mA).

Normally, the collector is operated at about one-half the supply voltage or 6 V. However, simply because the collector is at 4 V instead of 6 V does not make the circuit faulty. Some circuits are designed that way, so the transistor must be checked for leakage.

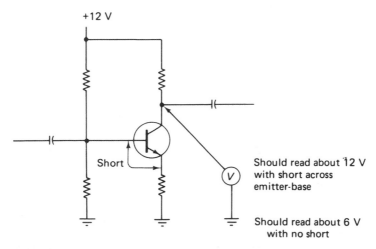

FIGURE 10-6 Checking for transistor leakage in amplifier circuit.

Now assume that the collector voltage rises to 10 V when the base and emitter are shorted. This shows that the transistor is cutting off, but there is still some current flow through the collector resistor, about 1 mA (2/2000 = 1 mA).

A 1-mA current flow is high for a meter. However, to confirm a leaking transistor, connect the meter through a 2-kΩ resistor to the 12-V supply, preferably at the same point where the collector resistor connects to the supply. Now assume that the indication is 11.7 V through the external resistor. This shows that there is some transistor leakage.

The amount of leakage can be estimated as follows: 11.7 − 10.5 = 1.2-V drop; 1.2/2000 = 0.6 mA. However, from a practical troubleshooting standpoint, the presence of any current flow with the transistor supposedly cut off is sufficient cause to replace the transistor.

10-3 EFFECTS OF CAPACITORS IN DISCRETE-COMPONENT AMPLIFIERS

This section discusses the effects of capacitors in discrete-component amplifiers from a troubleshooting standpoint.

10-3.1 Emitter-Bypass Capacitor

As discussed in Chapters 1 and 3, the emitter resistor (such as R4 in Fig. 10–7) is used to stabilize the transistor gain and prevent thermal runaway. With an emitter resistor in the circuit, any increase in collector current produces a greater drop in voltage across the resistor. When all other factors remain the same, the change in emitter voltage reduces the base–emitter forward-bias differential, thus tending to reduce collector current flow.

When circuit stability is more important than gain, the emitter resistor is not bypassed. When signal gain must be high, the emitter resistance is bypassed to permit passage of the signal. If the emitter-bypass capacitor (such as C2 in Fig. 10–7) is open, stage gain is reduced drastically, although the transistor d-c voltages remain substantially the same.

If there is a low-gain symptom in any amplifier with an emitter bypass, and the voltages appear normal, check the bypass capacitor. This can be done by shunting the bypass with a known good capacitor of the same value. As a precaution, shut off the power before connecting the shunt capacitor; then reapply power. This prevents damage to the transistor (because of large current surges).

Note that if an emitter bypass is shorted or leaking, the emitter voltage is not correct, thus localizing the problem. In this case, shunting a good capacitor across the suspected capacitor has little effect on gain. The suspected capacitor must be tested by substitution.

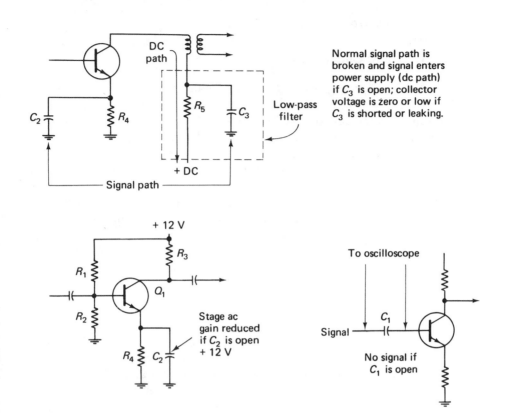

Normal signal path is broken and signal enters power supply (dc path) if C_3 is open; collector voltage is zero or low if C_3 is shorted or leaking.

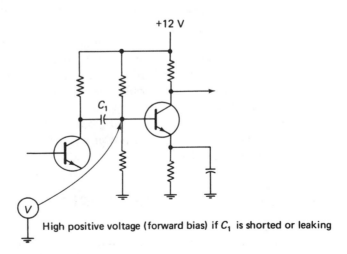

High positive voltage (forward bias) if C_1 is shorted or leaking

FIGURE 10-7 Effects of capacitor failure in solid-state amplifier circuits.

10-3.2 Coupling Capacitors

The function of C1 in Fig. 10–7 is to pass signals from the previous stage to the base of Q1. If C1 is shorted or leaking badly, the voltage from the previous stage is applied to the base of Q1. This forward biases Q1, causing heavy current flow and possible burnout of the transistor. In any event, Q1 is driven into saturation, and stage gain is reduced. If C1 is open, there is little or no change in the voltages at Q1, but the signal from the previous stage does not appear at the base of Q1.

From a troubleshooting standpoint, a shorted or leaking C1 shows up as abnormal voltages (and probably as distortion of the signal waveform). If C1 is suspected of being shorted or leaking, replace C1. An open C1 shows up as a lack of signal at the base of Q1, with a normal signal at the previous stage. If an open C1 is suspected, replace C1 or try shunting C1 with a known good capacitor, whichever is convenient.

10-3.3 Decouping Capacitors

The function of C3 in Fig. 10–7 is to pass operating signal frequencies to ground (to provide a return path) and to prevent signals from entering the power-supply line or other circuits connected to the line. In effect, C3 and R5 form a low-pass filter that passes dc and very low frequency signals (well below the operating frequency of the circuit) through the power-supply line. Higher-frequency signals are passed to ground and do not enter the power-supply line.

If C3 is shorted or leaking badly, the power-supply voltage is shorted to ground or greatly reduced. This reduction of collector voltage makes the stage totally inoperative or reduces the output, depending on the amount of leakage in C3.

If C3 is open, there is little or no change in the voltages at Q1. However, the signals appear in the power-supply lines. Also, signal gain is reduced and the signal waveform is distorted. In some cases, at higher signal frequencies, the signals simply cannot pass through the power-supply circuits. Since there is no path through an open C3, the signal does not appear on the collector circuit in any form. From a practical standpoint, the results of an open C3 depend on the values of R5 (and other power-supply components), as well as on the signal frequency involved.

10-4 BASIC IC AMPLIFIER TROUBLESHOOTING

This section describes some basic troubleshooting procedures for IC amplifiers, particularly op/amp/OTAs, but not limited to any specific type of amplifier. From a practical troubleshooting standpoint, it is often necessary to work with external feedback components. In the case of IC packages, the external components are the only ones that can be tested or replaced. In any event, trouble-shooting starts by isolating the problem to the external components or the amplifier. That is, the amplifier is tested as a separate function first. If the amplifier performs properly, the trouble is isolated to the external components, and vice versa.

10-4.1 Failure Patterns for IC Amplifiers

Major disasters are relatively rare in well-protected ICs, since input overloads never drive the circuit into saturation. Likewise, when such major failures occur, the failures are relatively easy to troubleshoot. The problems are easy to spot by normal signal tracing with waveforms or by voltage measurements at the transistor/IC terminals. For example, a major failure shows up as a normal input, but with no output at a particular stage (or at the input and output of an IC).

However, IC amplifiers are often plagued with such problems as hum, drift, and noise. The following paragraphs describe the most likely causes for such problems, with practical approaches for locating the faults.

10-4.2 Hum and Ripple Problems

In IC amplifiers any hum or ripple almost always comes from the d-c power supplies feeding the amplifier. A possible exception is when hum is picked up because of poor shielding or badly grounded leads.

The first step in locating a hum or ripple problem is to short the input terminals and monitor the output with a scope. If the hum or ripple is removed when the input terminals are shorted, the hum is probably being picked up by the leads or at the terminal. Look for loose shields, loose ground terminals, and cold-solder points where lead shielding is attached to chassis or feed-through terminals.

If the hum or ripple is not removed when the input terminals are shorted, the hum is probably coming from the power supply. Monitor the power-supply voltages at the point where the voltages enter the amplifier.

If the power supply is showing an abnormal amount of ripple, the problem is in the power supply. However, since the amplifier has considerable gain, the ripple (as monitored at the amplifier output) may be much greater than at the power supply.

10-4.3 Drift and Noise Problems

Drift and noise problems (particularly in IC op-amp/OTAs) are perhaps the most common complaint in IC amplifiers. There are several places to look when trying to track the causes of noise and drift.

Unstable Power Supplies.   Op-amp/OTAs are extremely sensitive to power-supply stability. For example, with IC op-amp/OTAs the typical dual-power-supply voltages (required for differential amplifiers, Chapter 6) are about ± 12 or ± 15 V. For satisfactory operation, the drift should be less than 1 mV/min (or less in some special amplifiers used with data-processing equipment).

Because of the low voltage involved, power-supply-stability measurements are best made with five- or six-place digital meters. Such meters can be connected to the monitoring point and checked at least once every minute, or over at least a 5-min interval. If the drift is less than 1 mV/min over this time interval, the power supply is probably satisfactory for typical op-amp use.

Noisy Zero-Correction Circuits.   Another possible source of output noise (particularly in op-amp/OTAs) is the zero-correction circuit, which takes voltages from both the positive and negative power supplies and provides a small d-c current to oppose the internally generated offset (such as the circuit shown in Fig. 7–33).

If zener diodes are used to regulate some portion of the zero-correction supply voltage (as is done by CR303 and CR304 in Fig. 7–33), the zeners should be checked carefully for drift and noise. Remember that any noise (or other signal) at the zero-correction circuit is injected into the amplifier at the point of highest gain (usually at the first-stage input).

Contaminated PC Boards.   Another frequent source of output drift (particularly in op-amp/OTAs) is contamination of PC boards, even with IC amplifiers. Although the IC package is sealed, the *summing point* is not sealed. (Typically, the summing junctions are placed on Teflon terminals near the IC package on the PC board).

When you remember that the input current at the summing junction of an op-amp/OTA is typically 10^{-11} A, it is easy to see why any contamination from fingerprints (providing leakage paths into the junction) can cause annoying output instability. Op-amp circuit boards should be handled as little as possible, and then only while wearing cotton gloves.

Great care should be taken never to touch the summing-junction terminals with bare hands. Boards suspected of being contaminated should be washed carefully with a clean degreasing solvent and dried with warm, dry air. (Never blow them dry with an air hose, as air lines invariably contain oil and water.)

Leakage in the Overload-Protection Circuits.   Another place to look for causes of output noise and drift is the overload-protection circuits. If an amplifier has been subjected to repeated serious overloads, there is a possibility of finding high and unstable reverse-leakage currents in the overload-protection circuits.

For example, in the circuit of Fig. 7–34, the reverse currents can be monitored by measuring voltages across R207 and R353. Substantial voltage across these resistors indicates considerable reverse leakage. If such voltages are measured in an amplifier with a noise symptom, try replacing the diodes, one at a time.

Unstable Choppers. Perhaps the most common source of output instability in chopper-stabilized amplifiers (Chapter 5) is instability of the chopper itself. Since choppers modulate low-level signals, it is virtually impossible to check choppers except by substitution (which is most practical since the choppers are generally sealed, or in IC form).

10–4.4 General Troubleshooting Hints for Op-Amp/OTAs

Because of the extremely high open-loop gain, troubleshooting op-amp/OTAs can be quite difficult. The basic test connections are shown in Fig. 10–8. The input is shorted, the drift output (if any) is monitored on a digital meter, and the hum and noise (if any) are monitored on a scope.

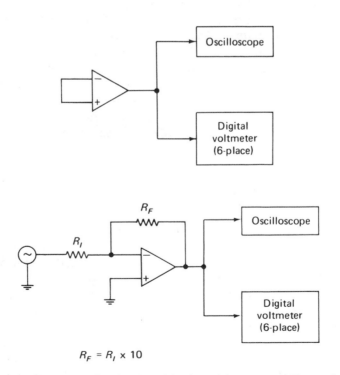

FIGURE 10-8 Basic troubleshooting connections for op-amp/OTA.

If the op-amp/OTA is used in a circuit where feedback is not by means of a resistor (such as an integrator where the feedback is through a capacitor), a resistor should be inserted in the feedback path. The feedback resistor should be 10 times the value of the input resistor. Thus, the op-amp operates with a gain of 10.

The same test connections can be used to check the range of zero-adjust circuits and the clipping level of the diode-protection circuits.

To check the zero-correction range, vary the zero control from one end of the range to the other while observing the amplifier output on the digital meter. (The meter should have a sensitivity of at least 1 mV.) If the zero control is provided with steps (instead of or in addition to a variable control), check that each step produces the same step size in output voltages.

Also check the output stability at each step. If the output voltage appears to be unstable at any particular step, look for poor contacts on the switch, poor solder connections, or a defective resistor connected to the switch contact.

Sometimes the range of the zero-adjust control is not sufficient to cause limiting of the amplifier (by means of protective-overload circuits). If this is the case, apply a small d-c voltage of known stability (preferably from a battery) to the input terminals of the amplifier. Check that limiting occurs at the amplifier output at the correct values for both polarities of the input voltage.

Unbalance between the positive and negative overload-breakdown voltage suggests either an open or shorted diode in the overload-protection loop or saturation taking place internally within the amplifier. In either case, leave the output driven lightly into saturation, and measure the d-c voltage appearing across *all elements* in the overload-protection circuit, plus the operating points of the various stages within the amplifier proper (unless the amplifier is an IC where internal stages cannot measured). This should locate the unbalance problem.

10-5 ELECTRONIC-VOLUME-CONTROL TROUBLESHOOTING

This section covers troubleshooting for the circuits discussed in Sec. 3–11.1 and shown in Figs. 3–25 and 3–26.

If the volume control does not operate, check for audio at pin 5 of IC604. If not, check the audio line to IC604. Remember that the electronic volume control is in the audio path between the amplifier input and output. So before you condemn the volume-control circuits, check for audio to IC604.

If there is audio at pin 5 of IC604, check for audio at pin 4 of IC604 while operating VOLUME UP and VOLUME DOWN. The audio volume should increase and decrease in 10-dB steps at pin 4 each time you press the corresponding VOLUME button. (If you hold the VOLUME buttons, the audio should increase or decrease steadily in 10-dB steps.)

If there is no audio at pin 4 of IC604, with audio at pin 5, suspect IC604.

If there is audio at pin 4, but there is no change when the VOLUME buttons are pressed, use a scope or logic probe to monitor the strobe (pin 12), data (pin 13), and clock (pin 14) outputs from IC901 while holding the VOLUME buttons.

Although you probably cannot decode the information on the control lines, the presence of pulse activity on the lines usually indicates that IC901 is good. If any one of the lines shows no activity with the VOLUME buttons operated, suspect IC901. On the other hand, if there is pulse activity on all three lines, but there is no change in the volume at pin 4 of IC604, with the VOLUME buttons held, suspect IC604.

If there is audio at pin 4 of IC604 and the volume changes, check for audio at pin 7 of IC604. If absent, suspect IC605.

If there is audio at pin 7 of IC604, check for audio at pin 8, and make sure that the audio changes in 2-dB steps when the VOLUME buttons are operated. If there is no audio, suspect IC604. If there is audio, but there is no change with the VOLUME buttons operated, suspect either IC901 or IC604. Note that IC604 is the most likely suspect if audio is good at pin 7, but it is possible that IC901 is not generating the correct code to produce 2-dB changes in volume.

If there is audio at pin 8, and the audio changes in 2-dB steps when the VOLUME buttons are operated, press MUTE and check that audio is cut off at pin 8. If not, suspect the MUTE switch circuit, IC901 or IC604.

You can also check for pulse activity on the data-code outputs from IC901 with a scope or logic probe (but you probably cannot decipher the code!). Press MUTE again and check that audio is restored at pin 8 and at the same level (if you have not pressed the VOLUME buttons during the mute condition). If not, suspect the MUTE circuit, IC901 or IC604.

Note that the *loudness network* is part of the electronic volume control. Also note that a failure symptom for the loudness functions is usually difficult to define in any amplifier. This is because the loudness function attenuates the midrange signals so that the ear hears what *appears to be the same level* across the audio range. (In some amplifiers, the loudness function boosts the treble and bass, but this is rare.)

Unfortunately, all ears are not the same, and not all loudness networks define midrange at the same frequencies. Usually, the customer complains that "there is no difference when I play the tape or recording with the loudness function on or off."

To troubleshoot such a symptom, apply an audio signal (say between 7 and 10 kHz) with loudness off. Then press LOUDNESS and check for a drop of about 20 dB in level at pin 4 of IC604. Repeat the test at 50 Hz and 20 kHz. There should be no substantial change in level (at pin 4 of IC604) at the low and high ends of the audio range (unless a boost circuit is used).

No matter what type of loudness circuit is used, if there is no change in audio level *at any frequency* when the loudness function is switched in and out, there is a problem in the loudness circuit. Start by checking the LOUDNESS switch circuit, IC901, IC604, and the network connected at pins 2 and 3 of IC604.

10-6 AMPLIFIER-OUTPUT TROUBLESHOOTING

This section covers troubleshooting for the circuits discussed in Sec. 3–14.1 and shown in Fig. 3–40.

Before you launch an attack on the audio-output circuits, make certain that there is audio at pins 3 and 5 of IC606. If not, check electronic-volume-control circuits as described in Sec. 10-5.

If audio is available, set the BASS, TREBLE, and BALANCE controls to mid-range. This simple act of faith has been known to cure a "no audio or weak audio in one channel" symptom.

If there is a total loss of audio in one or both channels, always look for speaker switches as a first step. Many "no audio" problems are solved instantly by setting a speaker switch to ON!

Note that the amplifier of Fig. 3–40 has *separate speaker switches* (S701/S702) in each channel and that there are two sets of outputs (A and B) for each channel. Of course, not all amplifiers have such a configuration, but the speaker switches (if any) should be checked first in all cases.

Once you are sure that all switches and controls are properly set, check for audio at pins 1 and 7 of IC606. If absent, but with audio at pins 3 and 5, suspect IC606. Also check the BASS and TREBLE controls since they are in the negative-feedback path of IC606 (which is essentially an op-amp).

If there is audio at pins 1 and 7 of IC606, check for audio (at about the same level) at pins 1 and 18 of IC701. If the audio level at pins 1 and 18 of IC701 is substantially different from the level at pins 1 and 7 of IC606, trace the audio path through the subsonic filter.

The setting of SUBSONIC FILTER switch S602 should have little or no effect on the audio level, except at very low frequencies (below about 20 Hz). If you notice a drastic change in audio level at different settings of S602, from about 1 kHz and up, look for problems in the subsonic filter circuit (such as leakage in C618).

If there is audio at pins 1 and 18 of IC701, but not at pins 10 and 13, suspect IC701. Before you pull IC701 (heat sink and all, Chapter 2), make sure that the 45-V supply is applied to various IC701 terminals.

The 45-V (both plus and minus) supply is applied through relay RY702. In turn, RY702 is turned on through Q701/Q702 when pin 34 of IC901 goes low (when the amplifier front-panel POWER switch is pressed). Most of the other ICs receive operating power when the power cord is plugged in (whether the POWER button is pressed or not).

Because of the heavy current drain and high heat dissipation, power amplifier IC701 is turned on only during play. This is typical for most modular and integrated amplifiers where the final power stage is a single IC in the range of 40 to 50 W. In any event, if the 45-V supply is absent at the terminals of IC701, suspect RY702, Q701, Q702, or IC901 (check pin 34 of IC901 for a low).

If there is audio at pins 10 and 13 of IC701, but the audio does not reach the speakers or headphones, suspect RY701, Q703, or Q704. Also check for a low at pin 33 of IC901. Pin 33 should go low at the same time as pin 34.

Remember that the power-output protection circuits described in Sec. 10-7 are designed to cut off the audio circuit in the event of an overload. Defective protection circuits can cut off the audio, *even without an overload.*

10-7 OUTPUT-PROTECTION TROUBLESHOOTING

This section covers troubleshooting for the circuits discussed in Sec. 3-14.2 and shown in Fig. 3-41.

The front-panel function display FL901 should flash on and off, and the speakers should be disconnected when any one of the following occurs: the power amplifier FL701 becomes overheated (the IC701 heat sink reaches 100 °C), the constant (no audio) d-c voltage applied to the speakers exceeds ± 1.7 V, or the speaker output line is shorted (or is at any impedance below that of the speakers).

Except for the low-impedance output, these conditions are difficult to simulate, making the circuits difficult to check. Also, if you do succeed in simulating any one of these conditions, and the protection circuits are not functioning properly, you can damage the equipment (for example, burn out the speaker voice coil and/or overheat IC701).

If you must check the circuits, try shorting the speaker lines (either L or R, or both) to ground temporarily (*very* temporarily). Check that the function portion of the front-panel display FL901 flashes on and off and that the speakers are disconnected. If not, temporarily short pin 3 of IC901 to ground, and check for a flashing display with the speakers disconnected.

If the display flashes, and the speakers are cut with pin 3 of IC901 shorted, but not when the speaker lines are shorted, suspect Q705, Q707, Q708, Q709, Q710, and Q902.

If the display does not flash and speakers are not disconnected with pin 3 of IC901 shorted, suspect IC901.

You can also check that the anode of D707 is at ground (unless the IC701 heat sink is at 100 °C or higher). If not, suspect that S703 is open.

You can also check the bases of Q707 and Q708. Both bases should be 0 V (ideally), but may be at some potential less than ± 1.7 V, without triggering the protection circuits. If the bases are at some value in excess of ± 1.7 V, pin 3 of IC901 should go low, and the display should flash. If not, suspect Q707 through Q710, or Q902.

10-8 AUDIO-OUTPUT AND MUTING
TROUBLESHOOTING

This section covers troubleshooting for the circuits discussed in Sec. 3-14.3 and shown in Fig. 3-42.

If there is no automatic search (AUTO MODE) operation for the tuner, first make sure that you are in the FM mode. (There is no AM auto mode.) Try pressing FM and AUTO MODE and make sure that the FM and AUTO MODE indicators are on (which might just cure the problem!).

Next make sure that the front-panel STEREO indicator is on when an FM station is tuned. It is possible that the station signal is not sufficient for good FM operation. If the STEREO indicator is on, check that pin 12 of IC201 is low and pin 26 of IC503 is high. If pin 12 of IC201 is not low, suspect IC201. If pin 12 of IC is low, but pin 26 of IC503 is not high, suspect Q406. If pin 26 of IC503 is high, but the FM tuning does not stop when the STEREO indicator turns on, suspect IC503. Also check the frequency-synthesis circuits (Sec. 10–13).

If there is no audio muting or the audio is muted at all times, troubleshooting the circuits may prove difficult. As in the case of most AM/FM tuners, our tuner is muted between stations or when stations are too weak to provide good operation. The muting function can also be removed so that the audio is not muted under any conditions.

If the tuner never mutes in any mode, there is an obvious problem in the muting circuits. On the other hand, if the muting circuits are defective and mute the audio under all conditions (or under the wrong conditions), this can lead you to believe that there are no stations of sufficient strength to unmute the audio or that the circuits ahead of the muting function are defective.

The most practical approach to any muting-circuit problem is to make some preliminary isolation steps. Start by checking for audio at pins 2, 6, and 7 of IC301. If there is no audio at pin 2, check the circuits ahead of IC301.

If there is audio at pin 2, but not at pins 6 and 7 of IC301, suspect IC301. If there is audio at pins 6 and 7 of IC301, trace the audio through to the L- and R-channel audio output terminals. If the audio drops off at pins 3 and 5 of IC401, it is possible that Q401/Q402 are turned on by a muting signal from Q408/Q409. Also, if audio is available at pins 1 and 7 of IC401, but not at the tuner audio output terminals, it is possible that RY401 has been turned off by a low at pin 15 of IC501.

If the audio path between IC301 and the audio output terminals is good, press AUTO MODE and check that pin 42 of IC503 goes high (and that the AUTO MODE indicator turns on). If not, suspect IC503 or the circuits between the AUTO MODE switch and IC503.

Next check that pin 16 of IC301 is low (in AUTO MODE). If pin 16 of IC301 is high, IC301 operates as a mono amplifier rather than an FM-stereo multiplex/decoder. However, IC301 should pass audio in either mode (auto or mono). Then check the base of Q404. If the base is low (zero volts), the audio should be unmuted. If not, suspect Q404/Q408. If the base is high (about 0.6 V), the audio should be muted. If not, suspect Q404, Q408, Q401, and Q402.

Note that in AUTO MODE the base of Q404 should go high only when an FM station of sufficient strength is tuned in. If not, try correcting the problem by adjustment of the muting level R202, as described in the service literature.

If this does not cure the problem, check that pin 12 of IC201 goes high and low, as you tune across the FM band. If not, suspect IC201. If pin 12 of IC201 changes status as FM stations are tuned in and out, but the base of Q404 does not change status, suspect Q406/Q407.

It is also possible that Q405 has not been turned off or on by a defect. (Q405 should be on only in mono mode.) If Q405 is on, Q406/Q407 and the signal at pin 12 of IC201 have no effect on the base of Q404. However, Q404 can be turned on to mute the audio by a high from pin 18 of IC503 (AM mute), a high from pin 14 of IC501 (system mute), or a high across R802 (temporary power-loss mute).

10-9   RECORD/PLAYBACK TROUBLESHOOTING

This section covers troubleshooting for the circuits discussed in Sec. 3–14.3 and shown in Figs. 3–43 through 3–47.

If audio is absent or abnormal during playback, but the tape transport appears to be normal (tape runs properly in both directions), first check the playback-gain adjustment as described in the service literature. Playback gain is adjusted by RT60 as shown in Fig. 3–44. Also, turn up the front-panel OUT-PUT control RV40 (Fig. 3–45). (This might just cure the problem!) If not, continue playing a known-good tape and monitor the audio at pin 7 of IC60. The output should be about 100 mV.

If there is no audio output from IC60, check circuits from the heads to IC60, including Q61 (off) and Q62/Q63 (on).

If there is audio from IC60, check for audio at pin 18 of Dolby IC300. If not, check the audio path from IC60 to pin 7 of IC300.

Make certain that pin 24 of IC500 and pin 11 of IC300 are low (about −7 V). If not, suspect IC500. If pin 24 of IC500 goes high, IC300 goes into the record mode rather than the playback mode.

If there is audio from pin 18 of IC300, check for audio at RV40. If absent, make sure that Q40/Q41 (Fig. 3–44) have not been turned on (mute condition) by a signal at pin 27 of IC500. If Q40/Q41 are not on, check the audio path from IC300 to RV40. Then check audio from RV40 to the LINE OUT jack and PHONES jack. If audio is available at RV40, but not at the jacks, suspect IC40.

If playback audio is present, but there is background noise that increases with time, or there is a decrease in playback audio at high frequencies with time, the heads may want degaussing. Some technicians never degauss the heads, while others degauss at each service (using a commercial head eraser). In between these extremes, some technicians degauss only when the classic symptoms (background noise and poor treble) occur. The author has no recommendations on head degaussing, except never degauss the heads with a cassette in place, particularly a customer's favorite tape, or an expensive shop-test tape.

While on the subject of test tapes, you will need certain tapes to test and adjust a cassette deck properly. Of course, always use the tapes recommended in the service literature. However, as a point of reference, our deck requires a *mirror tape* for tape-travel checks, an 8-kHz tape for azimuth adjustment, a 3-kHz tape for motor-speed adjustment, and a Dolby tape for Dolby NR checks (Sec. 10–10).

If audio is absent or abnormal during record, but the tape transport appears to be normal (tape runs properly in both directions), first check the playback circuits as just described. Always clear any playback problems before you check record. This applies to virtually all cassette decks. If there is no audio (or poor audio) during playback, the problem can be common to both record and playback. On the other hand, if the problem is only in record, you have quickly isolated the trouble to a few circuits (bias oscillator, record amp, or other).

If it is not possible to record audio (with good playback), start by checking for audio at pin 15 of IC300 (Figs. 3–46 and 3–47). Also try to cure the problem by turning up the front-panel RECORD control RV01 and by adjustment of record-level RT50 and bias-current RT400, as described in the service literature.

If audio is present at pin 15 of IC300, check the audio path from IC300 to the heads. Also make certain that pin 24 of IC500 and pin 11 of IC300 are high (about +6 V) to place IC300 in record mode. If there is no audio at the input to IC50, make sure that Q50/Q51 have not been turned on (mute condition) by a signal at pin 27 of IC500.

If there is audio at the heads, but the deck does not record, check the bias oscillator (Fig. 3–47). (As a general rule, if the tape can be erased, the oscillator and erase head are good. However, the bias signal may not be reaching the record/playback heads.) Check for an 85-kHz bias signal on both sides of C400 and RT400, as well as adjustment of RT400.

Remember that each type of tape requires a different amount of bias current. For example, the bias measured at C400 on our deck is about 3.4 V for normal tape, 5.2 V for chrome, and over 10 V for metal tape. This is determined by the TAPE TYPE switch S5 setting.

If there is audio to the heads and bias voltage is correct, but the deck does not record, suspect the heads. Before pulling the heads (a tedious job), make sure that Q1 is turned on and that Q62/Q63 are off, placing the heads in a condition to record. If not, suspect Q61, Q62, Q63, or Q70.

If there is no audio from pin 15 of IC300, trace the audio path from the LINE IN jack and MIC (microphone) jack (Fig. 3–46) through Q01, Q02, Q300, and IC01. Obviously, if you can record from LINE IN but not from MIC, suspect IC02.

Also remember that Dolby NR switch S1 controls Q300, which, in turn, controls the Dolby filter. However, even if the filter circuit or Q300/S1 fails, you will probably be able to record, even though the recording is poor.

10-10 DOLBY CIRCUIT TROUBLESHOOTING

This section covers troubleshooting for the circuits discussed in Sec. 3–14.5 and shown in Figs. 3–48 and 3–49.

Before you condemn any Dolby circuit, *make certain that the tape is being played back in the same mode as during record.* If Dolby was not using during record, do not use Dolby during playback (in a hopeless attempt to improve quality!). If Dolby C is used during record, play back in Dolby C, not Dolby B, and so on.

If there is no substantial difference in sound quality with and without ,Dolby (of the correct type, B or C), make the following checks.

During playback, check for correct voltage at pin 9 of IC300 in each mode (– 7 V for Dolby off, 0.6 V for Dolby B, and 6.3 V for Dolby C). If the voltages are not correct, suspect S1, S2, and Q300. If the voltages are correct at all modes, but there is no substantial difference in sound quality between Dolby and non-Dolby, suspect IC300.

Before you pull IC300, check that there is about 6.3 V at pin 11 of IC300 when record is selected and that pin 11 goes to about – 7 V when playback is selected. If not, suspect IC500.

10-11 CD AUDIO TROUBLESHOOTING

This section covers troubleshooting for the circuits discussed in Sec. 3–14.6 and shown in Fig. 3–50.

If audio is not available on the audio bus or at the rear-panel jacks, check for audio at the front-panel headphones jack. If audio is present at the headphone jack, suspect RY501 or Q501, and RY903 and Q907. Check for signals from pin 31 of IC901 to RY501 through Q501. Then check for signals from pin 30 of IC901 and pin 38 of IC402 to RY903 through Q907. If the signals are absent, suspect IC901 and/or IC402. If the signals are present, but the relays are not actuated, suspect Q501/Q907 and RY501/RY903.

If there is no audio at any output, including the headphones jack, start by checking for audio at pin 17 of IC403. If absent, suspect the signal-processing circuits. Next, check the sample/hold SHR and SHL signals from IC402 (pins 23/25). If SHR and SHL are present and there is audio at pin 17 of IC403, trace the audio signal from IC403 to the headphones and/or rear-panel output jacks. Note that the level for audio *at both* the rear-panel jacks and headphones is controlled by OUTPUT potentiometer R524.

Note that if relays RY501 and RY903 are not operating properly the head is cut off from the rear-panel jacks (but not the headphones). However, if the switches in IC506 are not responding properly to EMP (emphasis) signals from pin 41 of IC402 or if the signal is missing, the audio will pass, but may appear distorted. So if you get a "the audio appears distorted when I play certain discs" trouble symptom, start by checking the emphasis network and IC506 (as well as the emphasis signal from pin 41 of IC402).

10-12 AUDIO-OUTPUT SELECT TROUBLESHOOTING

This section covers troubleshooting for the circuits discussed in Sec. 4–6.1 and shown in Fig. 4–22.

If you get AM audio, but no FM audio, it is fair to assume that the circuits from R212 through to pin 2 of IC301 (and beyond to the audio-output terminals) are good. The first place to trace FM audio is at pin 6 of IC201 (TP3) and at the emitter of Q202.

If there is no audio at pin 6 of IC201 (or TP3), suspect the FM circuits. Check the FM PLL circuits as described in Sec. 10–13. If necessary, go through the FM adjustments described in the service literature.

If there is audio at pin 6 of IC201 (or TP3) but not at the emitter of Q202, suspect C209, C214, R209, R211, and Q202. Make certain that Q202 is turned on by FM B+. (The base of Q202 should be about 3 or 4 V.)

If there is FM audio at pin 2 of IC301, but no audio at the tuner-output terminals, suspect the muting circuits (Sec. 10–8) or possibly the audio control circuits.

If you get FM audio, but no AM audio, it is fair to assume that the circuits from R212 through to pin 2 of IC301 (and beyond to the audio-output terminals) are good. The first place to trace AM audio is at pin 13 of IC151, TP7, and at the emitter of Q201.

If there is no audio at pin 13 of IC151 or TP7, suspect the AM circuits. Check the AM PLL circuits as described in Sec. 10–13. If necessary, go through the AM adjustments as described in the service literature.

If there is audio at pin 13 of IC151, but not at TP7, suspect C203 and R165.

If there is audio at TP7, suspect C169, C213, R167, R210, and Q201, if the audio does not appear at the emitter of Q201. Make certain that Q201 is turned on by AM B+. (The base of Q201 should be about 3 or 4 V.)

If there is AM audio at pin 2 of IC301, but no audio at the tuner-output terminals, suspect the muting circuits (Sec. 10–8) or possibly the audio-control circuits.

10-13 FREQUENCY SYNTHESIS TUNER TROUBLESHOOTING

This section covers troubleshooting for the circuits discussed in Sec. 4–6.2 and shown in Fig. 4–24.

A failure in the PLL IC or in the FS circuits controlled by the PLL can cause many trouble symptoms. Unfortunately, a failure in other circuits can cause the same symptoms. For example, assume that you operate the TUNING UP/DOWN buttons and see that the front-panel frequency display varies accordingly, but the AM or FM station does not tune across the corresponding band (no stations of any kind are tuned in). This can be caused by a PLL failure or by a failure of the commands to reach the PLL IC503 (even though the commands are displayed); this creates special troubleshooting problems for AM/FM tuners with PLL.

Here is an approach that can be applied to most tuners, without regard to the exact method of PLL tuning. The most common symptoms for failure of the PLL are a combination of *no stations received* (on both AM and FM) and *noisy audio* (audio not muted when you tune across the broadcast band). Of course, not all tuners have both auto and mono modes as does our tuner, and not all muting circuits operate in exactly the same way. However, the following approach can be applied to the basic PLL problem.

10-13.1 Operating Control Checks

First make certain that the operating controls are properly set for a particular PLL function. For example, on our tuner, you must be in the auto mode (AUTO MODE button pressed, AUTO MODE indicator on) before the PLL will seek FM stations as you tune across the FM band with the TUNING UP/DOWN button. If you have selected mono (AUTO MODE indicator off) or the circuits have gone into mono because of a failure, the PLL tunes across the FM band in 200-kHz increments, whether the stations are present or not. Always look for some similar function on the tuner you are servicing.

When you are certain that the controls are set properly and that there is a true malfunction, the next step in troubleshooting is to isolate the problem to the tuning circuits or the PLL circuits. There are two basic approaches.

10-13.2 Frequency-Change Command Checks

With this approach, you apply a frequency-change command to the PLL and see if the error voltage and/or sample voltage changes accordingly. For example, to check the AM section of our tuner (Fig. 4-20), press the TUNING UP/DOWN buttons and see if the voltage at pin 22 of IC503 changes as the frequency display changes. If not, suspect IC503 or the circuits between the front-panel TUNING UP/DOWN buttons (key matrix) and IC503.

If the error voltage at pin 22 of IC503 changes, check that the frequency of the signal at pin 30 of IC503 (or pin 20 of IC151) also change as the frequency display changes. If not, suspect the tuning circuits D153, Q151, and/or low-pass filter Q507/Q508 (Fig. 4-20).

Next, press the TUNING UP/DOWN buttons while in the FM mode and see if the error voltage at the VCO input of MD101 changes. If not, suspect IC503 and/or IC507. You can also check that the PSC pulses at pin 28 of IC503 change (but this is usually more difficult to monitor).

If the error voltage applied to the VCO input of MD101 changes, check that the frequency of the sample-frequency output of MD101 and pin 27 of IC503 changes as the front-panel frequency display changes. If not, suspect MD101. If the frequency does change at MD101 and IC507, but not at pin 27 of IC503, suspect IC507. (It is also possible that IC507 is not receiving proper PSC pulses from pin 28 of IC503.)

10-13.3 Substitute Tuning Voltage

With this approach, you apply a substitute tuning voltage or error voltage to the tuning circuits and see if the circuits respond by producing the correct frequency. Although this sounds simple, here are some practical considerations.

First, you must make certain that the substitute tuning voltage is in the same range as the error voltage. For example, the error voltage in our tuner varies from about 1 to 20 V (at the tuning circuits). You can cover this range with a typical shop-type variable d-c supply. However, the shop supply can possibly load the tuning circuit with unwanted impedance, reactance, and so on.

Remember that if you apply a lower voltage the circuits will not respond properly. If you apply a voltage higher than the tuning-circuit range, the circuits can be damaged.

Although a number of tuners use circuits similar to our tuner (the MD101 tuner package is quite common), the tuning circuits are not the same for all tuners. (Some tuners combine the AM and FM functions in a single package.) Of course, if you are lucky, you can find the error-voltage range in the service literature, often in the adjustment chapter.

If the tuner is a *station-detect* function (most tuners do, at least in the FM section), you can use this feature together with a substitute tuner voltage to isolate PLL problems. Simply vary the substitute tuning voltage across the range and see if stations are detected.

For example, in our tuner (Fig. 4–20), you can check at pin 12 of IC201 and/or pin 26 of IC503 for a change of status each time a station is tuned in or out. Pin 26 of IC503 should go high and pin 12 of IC201 should go low each time an FM station of sufficient strength is tuned in. The status of the pins should reverse when the station is tuned out.

10-14   BASIC RF AMPLIFIER TROUBLESHOOTING

This section is devoted to troubleshooting an RF amplifier. The RF amplifier section of a radio transmitter is chosen as an example. It is assumed that you are familiar with basic transmitter tests and adjustments, such as AM/FM/SSB alignment, RF power-output checks, frequency checks, modulation checks, and so on. If not, you had better read the author's best-selling *HANDBOOK OF ADVANCED TROUBLESHOOTING* (Prentice-Hall, 1983), or you can follow the adjustment procedures found in the service literature (what a novel approach!).

It is assumed that you have localized a problem in a larger system to the transmitter by substitution. Now it is necessary to troubleshoot the transmitter circuits (particularly the RF amplifier circuits). This step-by-step troubleshooting problem involves locating the defective component in a solid-state radio transmitter.

10-14.1 General Instructions

The schematic for the transmitter is shown in Fig. 10–9. Note that the modulator portion of the transmitter is similar to an audio amplifier (Chapter 3), with one major exception. The modulator output is used to modulate the RF section (instead of driving a loudspeaker). A modulation transformer is used instead of an output transformer. Power for the final amplifier of the RF section is applied through the secondary windings of the modulation transformer. Thus any audio signal present at test point A is amplified by the modulator and serves to amplitude-modulate the RF unit.

Since the schematic diagram shows test points, it is not necessary for you to determine the location of the test point. However, *in practical situations, there is nothing that says you cannot add test points of your own.* For example, the diagram shows that there should be an RF signal of about 1 V at test point J (the collector of amplifier Q6). The same signal should also appear at the base of final amplifier Q7, even though no test point is indicated.

If, during the troubleshooting process, you find that there is a good signal at J but none at the base of Q7, you have traced a fault to that portion of the circuit (probably to the tuning network or possibly to the RF choke or Q7). In any event, *do not feel limited to the test points specified in service literature.*

Voltage and Resistance Information. No voltage or resistance information is available, except for the + 12-V supply voltage found on the schematic. However, you should be able to calculate the voltages found at the transistor elements.

The collectors of the three RF transistors (Q5, Q6, and Q7) are all connected to the + 12 V through RF chokes. Such chokes generally have very little d-c resistance and thus produce very little voltage drop (typically, a fraction of 1 V). Thus it is reasonable to assume that the d-c voltage at the collectors is about + 12 V (or slightly less).

The emitters of Q6 and Q7 are connected directly to ground (which is typical for NPN RF amplifiers). The emitter d-c voltage (and resistance) should be about 0 V. The same is probably true of the Q6/Q7 bases. The only resistance from the bases should be a few ohms produced by the RF chokes. It is possible that some d-c voltage might be developed across the chokes, but it is not likely.

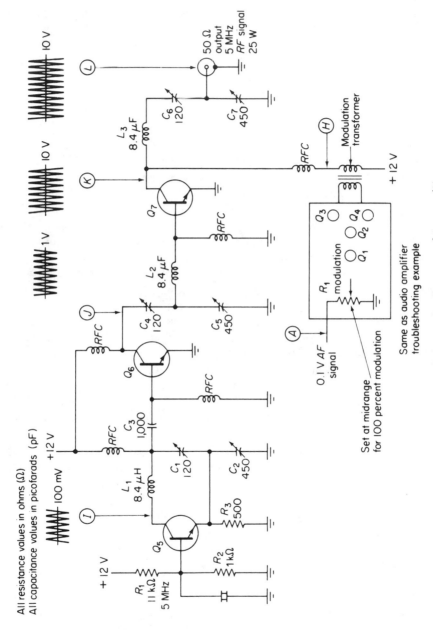

FIGURE 10-9 Schematic diagram of transmitter.

All resistance values in ohms (Ω)
All capacitance values in picofarads (pF)

Same as audio amplifier
troubleshooting example

Set at midrange
for 100 percent modulation

Modulation transformer

50 Ω
output
5 MHz
RF signal
25 W

The base of Q5 has a fixed d-c voltage applied through the voltage-divider network of R1 and R2. The ratio of R1 and R2 shows that the voltage drop across R2 is about 1 V, so the base of Q5 is at about 1 V with respect to ground. The emitter of Q5 should be about 0.5 V less than the base, or about +0.5 V with respect to ground. The resistance to ground from the emitter of Q5 should be equal to R3, or about 500 Ω.

Power Output. The schematic shows that the output is 25 W into a 50-Ω antenna. If a dummy load is used (as it should be during troubleshooting), use a 25-W (or larger) resistor with a value of 50 Ω. Do not use a wirewound resistor. (It is possible to use a 25-W lamp as the dummy, but the fixed resistance is generally preferred.)

Signal Frequencies. The schematic shows that the output frequency is 5 MHz and that of the crystal is also 5 MHz. This means that the *three RF stages are tuned to the same frequency.* In some transmitters, one (or more) of the RF amplifier stages are tuned to a harmonic of the fundamental frequency. Such stages act as *frequency multipliers,* as discussed in Chapter 4.

Generally, this is of concern only when you are trying to measure the frequency at each stage. From a practical standpoint, *it is usually possible to troubleshoot a transmitter without knowing the frequencies at all stages.* You are more concerned that RF is present and is of the correct amplitude. Even when making final tuning adjustments, *you tune for a peak indication* at each stage, rather than to a specific frequency.

Tuning Adjustments. Capacitors C1, C2, and C4 to C7 are tuning *adjustments* rather than tuning *controls* (or operating controls). These capacitors are accessible for adjustment (only after the covers are removed) and are not touched during routing operation.

Operating Control. The only operating control is R1, the modulation potentiometer. The schematic indicates that the RF output is modulated 100% when R1 is set to the approximate midrange, and there is a 100-mV audio signal at the modulator input (test point A).

With this "wealth of information" to draw upon, you are ready to plunge into the troubleshooting by determining symptoms.

10-14.2 Determine the Symptoms

In some troubleshooting work, you will find that the equipment has never been properly tuned, adjusted, or otherwise put into operation. Some troubleshooters approach each new problem with that assumption. However, in the majority of troubleshooting problems, you will find that the equipment was working satisfactorily for a time before the trouble occurred.

This is the case with our example. The transmitter has been working for many months, but the operator reports to you that there is now "no output. The transmission cannot be heard by any stations tuned to the operating frequency (5 MHz)." What is your first step?

You could start by checking each circuit to determine which circuit is not operating. This rates a *definite no.* You will eventually locate the faulty circuit with this procedure. However, it is not a logical step in troubleshooting and probably requires several unnecessary tests before the trouble is isolated to a circuit.

You could perform the tuning procedure to determine which circuit group does not perform properly. This is about one step above checking each stage. The procedure is not a logical approach. Although you *might* locate the trouble areas with this procedure, it would probably require many unnecessary steps. *Each step in troubleshooting should provide the most information with the least amount of testing.*

You should check the output. Notice that we did not tell you how to check the output at this time. You have two basic choices for checking the output: with test equipment and without test equipment. You can check the output by tuning a receiver to the operating frequency, turning on the transmitter, and trying to transmit a signal (probably voice into a microphone at the modulator input). This is a quick and easy test, but it does not prove much.

Even if you hear the voice transmission on the receiver, it is possible that the transmitter output is low. A weak transmission might be picked up by a receiver nearby but could not be heard over the normal communications range. If you do not hear the transmission on the receiver, it only proves that the operator was right (they are, occasionally), but it does not give you any clue to which section of the transmitter is at fault.

A more practical method for determining the symptoms is to monitor the output of the final amplifier stage Q7. Note that the transmitter is not provided with any built-in operating indicators. There is no RF output indicator (lamp or meter), no modulation meter, no SWR meter, and so on. You must use external test equipment. Before making any such tests, disconnect the antenna and connect a dummy load (no matter what combination of test equipment you use).

You can monitor the RF output at test points K and L with a meter or scope, using an RF probe (Chapter 9). Remember that an RF probe only proves the presence (or absence) of RF signals. The most satisfactory test is to use a scope connected for a *modulation check.* Remember that the vertical amplifier of the scope must be capable of passing the 5-MHz RF signals.

Modulate the transmitter with an audio tone (for a complete output test). You can use voice, but the results are inconclusive. A better test is to set R1 to midrange and apply a 100-mV audio tone (say, at 1 kHz). As shown on the schematic, this should result in a 100% modulation (the modulation envelope should drop to zero, or near zero, between peaks).

The percentage of modulation is controlled by R1. Although no specifications are given on the schematic, it is reasonable to assume that, with R1 set to full on, the percentage of modulation is something in excess of 100%. This can be verified during the output check.

Now assume that there is no RF indication on the scope (no vertical deflection whatsoever, with R1 set at any position). In which circuit group do you think the trouble is located?

You could choose the modulator circuits. If you choose the modulator, you will probably not succeed as a troubleshooter, or you are not paying attention, or you simply do not understand transmitters. The vertical deflection on the scope is produced by the RF signal; the shape of the signal is determined by the modulation. Thus, with no vertical deflection, there is no RF, and the trouble is traced to the RF circuits. The setting of R1 has no effect on the presence or absence of RF. (One possible exception to this is if you consider the modulation transformer a part of the modulator circuits, and the secondary winding is open.)

You should choose the RF circuits. The next step is to isolate the trouble to one of the circuits (oscillator Q5, buffer-amplifier Q6, or power amplifier Q7) in the defective circuit group. What is your next most logical test point?

You could check the supply voltage for the RF circuits. This approach has some merit. Note that Q5 and Q6 get collector voltage directly from the +12-V supply line, whereas Q7 gets collector voltage through the modulation transformer winding. To make a complete check of supply voltage for the RF circuits, you must measure the voltage at each of three collectors and at the base of Q5. If the collector voltages are absent or abnormal, suspect the corresponding RFC. If the Q5 base voltage is bad, suspect R1 or R2. If the voltage at any one collector is present and correct, the power supply is good.

You could check for RF at test point I. This is not a bad choice. If there is no RF at I, the trouble is traced to oscillator Q5 and associated parts. This would be a lucky guess. If there is RF at I, you know that Q5 is oscillating, but you eliminate only one circuit as a possible trouble area, so this is not the most logical choice.

You should check for RF at test point J. This is the most logical choice and can be done with a meter and RF probe (or scope). If there is a good RF signal indication at J, both Q5 and Q6 can be considered as operating properly, and the trouble is traced to the power amplifier Q7.

You also eliminate the power supply (which must be good if Q5 and Q6 are functioning normally). However, you do not eliminate the lead between the power supply and the collector of Q7 (which is a separate path from that between the supply and the Q5/Q6 collectors).

If there is no RF indication at J, the trouble is traced to Q5 or Q6. In this case, the next step is to check for RF at I. If RF is present at I but not at J, suspect Q6. If RF is absent at I, suspect Q5.

Now assume that there is a good RF indication at J. You can "bracket" the power amplifier Q7 and associated parts as the faulty circuit group (because the input signal is good and the output is bad).

10–14.3 Locate the Specific Part

At this point, a visual inspection of Q7 and the related circuit is in order. But let us assume that there is no *apparent sign* of where the trouble could be located. Just to make it difficult, there is no sign of overheating, and all parts (as well as wiring) appear proper. What is your next step?

You could make an in-circuit test of Q7. This is difficult because Q7 is operated as class C and does not respond to the usual in circuit forward-bias test (such as the leakage tests described in Sec. 10–2). There is no forward bias applied to Q7, so you cannot remove the bias. If you attempt to apply a bias, the voltage relationship between emitter and collector will probably not change (because there is no d-c load). An in-circuit transistor tester might prove that the transistor is good at audio frequencies. However, in-circuit testers are generally useless at radio frequencies.

You could make a substitution test of Q7. This is more certain than the in-circuit test, and it is possible that you may substitute Q7 before you have located the fault. However, there are more convenient tests to be made at this time (particularly if Q7 has a heat sink, Chapter 2).

You could check the resistance at all elements of Q7. Although you will probably check resistance and continuity before you are through, resistance checks at this time prove little. The resistance-to-ground at the emitter of Q7 should be zero. The base resistance-to-ground should also be zero. (The RF choke may show a few ohms on the lowest ohmmeter scale.) Only a high resistance-to-ground reading at the base and emitter is of any significance.

You do not know the correct resistance-to-ground for the collector of Q7. Of course, you could guess. The resistance of the Q7 collector should be substantially the same as that for the Q5/Q6 collectors, plus any resistance in the modulation transformer winding. If the Q7 resistance is slightly higher than the Q5/Q6 collector resistance, you can *assume* that the value is correct.

You should check the voltage at all elements of Q7 first. The voltage at the base and emitter should be zero (we are speaking of d-c voltage, not RF signal voltage). The d-c voltage at the Q7 collector should be about +12 V. If the voltages are all good, you can skip the resistance-to-ground measurements. However, you may still have to make continuity checks if the voltages are abnormal. Let us examine possible faults that are indicated by abnormal voltages.

If there are large d-c voltages at the base or emitter, this indicates that the elements are not making proper contact with ground. For example, if there is a high-resistance solder joint between the Q7 emitter and ground, it is possible for a d-c voltage to appear at the emitter. Or if the emitter–ground connection is completely broken, the emitter is floating and can show a d-c voltage.

If the collector shows no d-c voltage, the fault is probably in the RF choke or the modulation transformer. This requires a continuity check. Start by checking the d-c voltage at test point H. If the voltage is correct at H, but not at K (the collector of Q7), suspect the RF choke. If the voltage is absent at H, suspect the transformer.

10-14.4   RF Amplifier Troubleshooting Summary

To summarize this troubleshooting example, let us assume that the actual trouble is caused by an open L3 coil winding. Assume that the transmitter has been subjected to excessive vibration, and the L3 winding is broken from the coil terminal, underneath where you cannot see the break (that is where they always break!).

This trouble does not affect d-c voltage or resistance at the transistor elements, and substitution of Q7 does not cure the fault! These are the kinds of problems you find in *real troubleshooting*. Everything appears to be good, but the equipment does not work.

To solve such problems, you must make point-to-point continuity checks, starting with the suspected stage (Q7). In this case, if you check from point K to C6, you will find the open coil winding.

INDEX

ABW-7553

37.0   2/18/93

DATE DUE